Genome and Genomics

K. V. Chaitanya

Genome and Genomics

From Archaea to Eukaryotes

Springer

K. V. Chaitanya
Department of Biotechnology
GITAM University
Visakhapatnam, Andhra Pradesh, India

ISBN 978-981-15-0701-4 ISBN 978-981-15-0702-1 (eBook)
https://doi.org/10.1007/978-981-15-0702-1

This Springer imprint is published by the registered company Springer Nature Singapore Pte Ltd.
The registered company address is: 152 Beach Road, #21-01/04 Gateway East, Singapore 189721, Singapore

Knowledge of Sequences Could Contribute Much to Our Understanding of Living Matter

Frederick Sanger

Dedicated to My Teachers

Preface

A genome encodes all the necessary information for the function of both single cell and highly complex organisms. Genome comprises a cluster of genes, regulated in a wide variety of cells, whose division makes an organism. Apart from the genes, a genome is also composed of noncoding regions, regulatory regions, etc. whose coordinated function will make the process of life. A lot of advancement has been made in the field of genomics with the genomes of all model organisms and economically important organisms being sequenced and deposited in the databases.

This book addresses the new tools, technologies, and approaches that were made to sequence a variety of genomes belonging to various organisms for both the students and experienced, practicing biologists. The achievement of long-sought decoding a genome sequence was possible through the development of instrumentation and computational technologies with user-friendly softwares and tools, which has irrevocably changed the perspective and provided a new direction for the a better understanding of biology. The development of other omics technologies such as proteomics, transcriptomics, and metabolomics has also provided a comprehensive scope of understanding a living system in detail.

Genome and Genomics: From Archaea to Eukaryotes is the most updated book, which mentions about the components of the genomes of all three major life forms along with the organelles. This book covers the concepts of the genomes, how various components in the genome are operating for the life of an organism, how the genomes and different life forms have evolved, what are other "omics" which provides a better understanding of genome functions, and what are the major applications of the genomics in providing a healthy, hunger-free, and disease-free human society.

This book is divided into nine chapters. Chapter 1 describes the variations in the viral genomes and their evolution. In Chap. 2, archaeal genomes and their relationship with the prokaryote and eukaryote genomes were discussed. Chapter 3 explains about the bacterial genomes. Chapter 4 mentions the organellar genomes and their evolution from the bacteria. Chapter 5 describes the eukaryotic genomes with model organisms including the human genome. Chapter 6 explains the sequencing technologies that are applied for sequencing the genomes of various organisms including fossils, their annotation, and assembly mechanisms. The role of other omics technology that has helped in the better understanding of the life processes such as proteomics, transcriptomics, metabolomics, and exposomics was discussed in Chap. 7. Applications of genome sequencing were described in Chap. 8. Chapter 9 consists of the genome databases and their URLs.

I owe my profound gratitude to Late Dr. M.V.V.S. Murthy, Founder President, GITAM Deemed to be University whose vision, dynamism, and motivation have deeply inspired me. I am indebted to Dr. Utpal Nath, Associate Professor, MCB Department, Indian Institute of Science, Bengaluru, Professor Ch. Ramakrishna, Professor Sk. Khasim Beebi, and Professor T. Sekhar, GITAM University, for their unstinting support. I thank Dr. Nageswara Rao Reddy for sparing his time to read the drafts and provide invaluable suggestions and comments. I thank my wife Lalita and my children Abhiram and Aamukta for their patience and constant support.

I express my deep sense of gratefulness to all authors, whose works have been consulted for writing this book. I shall highly appreciate any valuable suggestions for further improvement of this work.

Visakhapatnam, Andhra Pradesh, India K. V. Chaitanya

Contents

About the Author

K. V. Chaitanya is an Associate Professor at the Department of Biotechnology, GITAM University, Visakhapatnam. He completed his Ph.D. in Life Sciences at Pondicherry University and then received a fellowship from DBT to pursue post-doctoral studies at the Indian Institute of Science, Bengaluru. He has over 15 years of experience in research in the fields of genomics, molecular biology, and plant biotechnology. He has worked in various capacities in internationally respected academic and research institutes and has published a number of articles in leading international journals. He has published one book on cell and molecular biology and has filed five patents. Dr. Chaitanya has received numerous academic awards and fellowships.

Structure and Organization of Virus Genomes

1

Abstract

This chapter provides an in depth study on the structure, composition, and organization of viral genomes, their classification into double stranded and single stranded DNA viruses, positive and negative stranded RNA viruses with and their genome diversity. Segmentation and re-assortment of viral genomes have been discussed along with the multipartite virus genomes. Genome details of 13 different viruses have been provided as type studies for better understanding of these topics. Concepts of viral genome evolution have also been discussed.

1.1 Introduction

A Virus is a small electron microscopic parasite, incapable of reproducing by its own, survives by directing the host cell machinery for the production of more viruses, which emerges from their respective host cell through lysis. Most of these viral organisms contain either double stranded or single stranded DNA as well as RNA in their genomes, which may be either single stranded or double stranded. After the purification and partial crystallization of Tobacco Mosiac Virus in 1935 by Wendell Stanley, the study of viruses has inspired many scientists, which lead to identification and characterization of plant, bacteria, archaea, and animal viruses. Since viruses are capable of infecting a large number of various cell types, genetically modified viruses are being considered for the gene therapy. All these factors and applications make the virus an important organism for its capability to infect any living organism on this planet.

Viruses are small submicroscopic, obligate intracellular parasites, which contains either DNA or RNA as genome protected by a virus-encoded protein coat called capsid. Viruses are mobile genetic elements, depends on metabolic and biosynthetic machinery of host cells for their propagation. Viruses cannot carry out their life sustaining functions outside the host cell. They cannot synthesize proteins as they lack ribosomes and uses the host cell ribosome machinery for translating their

K. V. Chaitanya, *Genome and Genomics*,
https://doi.org/10.1007/978-981-15-0702-1_1

mRNA into proteins. Viruses can neither generate nor store ATP but derive the necessary energy and other metabolic functions by parasitizing the host cell for the basic materials such as amino acids, nucleotides, and lipids. Even though the viruses are speculated as a form of proto-life, their inability to survive outside living organisms does not make them as living organisms in the strict sense and it is highly unlikely that they preceded cellular life during the evolution of the earth. Some scientists even speculate that viruses have begun as rogue segments of genetic code which got adapted to a parasitic form of existence. Virions are the complete virus particles that are produced by the assembly of the pre-formed viral components, whose main function is to deliver its genome into the host cell for its expression (transcription and translation). The icosahedral virion belonging to the *icosahedral* or *quasi-spherical* structural class of viruses is made of 20 identical triangular faces, each face is constructed with three identical capsid protein units making 60 subunits per capsid, with five subunits symmetrically contacting each of the 12 vertices, thus making all proteins in equivalent interaction with each other. The viral genome is packed inside a symmetric protein capsid, composed of either a single or multiple proteins, each of them is encoding a single viral gene. Due to this symmetric structure, viruses could encode all the necessary information for constructing a large capsid using a small set of genes. A capsid along with the enclosed nucleic acid is called a nucleocapsid. In enveloped viruses, the nucleocapsid is surrounded by a lipid bilayer and is studded with a layer of glycoproteins. Often, the nucleic acid is associated with a protein called nucleoprotein. Viruses possess a great diversity with respect to their size. *Mimivirus* is the largest virus reported with 400 nm in diameter, bigger than the *Mycoplasma* bacteria, which is ~200 to 300 nm long. They also exhibit a wide diversity in shapes and forms, such as spherical, rod-like, etc.

There are few subviral entities which are highly similar, pathogenic and are possessing the properties similar to that of viruses. These entities are viroids, virusoids, and prions. Viroids are a short stretch of highly complementary, single stranded (200–400 nucleotides), circular RNA molecules possessing a rod-like secondary structure without any capsid or envelope. These viroids are associated with plant and human diseases such as hepatitis D. They are obligate intracellular parasites, with replication strategies similar to viruses. Virusoids are a satellite, circular single-stranded RNAs (1000 nucleotides) dependent on plant viruses for replication and encapsidation. They are packed into virus capsids as passengers. Their genome encodes only structural proteins. Prions are anomalous infectious agents that cause fatal neurodegenerative diseases mediated by contemporary mechanisms. Prions consist of a single type of protein molecule without any nucleic acid component. The protein is a modified isoform of prion protein (PrP) designated PrPSc. The normal cellular PrPC is converted into PrPSc by a structural transition of its α-helical and coil structure is refolded into β-sheet. The prion protein and its encoding gene are often found in normal uninfected cells and are associated with viral diseases such as Creutzfeldt–Jakob disease in humans, scrapie in sheep and bovine spongiform encephalopathy (BSE) in cattle.

1.2 General Features of Viral Genomes and their Classification

It has been estimated that there are 10^{31}–10^{32} viruses in the earth's atmosphere, which exceeds the number of host cells fairly by an order of magnitude. As a consequence, every organism on the planet or even every living cell is under constant attack from viruses, even viruses are highly responsible for the greatest selection pressure on the living organisms. In spite of their small size, viruses play an important role as obligate intracellular parasites, modulating their host cells for energy and reproduction leading to adverse effects. The main emphasis of virology is focused on the identification and control of pathogenic viruses that invade humans, domestic animals, and plants. But, origin and organization of viruses, their evolution is the deep questions which are fundamental to molecular virology. The comparative genomics has allowed closely related viruses to be compared and classified. In addition, the sequencing of eukaryotic genomes has revealed that 5–10% of their DNA encodes information for these organisms. A large fraction of the remainder is thought to be composed of mobile retrovirus-like elements (retro-transposons), which may have played a considerable role in shaping these complex genomes. Bacterial genomes do not have such extra genetic material. But, the genomes of certain bacteriophages have a close resemblance with bacterial plasmids in their structure and in the way of their replication, revealing that the relationship between viruses and other living organisms is perhaps more complex than what was previously thought.

1.2.1 Classification of Viral Genomes

Currently, over 4000 viruses have been described, classified into 71 families. Even though viruses possess small genomes, they exhibit enormous diversity compared with plants, animals and even bacteria. With respect to the genome, viruses are broadly divided into DNA viruses and RNA viruses. Both DNA and RNA viruses can either single stranded or double stranded, with a circular, linear or segmented arrangement. DNA and RNA viruses are distinguished by their features, such as monopartite or multipartite. In monopartite, their genome is having a single nucleic acid molecule. All double stranded DNA genomes contain only a single nucleic acid molecule and few of the viruses with single stranded genomes are reported to have multiple segments. In contrast, RNA viral genomes are generally multipartite, with more frequency for single stranded RNA viruses. Additionally, single stranded virus genomes may be either positive sense (+) where the RNA present in the genome will of the same polarity as mRNA and will encode the genes or negative sense (−), where the entire ssRNA genome must be copied and the copied strand is transcribed. Some single stranded viruses are ambisense (a mixture of + and − sense). Bacteriophages such as MS2, Qb, and Mimivirus belonging to family Leviviridae consists of ambisense viral genomes. Size of the DNA viruses is larger than that of RNA viruses. Few DNA viruses can be as large as 305,000 nucleotides. DNA viruses are called large viruses and RNA viruses are small. Size of a few single stranded RNA genomes is up to 31,000 nucleotides. The small size of the single-stranded virus might be limited due to the fragility of the RNA, which provides the tendency of the large RNA strands to break and also due to the fact that RNA viruses are more susceptible to mutations than DNA viruses. Single stranded DNA and RNA viruses are fragile than double stranded viruses.

The genomes of the largest double-stranded DNA viruses such as herpesviruses and poxviruses are quite complex, resembling their host cells. Genomes of Polyomavirus are complexed with cellular histone proteins, which form a chromatin-like structure inside the virus particle. After infecting the host cell, these genomes behave like miniature satellite chromosomes inside the host cell following the dictates of cellular enzymes and the cell cycle. mRNAs of Vaccinia virus were polyadenylated at their 3¢. The genome of Adenovirus consists of split genes with non-coding introns, protein-coding exons, and spliced mRNAs. In order to streamline the replication cycle, the Adenovirus takes control of many biological processes in the cell, such as alternative RNA splicing and polyadenylation for expanding the coding potential of the limited viral genome.

Phage genomes are simple and least complex. Introns in prokaryotes were first discovered in the genome of T4 phage in 1984. The genome of T4 phage is 160 kbp double-stranded DNA, highly compressed with promoters and sequences that control translation are nested within the coding regions of overlapping upstream genes. It consists of three self-splicing group I introns, located in the genes that codes for thymidylate synthase (td), aerobic ribonucleotide reductase (nrdB) small subunit and the anaerobic ribonucleotide reductase (nrdD). Like any other group I introns, the td and nrdD introns each contain an open reading frame encodes for a homing endonuclease that renders the introns mobile and they can be inserted into new phage genomes. The nrdB intron is non-mobile due to a deletion in the intron-borne homing endonuclease gene. It was speculated that the presence of in these introns may confer a selective advantage to the phage by offering the possibility of regulating the expression of the intron-containing genes by the regulation of splicing. It was also believed that horizontal transfer of the introns between phages through homing has played a significant role in the evolution of group I introns in T-like phages.

The size of viral genomes also depends on the type of host cell. Viruses with prokaryotic host cells tend to replicate quickly to keep up with their host cells, which are reflected in the compact nature with overlapping genes of many bacteriophages, leading to the minimum genome size. Viruses with eukaryotic cells as hosts show tremendous compression while the core is getting packed into the capsid so that only optimum amount of genome can be packed. Viruses pack their genomes into protective protein contained capsids. Packaging strategies of viruses are related to the type of genome. Viruses with double stranded DNA genomes use a molecular motor with a spool like structure to pack their genomes into the capsid. In contrast, viruses with single stranded DNA or RNA genomes employ a co-operative mechanism, in which package and assembly of the genome into the capsid will occur simultaneously, enhancing the assembly efficiency of the capsid. In bacteriophage MS2, genomic RNA sequence will form a short term loop, for its interaction with the

Table 1.1 General features of viral genomes sequenced

S. No.	Class	Sequenced genomes	Size (Nt)	Proteins
1	DsDNA	414	4697–335,593	6–240
2	SsDNA	230	1360–10,958	6–11
3	DsRNA	61	3090–29,174	2–13
4	SsRNA (+)	421	2343–31,357	1–11
5	SsRNA (−)	81	8910–25,142	5–6

capsid proteins. This interaction will trigger conformational changes that convert symmetric protein dimers into asymmetric form, needed to build the capsid. Some bacteriophages of the family Myoviridae, such as T4 contain relatively large genomes, up to 170 kbp. The largest viral genome currently is known that of Mimivirus, ~1.2 Mbp, consisting of 1200 open reading frames, with only 10% of them showing the similarity with proteins of known function. Among the eukaryotic viruses, herpesviruses and poxviruses possess relatively large genomes, up to 235 kbp. These genomes contain genes involved in their own replication, particularly enzymes concerned with nucleic acid metabolism. Therefore, these viruses can partially escape the restrictions of the host cell biochemistry by encoding additional biochemical apparatus with the penalty of encoding all information necessary for a genome to pack into the capsid, which also causes upward pressure on its size. Whatever might be the composition of a genome, all viruses are obligate intracellular parasites, which can replicate only inside the appropriate host cells, their encoded genomes must be recognized by a specific host cell, which is getting parasitized. For that, the genetic code employed by the virus must either match or recognized by the host cell. Similarly, the signals that regulate the expression of viral genes must be inappropriate to the host. The general features of viral genomes sequenced are displayed in Table 1.1.

1.3 Size, Structure, and Composition of Double Stranded DNA Virus Genomes

Cellular life forms of all categories possess genomes with double stranded DNA and utilize the same as a standard scheme for their replication and expression. In contrast, viruses and other pathogenic and selfish elements exploit all possible interconversions of nucleic acids, with their genomes that can be either DNA or RNA, which are single-stranded or double-stranded. Viral genomes that have been sequenced and annotated are compared with genomes of cellular life forms, which were small with unknown gene functions. But, in the past few years, discovery of giant viruses has rapidly expanded the size of viral genomes whose range spans up to 3 orders of their magnitude, from ~2 kb to ~2 Mb. Surprisingly, the genomes of giant viruses are larger than the genomes of many bacteria and archaea, obliterating the gulf between cells and viruses in terms of genome size and complexity. A number of viral groups possess double-stranded DNA as their genomes are of considerable complexity, classified into small and large genomes depending on the type of replication. Small genomes use a host DNA polymerase for replication, In contrast, large genomes encode a virus-specific DNA polymerase, responsible for their genome replication. These viruses are genetically similar to the host cells they infect. Large DNA viruses encode more proteins than small DNA viruses.

There are two major viral groups representing the double stranded DNA genomes ie. members of the Adenoviridae and Herpesviridae families. Human adenovirus is one of the most common pathogens that causes minor, self-limiting illness to most of the patients. Adenovirus is one of the very well understood viruses, whose basic biology has been extensively studied over the past 60 years. Human Adenovirus has been isolated in the early 1950s from the adenoid tissue causing respiratory infections. This virus has a remarkable capacity to spread among patients contacting from as few as five virus particles. Adenovirus-induced acute respiratory disease is the most common infection in confined populations of daycare centers, hospitals, retirement homes, and military training venues, accounting for ~8% of all childhood respiratory tract infections, which can lead to bronchitis, bronchiolitis, or pneumonia, requiring hospitalization in ~25% of diagnosed cases. It also causes other localized diseases, such as colitis, hemorrhagic cystitis, hepatitis, nephritis, encephalitis, myocarditis and disseminated disease with multiorgan failure. A few such diseases can be more serious in pediatric and geriatric populations, especially in the individuals with suppressed immune systems, such as transplant recipients or patients suffering from AIDS. Many aspects of Adenoviral life cycle have been completely elucidated in great detail, which has allowed the development of adenoviral vectors that are highly efficient in delivering genes into the mammalian systems, especially human cells for the transgene expression and for delivering therapeutic genes in human gene therapy. Adeno viruses specifically consist of a nucleoprotein core with a 30–40 kb linear double-stranded DNA, surrounded by an icosahedral, non-enveloped capsid of 70 to 100 nm diameter. These viral genomes contain 30–40 genes. The terminal sequence of each DNA strand has an inverted repeat of 100–140 bp. The denatured single strands form a '**panhandle**' like structures, which are important for the DNA replication. A 55 kDa protein known as the terminal protein is covalently attached to the 5¢ end of each strand. During the genome replication, this protein might acts as a primer, for initiating the synthesis of new DNA strands. The expression of the genes is rather more complex in Adenoviruses with clusters of genes expressed from a limited number of shared promoters. Multiply spliced mRNAs and alternative splicing patterns are used to express a variety of polypeptides from each promoter.

Apart from the differences in host and tissue tropism, a very less amount of variation is found in the Adenoviral genomes and their structural parameters. Human adenoviral serotype 5, is one of the highly characterized adenovirus, consisting of ~36 kb genome. Its coding region is divided into early (E1–E4) and late (L1–L5) transcripts based on their stage of expression. Essential early E1 region is deleted in most adenoviral vectors, rendering their incapability of replicating in most cell lines. Numerous studies have shown that after the deletion of E1, Adenoviral vectors are more ideal for the in vivo and in vitro studies requiring short-term transgene expression. The Second generation Adenoviral vector constructs are made after deletions in the essential E2 or E4 regions, facilitating the prolonged transgene expression. Helper-dependent Adenoviral vectors (hdAd) are generated by deleting all genes that code for viral proteins for increased cloning capacity. Removal of protein coding sequences in these vectors allows the overall reduction in their genome size from 36 kb for the wildtype to 30 kb in E1/E3-deleted Adenoviruses and ~500 bp for helper-dependent Adenoviral vectors.

1.3.1 Analysis of Adenovirus 2(AD2) Genome

The genome of adenovirus 2(Ad2) was the first adenoviral genome to be fully sequenced, which is of the size ~36 kb, encoding over 40 proteins. Adenoviral coding regions are designated as early or late depending on their expression before or after the DNA replication. The early genes E1A, E1B, E2, E3, and E4 are the first ones to get transcribed. They encode for the proteins involved in activating transcription of other viral regions, altering the cellular environment to promote viral production. E1A proteins also induce mitogenic activity in the host cell, which stimulates the expression of other viral genes. E2 proteins regulate viral DNA replication, while E3 and E4 proteins are involved in altering the host immune responses and cell signaling. Activation of the major late promoter (MLP) followed by the start of viral DNA synthesis, allowing the expression of late genes encoding primarily virion structural proteins. L1–L5 of the late regions are transcribed from an alternatively spliced transcript. The regions encoding the L4-22 K and L4-33 K proteins are initially expressed at low levels from a novel promoter located within the L4 region and these proteins functions in fully activating the major late promoter (MLP). Four small proteins including the structural protein IX (pIX) and the IVa2 protein produced at intermediate/late stages of infection, helps in packing of viral DNA into immature virions. The late products VA RNA I and II inhibits the activation of the interferon response, impede cellular micro-RNA processing and also influence the expression of host genes. There are 100 bp inverted terminal repeats (ITRs), located at both ends of the genome, which act as the sites for the origin of replication, with the ~200 bp viral packaging sequence positioned next to the left ITR (Fig. 1.1). Even though Adenovirus is being studied in great detail for more than 60 years, our knowledge of the genes encoded by this virus is still expanding. In 2007, a new open reading frame (ORF) has been identified to be located between the fiber ORF and E3, which is termed as U exon. The U exon protein (UXP) is expressed from a unique promoter during later stages of infection and is hypothesized to play a significant role in the transcription.

Most of the transcription processes in the adenovirus result in more than two alternatively spliced mRNAs. Keeping the compactness of the adenovirus genome, regulatory events that take place at the level of RNA processing are of great importance for controlling the lytic cycle of the virus. Splicing of large introns results in the production and accumulation of shorter mRNAs, at the later stages of viral infection. Except for E1A, L4-33Kand the U exon protein, viral introns do not

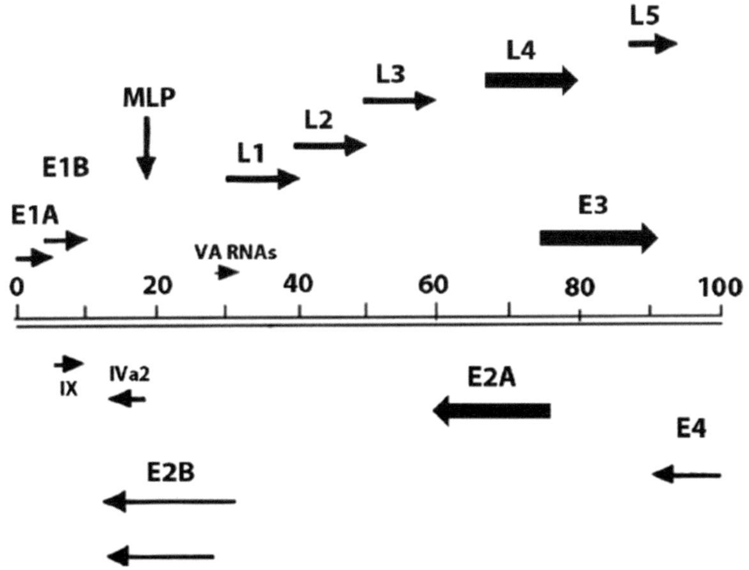

Fig. 1.1 Organisation of the Adenovirus genome (Courtesy Russel WC, School of Biology, University of St Andrews, UK)

interrupt the ORF of the gene. Further, Adenoviral genes contain very few introns compared to that of cellular genes. Viral mRNAs mature by removing one to three introns. As the virus has to compress much of its genetic information into a small genome, the selection pressure appears to have favored few and small introns. Deep cDNA sequencing of Adenovirus genome has identified many novel alternatively spliced transcripts, suggesting that there may be numerous new or altered polypeptides produced by this virus in the infected host cell, reflecting that this genome still has many secrets that remain to be uncovered.

1.3.2 Herpesviridae

Herpesviridae belongs to a large family of ~100 members with at least one for each of the animal species that have been examined till date. There are eight human herpesviruses, all of them share a common overall genome structure, but differs in the fine details of genome organization and at the level of the nucleotide sequence. Herpesviridae is a family of enveloped, DNA viruses with complex genomes. They replicate in the nucleus of a wide range of vertebrate and invertebrate hosts, such as humans, horses, cattle, mice, pigs, chickens, turtles, lizards, fish, and invertebrates such as oysters. There are eight human Herpesviruses divided into three subfamilies. Herpesviruses possess very large complex genomes composed of large complex virus particles up to 235 kbp of linear, double-stranded DNA and ~35 virion polypeptides. Members of Herpesviridae are highly diversified in terms of their genome sequence and protein synthesis but show a great similarity in terms of structure, genome organization and almost all of their genomes encode the enzymes involved in nucleic acid metabolism, DNA synthesis, and protein processing. Few Herpesviral genomes consist of two covalently joined sections named as a unique long (UL) and a unique short (US) regions, each bounded by inverted repeats. These repeats allow structural rearrangements of the unique regions, facilitating the existence of genomes as a mixture of four functionally equivalent isomers. These genomes also reported contain multiple repeated sequences and depending on their number, size of the genome may vary up to 10 kbp.

Integration of viral genome into the chromosomes of the host cell is mandatory for the successful completion of the virus lytic cycle. In contrast, Herpesviruses maintains their genomes as extrachromosomal circular episomes in the nuclei of infected cells without the need for integration. There are also reports of chromosomal integration of Herpesviral DNA, suggesting that Herpesviruses are also capable of integrating into the host's chromosomes under few circumstances. It was hypothesized that the replication of non-integrated Herpesviral DNA occurs through the rolling-circle mechanism, yielding long DNA concatemers that are subsequently cleaved into single genome equivalents during nucleic acid encapsidation. In addition, Human Herpesvirus 6 (HHV-6) is found to be integrated into the germlines of approximately 1% of the world's population. But how the replication of linear CIHHV DNA occurs still remains unknown. Chromosomal insertions of α-Herpesvirus DNA, Herpes simplex viruses and Equine Herpesvirus have been detected following infection with defective interfering particles or transfection of sheared or subgenomic viral DNA fragments. The integrated viral genome consists of mostly subgenomic fragments without any possibility for the production of infectious viral particles to occur. Many of the cells carrying integrated viral DNA displayed a transformed phenotype, fueling the hypotheses on the oncogenic nature of these viruses.

1.3.2.1 Analysis of Herpes Simplex Virus Genome

The prototype member of the Herpesviridae family is Herpes Simplex Virus (HSV), whose genome is ~152 kbp, composed of double-stranded DNA. The complete nucleotide sequence of the Herpes Simplex Virus has now been determined. This virus contains about 80 genes, densely packed and with overlapping open reading frames. Each gene is expressed under its own promoter. HSV-1 is perhaps the most intensively studied complex virus genome. Before the development of nucleotide sequencing, the HSV genome has been extensively mapped by conventional genetic analysis and mutant analysis. The HSV-1 genome is composed of a single linear double stranded DNA of ~152,261 base pairs in length. The genome is divided into two segments called Unique Long (UL) and Unique Short (US). Small regions of repeated sequence occur at the genome ends between the L and S segments. As DNA is replicated, the inversion of L and S segments takes place at a very high rate, creating four genome isomers, occurring at equal frequencies in most of the HSV-1 wild type populations. This virus genome synthesizes 75 genes encoding for known proteins. Among them, 69 genes are found to exist in a single copy and three genes in two copies each. Among the 75 genes with identified functions, 43 are considered to be the core genes common to α, β and γ Herpesviruses located in the UL segment of the genome. These genes are involved in many vital functions including the entry of the core DNA into the host cell and its replication, assembly into the capsid and its dispersal. All genes located in the unique short segment are non-core and are highly divergent. These are mainly found at the ends of the segment. Proteins encoded by the non-core genes are involved in both lineage and species-specific functions such as transcriptional transactivation, immune evasion, and host cell recognition.

Type 1 Herpes simplex virus is a member of the α-Herpesvirinae subfamily of the Herpesviridae family, whose infection results in cold, ocular, genital sores and encephalitis. Several strains of HSV-1 have been isolated varying in virulence, which might be due to base substitutions resulting in amino acid or *cis*-regulatory changes. One of the HSV-1 strains, KOS isolated from a human labial lesion and is frequently used as a marker to investigate the HSV-1 gene function and pathogenesis. KOS is less virulent compared with other HSV-1 strains, such as McCrae and 17, which has raised the curiosity for the comparative genomics. KOS genomic DNA was isolated from infected African green monkey (Vero) kidney cells and an unpaired 42-bp Illumina library was generated and run at Genome Technology Access Center, Washington University. Since viral DNA was isolated from Vero cells, potential contaminating host reads that matched the *Rhesus macaque* and/or human genomes were removed using Bowtie. The remaining reads of 16,494,831 bp were assembled into contigs using the Velvet de novo assembler against the reference HSV-1 strain 17 genome (GenBank accession number NC_001806) with SeqMan Pro (DNASTAR, Inc.). Because the HSV-1 genome includes two sets of inverted repeat regions TRL/IRL and IRS/TRS, contigs assembling into one of the repeat units were reverse complemented and also placed into the other repeat unit. The final KOS genome is a linear double stranded DNA of 152,011 bp, consisting of 80 genes and has 13 gaps, exclusively at VNTR regions, totaling up to 1582 bp in length. In the Gen Bank annotation, the sequence and length of each VNTR were copied from strain 17. Using Bowtie for aligning the filtered reads against the de novo assembly, the average sequence coverage per base pair for the KOS genome was determined to be 4257 (Fig. 1.2).

To identify nucleotide variants between the genomes of strains KOS and 17, the genomes were aligned using fast statistical alignment (FSA) and applied custom Perl and R scripts. KOS differs from strain 17 by 1024 SNPs, 320 of them are non-synonymous changes in 65 of 77 HSV-1 open reading frames. The two genomes also differ by 172 indels, most of them are insertions or deletions of single bases in non-coding regions. However, 26 indels are in frame additions or deletions of codons. Further analyses are in progress for the comparative genomics of KOS with other genomes of HSV-1 strains. Such studies will increase the scope for better identification of the genetic attributes of KOS and its contributions to its pathogenesis.

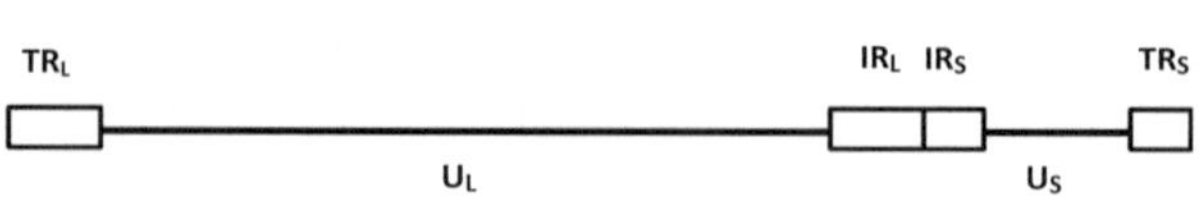

Fig. 1.2 Organisation of the Herpes Simplex Virus-1 genome

1.4 Genomes of Single Stranded DNA Viruses and their Mosaicism

Viruses with single stranded DNA genomes infect hosts that belong to all three domains of life and are considered to be economically, medically and environmentally important pathogens. Recent studies have shown that these single stranded DNA viruses exist in great numbers in highly diverse habitats, ranging from extreme geothermal springs to the gut of humans and other animals. International Committee on Taxonomy of Viruses currently classified single stranded DNA viruses into 10 different taxa. However, several viruses that can be classified into additional groups have been isolated and many of their genomes were sequenced. All single stranded DNA viruses are pathogenic on eukaryotes, possess non-enveloped, icosahedral capsids, along with Microviridae family members, which infects bacteria. Single stranded DNA viruses pathogenic on other prokaryotes have filamentous (Inovirus), rod-shaped (Plectrovirus), coil-shaped (Spiraviridae), or pleomorphic (proposed family"Pleolipoviridae") virions (Table 1.2).

Single stranded DNA viruses are the group comprising of smallest viruses and their genomes are as small as 1–2 kb, encoding two proteins; one for capsid formation and the other for genome replication. Such irreducible simplicity of single stranded DNA viruses epitomizes their essence of being a virus and makes them an attractive model for investigating virus origins and evolution. Numerous metagenomic studies have revealed a high range of genetic diversity existing in single stranded DNA viruses in the environment, suggesting a highly dynamic interaction between these viruses and their respective hosts. Also, single stranded DNA viruses with the smallest genomes and simplest proteomes were found to be widespread in cellular chromosomes, providing new important insight into the evolution of these viral.

Table 1.2 Morphological diversity of single stranded DNA viruses

Host virus taxon	Virion morphology	Genome topology	Genome size
Microviridae	Icosahedral	Circular	4.4–6.1
Inoviridae			
Inovirus	Filamentous		5.8–12.4
Plectrovirus	Rod-shaped		4.5–8.2
Pleolipoviridae	Pleomorphic	Circular	7–10.6
Spiraviridae	Coil-shaped	Circular	24.9
Anelloviridae	Icosahedral	Circular	2–4
Bidnaviridae	Icosahedral	Linear, segmented, 6.5 per segment	
Circoviridae	Icosahedral	Circular	1.7–2.3
Geminiviridae	Icosahedral	Circular, segmented 3 per segment	
Nanoviridae	Icosahedral	Circular, segmented 0.98–1.1per segment	
Parvoviridae	Icosahedral	Linear	4–6.3

1.4.1 Genomes of Bacteriophages

Bacteriophages are the smallest viruses with simple genomes. Since their discovery in 1915 and 1917 by Fredrick Twort and Felix d'Herelle respectively, bacteriophages have been studied in many laboratories and are being used in a variety of practical applications. The Density of phage viruses present in the oceans is 10^6–10^7 particles per ml. It was estimated that the total population of the bacteriophages is 10^{31} particles and the ratio of environmental virus and bacteria are 5–10:1, after the validation of 10^{30} bacterial cells in the biosphere. Altogether, the prokaryotic population is highly dynamic, with an estimated number of ~10^{23} global infections per second. It has been hypothesized that oceanic bacteriophages infect bacterial cells at the rate of 10^{29} phage infections per day, which releases over 10^{11} kg of carbon from the biological pool per day. Over the past three decades, research on bacteriophages has revealed their abundance in nature, genome diversity, impact on the evolution of microbial diversity, their utilization in control of infectious diseases and their influence in regulating the microbial balance in the ecosystem has been explored, leading to a resurgence of interest in the phage research. Research on phages has played a pivotal role in the most significant discoveries, that were made in biological sciences right from the identification of DNA as the genetic material, in the elucidation of the genetic code, leading to the development of the molecular biology. Research on phages has continuously broken new grounds in our understanding of the basic molecular mechanisms of gene expression and their structure. In recent times, phage genomics has revealed novel biochemical mechanisms for replication, maintenance, and expression of the genetic material and is providing new insights into the origins of infectious diseases, utilization of phage gene products and even whole phage as an agent for the gene therapy.

In addition to the killing of bacterial cells, temperate phage genomes also carry toxins and other critical virulence factor genes that are important for many bacterial pathogens to infect human beings. Phages also contribute to the diversity of the bacterial community by serving as vectors for the transduction of different genetic alleles, such as antibiotic resistance genes, between bacterial cells. Phages also have great medical and nanotechnological potential. Strategies for using tailed phages for detecting bacteria, curing bacterial diseases through phage therapy or decontaminating surfaces have been implemented for almost 100 years in Russia and Georgia. These phages are currently being used to treat agricultural diseases as well as in the prevention of food contamination in western countries. Phage virions are being developed as nanocontainers for specific chemical cargoes that can be delivered to specific targets.

Small size and the simplicity of isolation have made bacteriophages as the primary choice for the complete genome sequencing. Phage φX174 is the first organism with the complete genome sequence of 5386 bases of single stranded DNA and λ phage genome is the first organism with double stranded DNA of 48,502 bp, followed by phage T7 genome of 39,936 bp. dsDNA tailed mycobacteriophage L5 is the first among non-*E. coli* phage genomes to be fully sequenced. Further, the sequencing of the bacteriophage genomes are propelled exponentially with two main objectives;

1. To understand the relationship between the phage genomes the evolutionary mechanisms that shaped these bacteriophage populations.
2. For increased utilization of bacteriophages in the development of tools, utilities, and techniques related to genetics and biotechnology.

Phage genomes display a considerable amount of variation in their size, varying from *Leuconostoc* phage L5 (2435 bp) to *Pseudomonas* phage 201 (316,674 b). Tailed phages with double stranded DNA genomes vary in their size from >10 kbp to <15 kbp, consistent with their overall virion structure and gene assembly, which encompass up to 15 kbp of the genome space. *Siphoviruses* of the genome size 1.5–6 kbp are characterized by a long flexible non-contractile tail with a tape measure protein gene, whose length corresponds to the phage tail length Many phages with the morphologies similar to *Siphoviruses* have genomes longer than 20 kbp. Contrastingly, *Myoviruses* with contractile tails are the phages with larger genomes of >125 kbp and the *Bacillus* phage SPBc2, is the largest Siphoviral genome of the length 134,416 bp. The main reason for the absence of large *Siphoviruses* is still unknown.

1.4.2 Phage Genome Sequence Diversity

Bacteriophages are estimated to be the most widely distributed biological entity of the biosphere. They are found in all habitats of the world, where bacteria proliferate. Most of the viral population is dominated by bacteriophages, with double stranded DNA tailed phages, or Caudovirales, accounting for 95% of all the phages, possibly making up the majority of phages on the planet. However, phages belonging to other groups also occur abundantly in the biosphere, such as phages with different virions, genomes, and lifestyles. Two key approaches were made for studying the viral diversity are metagenomics of total concentrated phage samples collected from the environment and a genome-by-genome strategy of individually isolated phages. These two approaches are compatible, having distinct outcomes. Metagenomics generates a large amount of sequence data, which provides a good insight into their diversity. Sequencing and analysis of individually isolated phages generate small data sets, which are structured into whole genomes. As phage genomes are architecturally mosaic, the availability of complete genomes contextualizes the complexities of their relationships. The nucleotide sequences of phage genomes with non-overlapping hosts rarely share sequence similarity, as noticed in the published genomes of four *Streptomyces* phages and available collection of 50 mycobacteriophage genomes. Phages infecting a common bacterial host are in genetic contact with each other, and they share common nucleotide sequences. Genomes of over 30 phages with common host have been isolated and sequenced from *Pseudomonas*, *Staphylococcus,* and *Mycobacterium* containing related sequences, with a few exceptions. Most of these phages share a very low or no sequence similarity, as illustrated by the nucleotide sequence comparisons of *mycobacteriophages* and *Pseudomonas phages*.

1.4.3 Genome Mosaicism of Phages

Phages were evolved not only by the accumulation of mutations but also through the recombination events, during which they exchanged genetic material with other phages. These events have been suggested to explain the mosaic structure of the phages, arisen by comparison of two or more phage genomes. During the comparison of the genomes, nearly identical sequences alternate with merely similar sequences or completely divergent sequences. Such type of exchanges in bacteriophages was obtained by heteroduplex mapping in the early 1990s. Since then, numerous mosaics have been identified by sequence comparison, and the mosaic structure of bacteriophages is now a well-documented phenomenon. This mosaicism is also found to be ubiquitous among bacteria, where the genes are acquired through horizontal genetic exchange mostly through transduction, transformation, and conjugation. But, the extent of mosaicism is highly remarkable in phage genomes as evidenced by the increasing number of genomes available for comparative genomics analysis.

The mechanism of genome mosaicism in bacteriophages can be understood at two levels; 1. by comparing nucleotide sequence through DNA heteroduplex mapping, 2. by comparing their DNA sequences. There are two models which explain the recombination mechanisms that are responsible for these patterns. Model 1 describes the role of short conserved boundary sequences that are located at gene junctions in targeting various exchange events that are catalyzed by homologous recombinations, by using the recombinases synthesized by either host-or phages. Model 2 attempts to explain that the homologous recombination events are not specifically targeted and occur randomly with the preference of a few short sequences so that most of the events results in non-functional genomic trash. Comparison of the predicted amino acid sequences encoding phage gene products is an alternative manifestation of mosaicism. This is an informative approach, since many phages including those that infect common hosts may not share any nucleotide sequence information. In that case, protein sequence data reveals genes that share much older ancestry.

1.4.4 Genomes of Enterobacteria Phage M13 and λ Phages

M13 Enterobacteria phage infects *E. coli*. The genome of M13 phage consists of 6.4 kb single-stranded, (+) sense, circular DNA, which encodes for 10 genes. Unlike most icosahedral virions, the capsid of M13 phage is filamentous, which can be expanded by the addition of further protein subunits. Hence, the genome size can also be increased by the addition of extra sequences in the nonessential intergenic region without becoming incapable of being packaged into the capsid (Fig. 1.3).

In λ phage, the packaging constraints are much more rigid with DNA of ~46–54 kbp of the normal genome size can be packaged into the virus capsid and the substrate packaged into the phage heads during assembly consists of long concatemers of phage DNA that are produced during the later stages of

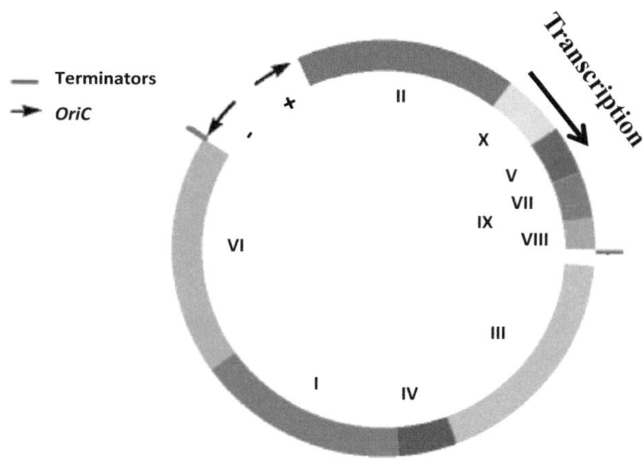

Fig. 1.3 Genome map of M13 phage

vegetative replication. The DNA is apparently reeled into the phage head and after the incorporation of the complete genome, DNA is cleaved at a specific sequence by a phage-coded endonuclease, leaving a 12-bp 5¢ overhang on the end of each of the cleaved strands, known as the *cos* site. Hydrogen bond formation between these 'sticky ends' can result in the formation of a circular molecule (Fig. 1.4).

In a newly infected cell, the gaps on either side of the *cos* site are closed by DNA ligase, and resulting circular DNA undergoes vegetative replication and integration into the bacterial chromosome.

1.4.5 The Genome of T4 Phage

Bacteriophages T2 and T4 are the model organisms playing an instrumental role in the development of modern genetics and molecular biology since the 1940s. They were involved in the development of many salient concepts related to biological sciences, including the recognition of nucleic acids as genetic material, identification of a gene through structural, mutational, recombinational, and functional analyses, in the demonstration triplet genetic code, in the identification of mRNA and establishing the importance of recombination in the replication of DNA, in the light-dependent and light-independent DNA repair mechanisms, restriction and modification of DNA, self-splicing introns in prokaryotes, etc. The main advantage of using T4 phage as a model system is its capability of totally inhibiting its host's gene expression, permitting the investigators to identify the differences between host specific and phage specific macromolecular syntheses. Analysis of the T4 capsid assembly and functioning of its nucleotide-synthesizing complex, replisome, and recombination complexes has led to important insights into

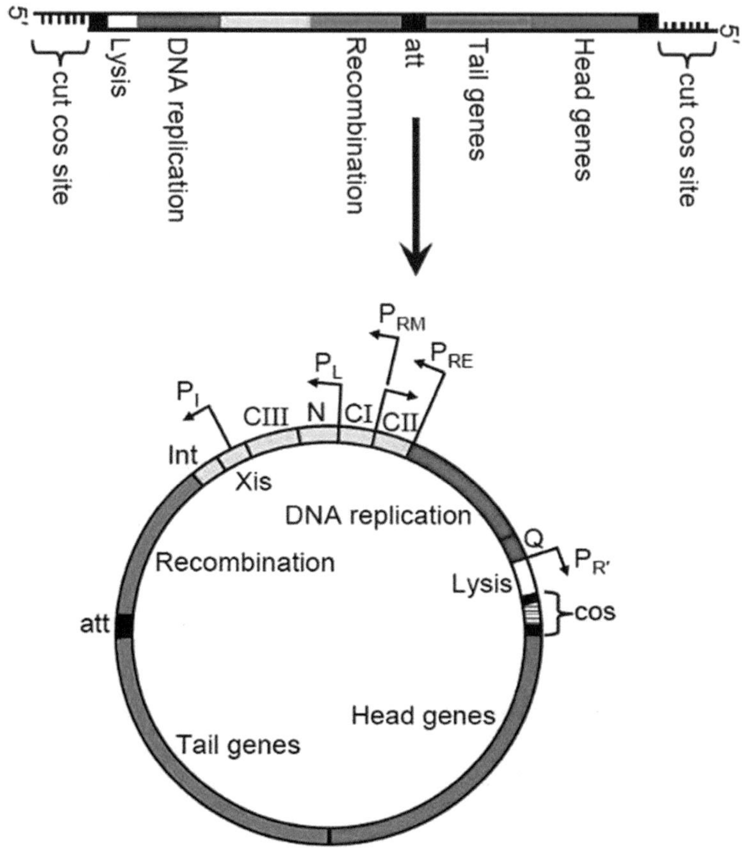

Fig. 1.4 Genetic map of λ phage

macromolecular interactions, substrate channeling, and co-operation between phage and host proteins within such complexes.

The genome of T4 phage is considered as the best avenue for understanding and evaluating the complete genome of a well organized biological system. On the basis of all available information, T4 phage genome comprises of ~300 probable genes, packed into a 168,903 bp genome. This genome comprises 289 expressing genes, 8 tRNA genes, and a minimum of 2 genes that encodes small, stable RNAs with unknown function. Genes 16, 17, and 49 contains multiple coding regions that encode more than one protein. T4 phage genome is four times higher than that of Herpesviruses and yeast, two times higher than *E. coli*. A very small number of genes contains non-coding regions of ~9 kb, accounting for 5.3% of the genome. Regulatory regions in this phage genome are compact, occasionally with overlapping coding regions. Another significant feature of this genome is the overlap of one gene's termination codon with the start codon of the next one. T4 phage has several groups of nested genes. It was found that only 62 genes in this organism are absolutely essential under standard laboratory conditions (rich medium, aeration, 30 to 37 °C). Mutants generated by altering a few other genes produced very small plaques under similar standard laboratory conditions. Many of the 62 essential genes are larger than an average T4 gene, occupying half of the genome. Essential genes encode proteins of the replisome and nucleotide precursor complex, transcriptional regulatory factors, and proteins involved in the structure and assembly of the phage particle. The genome of T4 phage illustrates another rare molecular feature of certain linear viral genomes, terminal redundancy. Replication this phage genome produces long concatemers of DNA, which are cleaved by a specific endonuclease, gets incorporated into the particle with the length exceeding its complete genome due to the repetition of some genes at each end of the genome. Resulting T4 phage genome containing reiterated information is packed into the phage head.

Three T4 phage genes that encode for thymidylate synthase (td), subunit of the aerobic ribonucleotide reductase (nrdB) and the anaerobic ribonucleotide reductase (nrdD) are found to contain introns that are later spliced out of these transcripts. A possibility of an unusual relationship between the nucleic acid sequence and protein sequence occurring through translational bypassing is demonstrated in gene 60 of the T4 phage genome. A 50 bp mRNA segment in the coding region of this gene is not translated by the regular mechanism. This mRNA segment is the only known and unique high-efficiency translational bypass site in the entire T4 phage genome.

DNA in the genome of this phage contains only 34.5% GC, compared with its host genome of *E. coli* consisting of 50% GC. In the genome, 18 of the known or predicted genes containing less than 60% AT and 4 predicted genes have less than 58%. Capsid proteins, which are the most widely conserved among the T4-related phages have the lowest AT contents. Gene 23, which encodes for the major head protein, has the lowest AT content of 55%. A substantial decrease in the pairing of G against C in the coding strands of translated regions has been identified. 4 genes having more than 20% C in the coding strand, while more than 130 genes have more than 20% G and 37 genes have more than 22% G. A and T are equally divided between the coding strands. However, some AT bias has been identified in the T4 phage genome, which is stronger in the third position of codons, as expected in genomes with a high amount of AT-rich regions.

1.5 Positive and Negative Stranded RNA Viral Genomes

The ultimate size of the RNA viral genomes is affected by the fragility of RNA and the tendency of their long strands to break. In addition, RNA genomes tend to have higher mutation rates than those composed of DNA because they are copied less accurately. This tendency might have tended to drive RNA viruses towards smaller genomes. Genomes of RNA viruses encode for a limited number of proteins. RNA viral genomes are broadly divided into double stranded RNA, positive and negative strand single stranded RNAs, monopartite and multipartite RNA viruses. One of the primary proteins encoded by all these RNA genomes is RNA dependent RNA polymerase, essential for their replication. A major difference between + and − strand ssRNA viruses is that the RNA polymerase can be immediately translated from ss(+) RNA, whereas it is contained in ss(−) RNA. In monopartite ssRNA viruses, the genome encodes a single polyprotein, which is further processed into a number of small molecules, critical for the completion of the viral life cycle. In multipartite ssRNA genomes, each segment encodes for a single gene.

1.5.1 Positive Stranded RNA Viral Genomes

Among the viruses possessing RNA genomes, single stranded plus sense (+) RNA genomes represent an important subgroup including many pathogenic plant, animal and human viruses. These genomes contain cis-acting RNA elements that direct different viral processes, such as protein translation, genome replication, and subgenomic mRNA transcription mRNAs. Single stranded RNA genomes vary in size from coronaviruses (~30 kb) to those of phages such as MS2 and Qb(~3.5 kb). Although members of distinct families, most (+) sense RNA viruses share common features in terms of their genomes. Importantly, purified (+) sense virus RNA is capable of infecting the host cells in the absence of any viral proteins.

1.5.1.1 Picornavirus Genome

Picornaviruses are the etiologic agents of numerous diseases with medical and veterinary importance such as Poliomyelitis, common cold, flu, hepatitis, foot-and-mouth disease all are caused by picornaviruses. These viruses have a single-stranded RNA genome of positive polarity in the size of 7200 nucleotides in human rhinoviruses to 8500 nucleotides in foot and mouth disease virus, containing a number of features conserved in all picornaviruses. There is a long 5′ untranslated region (UTR) of 600–1200 nucleotides, which is important for translation, virulence, possibly encapsidation as well as a shorter 3′ UTR of 50–100 nucleotides, necessary for the (−) strand synthesis during replication. 5′ UTR contains a 'clover-leaf' secondary structure known as the internal ribosomal entry site (IRES). The rest of the genome encodes a single polyprotein between 2100 and 2400 amino acids. Both ends of the genome are modified, with the 5′ end by a covalently attached small, basic VPg protein of the 23 amino acids length and the 3′ end by poly(rA) tail. Genome replication occurs in a process that uses the (+) strand as a template for the (−) strand synthesis, which, in turn, is used as a template for the production of an excess of (+) strands. Initiation of both (+) and (−) strand RNA synthesis is thought to be primed by a uridylylated form of VPg, VPg-pUpU.

1.5.1.2 Toga Virus Genome

Togaviridae consists of two genera, alphaviruses, and rubiviruses. Alphavirus genus has 27 members, many of which can be transmitted *via* insect vectors. Rubella virus is the only known member of the rubivirus. The genome of togavirus consists of a single stranded, non-segmented, + sense RNA of ~11.7 kb. Capsids of these viruses

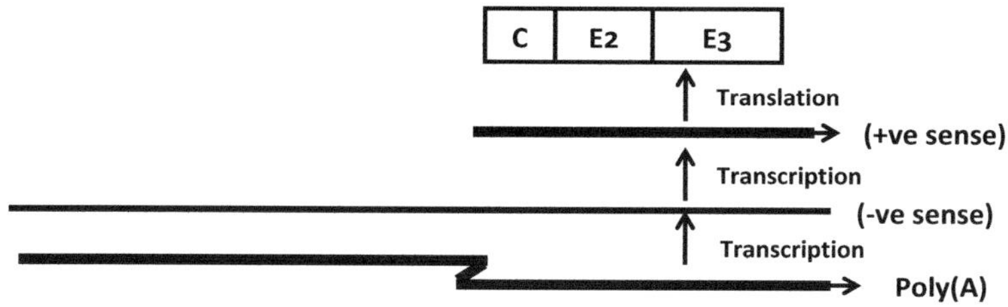

Fig. 1.5 Translation of togavirus subgenomic RNA

are composed of 240 copies of a single capsid protein of ~264 amino acids. The envelope contains 2 virus-encoded glycoproteins, E1 and E2. The genome resembles an mRNA with a cap at the 5′ end and polyadenylation at the 3′ end. The 5′ of the genome encodes the nonstructural proteins required for transcription and replication and the 3′ encodes the structural proteins (capsid, E1, and E2). Translation of the non-structural proteins will be done from genomic RNA, resulting in the synthesis of a polyprotein, which is cleaved into the matured proteins. A subgenomic mRNA is derived from the 3′ end of the genome synthesized using an internal promoter on the (−) strand. Translation of the subgenomic mRNA gives three structural proteins, capsid, E1, and E2 respectively (Fig. 1.5). The capsid protein is synthesized on free cytoplasmic ribosomes. It is cleaved off co-translationally, exposing a signal sequence which directs the ribosome to the ER membrane where the translation of the remaining E1 and E2 proteins gets completed. E1 and E2 are synthesized in association with the rough ER membrane and are then processed through the Golgi apparatus before being transported to the plasma membrane.

1.5.1.3 Flavi Virus Genome

The family *Flaviviridae* consisting of three genera, (1) Flaviviruses (yellow fever virus), (2) Pestiviruses (Bovine viral diarrhea virus) and (3) Hepatitis C Viruses. Many of the Flaviviruses are transmitted *via* insects. These viruses are enveloped and have a + strand genome of 10.5 kb capped at the 5′ and non-polyadenylated 3′ end. 5′ end of the genome encodes for structural proteins and 3′ end encodes for non-structural proteins. The entire viral genome is translated as a single polyprotein, which is cleaved into the mature proteins. No subgenomic RNA is formed.

1.5.1.4 The genome of Coronavirus

The family *Coronaviridae* is comprised of 2 *genera*, coronaviruses, and toroviruses. Inclusion of arteriviruses into this family is recent and is not still widely accepted. Coronaviruses are enveloped with a large 27–30 kb + strand RNA genome, which is capped at the 5′ end and is polyadenylated at the 3′ end. Approximately the first 60% of the genome from the 5′ end consists of two overlapping open reading frames (ORF1a and ORF1b) that encode the viral RNA-dependent RNA polymerase, proteases, and other non-structural proteins, along with a 60–80 nucleotide-long "leader" RNA, followed by 200–500 untranslated nucleotides. ORF 1b is translated *via* ribosome frame-shifting. The replication or gene expression in these viruses is done by the translation of ORF 1a or 1b to make the viral polymerase and other non-structural proteins, the transcription of (−) strand RNA using the viral polymerase and the synthesis of both full-length viral RNA and subgenomic mRNAs, using (−) strand RNA as template.

All subgenomic mRNAs contain similar leader sequence, derived from the 5′ end of the genome, followed by different regions of the genome, forming a nested set of transcripts, each of them is polyadenylated at the same site. Each of these transcripts is monocistronic and contains a single translation unit, which starts at the first AUG after the leader. The mechanism of the nested transcript synthesis is not completely clear, but it is not *via* RNA splicing since it occurs in the nucleus and coronaviruses replicate in the cytoplasm. Translation of the subgenomic mRNAs yields structural proteins along with some additional non-structural proteins. The new full-length + strands can either be translated or packaged into new virions. These viruses have a unique crown-like structure as identified in the electron microscope. The envelopes contain two viral glycoproteins; M protein (membrane protein), which binds the viral nucleocapsid to the viral envelope during budding and S (spike protein), which facilitates receptor binding and cell fusion. Entire (+) strand is coated with a nucleocapsid protein, directed to intracellular membranes bearing the M protein. The new virions bud through these membranes and are transported to the cell surface through the Golgi smooth-walled vesicles, which then fuse with the plasma membrane, releasing the virus from the cell, without lysis.

1.5.2 Negative Stranded RNA Viral Genomes

The life cycle of negative-strand RNA viruses differs from that of the other RNA viruses in many ways. Specifically, the genome of (−) RNA viruses is not infectious, and infectious virus particles must also deliver their own RNA-dependent RNA polymerase into the infected cell to start the first round of virus-specific mRNA synthesis. These viral genomes are more diverse than the (+) stranded viruses, possibly because of the difficulties in expression. These organisms tend to have large genomes encoding more genetic information. Because of this, most of these viral genomes are segmented. None of these genomes are infectious in its purified RNA. Although the gene encoding for RNA-dependent RNA polymerase has been found in some eukaryotic cells, most of the uninfected cells do not contain enough RNA-dependent RNA polymerase activity to support virus replication. As the (−) stranded RNA genome cannot be converted into mRNA without the activity of viral polymerase packed in each particle, these genomes are effectively inert.

The non-segmented (−) stranded RNA viruses belongs to the order Mononegavirales having linear, single stranded (−) sense RNA as their genetic material. This order includes four families: Rhabdoviridae, Paramyxoviridae, Filoviridae and the Bornaviridae, comprising a wide variety of human, animal and plant pathogens such as rabies virus, measles virus, canine distemper virus, Rinderpest virus and human, bovine respiratory syncytial viruses as well as the lethal Ebola and Marburg viruses and the recently described Nipah and Hendra virus. These four virus families hold one common factor, that the genetic information in their (−) sense RNA genomes is expressed *via* transcription of a series of discrete monocistronic mRNAs. Transcription originates from a single polymerase entry site near the 3′ end of the genome and is obligatorily sequential, but attenuation occurs specifically at each gene junction, resulting in a progressive reduction in the transcription of genes that are located further from the promoter. This simple method of regulation supports the order of core genes in the Mononegavirales, which are highly conserved among these four families and the genes whose products are required in large amounts are located proximally to the promoter, whereas those needed in catalytic amounts are distally located. The complete genome sequences of almost all genera belonging to all four families of the Mononegavirales have been determined, which ranges in size from the ~8.9 kb of Borna disease virus to ~18.9 Kb of Ebola virus genome, which is twice the size and contains 5–10 genes. The genomes of all four virus families contain 4 core genes encoding for a nucleocapsid protein gene (N), phosphoprotein gene (P), matrix protein gene (M) and an RNA-dependent RNA polymerase gene (L). There is a single additional G gene in Rabies virus encoding for the transmembrane attachment and glycoprotein entry. In a few members of this order, three additional genes encoding for transmembrane glycoproteins were also identified. Additionally, a small hydrophobic gene encoding for a protein of unknown function was found in the genus Rubulavirus. Many members of this order encode nonstructural proteins, which are encoded either individually as separate genes or by multiple ORFs within a single gene. The pneumoviral genomes have separate genes encoding for two non-structural proteins involved in evading the host response. These genes are not found in the genomes of avian pneumoviruses. Some rhabdoviruses have a gene between the G and L genes encoding a small nonstructural protein. In the Paramyxovirinae the P gene coding capacity is extended to give rise to a surprising number of polypeptides by utilizing multiple overlapping open reading frames on a single transcript or by co-transcriptional editing. In Pneumovirinae, an M2 gene encoding for an additional transcription factor M2–1 protein has been reported with a second overlapping ORF encoding the M2–2 protein. Some of the viruses are not strictly (−) sense but are ambisense, ie., they are partly (−) sense and partly (+) sense. Ambisense coding strategies occur in both plant viruses (Tospovirusgenus of the bunyaviruses) and animal viruses (the Phlebovirusgenus of bunyaviruses and arenaviruses).

1.5.2.1 Genome of Bunyavirus

Viruses in the family Bunyaviridae possess three distinct linear, single-stranded, negative or ambisense RNA segments in their genome named as small (S), medium (M) and large (L). RNA of the S segment is 0.9 kb, codes for the nucleocapsid protein and a non-structural protein NSs, which interferes with innate immunity. M segment RNA is 5.7 kb, encodes for a polyprotein, eventually giving rise to two glycol proteins Gn and Gc. The large segment (L) RNA is 8.5 kb, codes for the transcriptase, replicase p, and the large RNA-dependent RNA polymerase proteins.

In addition to polymerase activity, Bunyaviral L proteins have an endonuclease activity which cleaves cellular messenger RNAs for the production of capped primers used to initiate transcription of viral messenger RNAs which is known as cap snatching. Exceptionally, in the members of the Phlebovirus and Tospovirus genera, S segment is rather larger than that of M and L. All the three segments of the genome have the same basic structure with the coding region flanked by untranslated regions (UTRs) at the 5′ and 3′ ends. In common with all (−) sense RNAs, the 5′ ends are not capped and the 3′ ends are not polyadenylated. Even though bunyaviruses are categorized into (−) strand viruses, some of the members have genome segments with an ambisense coding strategy, such as Phlebovirus and Tospovirus with 5′ end of each segment is (+) sense, but the 3′ end is (−) sense.

Bunyaviral genomes are remarkably flexible. UTRs of the three segments can be exchanged or can be drastically shortened, the ORFs within an ambisense segment can be swapped around, the genomes can be lengthened through insertions of epitope tags and additional ORFs and the tripartite genome can even be converted into two-segmented or four-segmented versions. The promoters for replication and transcription of the genome segments are located in the terminal sequences of the UTRs that are largely complementary and form panhandle like structures.

1.5.2.2 Genome of Rhabdovirus

Rhabdovirus is a member of a special class of viruses with linear single-strand (−) RNA genome, which is completely non-infectious and is complementary to functional, virus-specific, (+) sense messenger RNAs (mRNAs). These viruses are bullet-shaped, ubiquitous in nature with a uniquely broad and highly diversified range of host system comprising of vertebrates, invertebrates, and plants. Two most popular and frequently studied rhabdoviruses are the animal pathogenic *Vesicular Stomatitis Virus* (VSV) and the human pathogen *Rabies Virus*. Their genomes are non-segmented and are up to 11 kb. There are 60 nucleotides UTR at the 5′ end and a leader region of approximately 50 nucleotides at the 3′ end of the genome. It contains five genes nucleoprotein (N), phosphoprotein (P), matrix protein (M), glycoprotein (G) and polymerase (L). Each gene is terminated with a conserved polyadenylation signal with short intergenic regions between these five genes. There are two structural components in Rhabdoviral genomes: a helical ribonucleoprotein core (RNP) and a surrounding envelope. In the ribonucleoprotein, genomic RNA is tightly encased by the nucleoprotein. Two viral proteins, the phosphoprotein and the large protein (L-protein) are associated with the ribonucleoprotein. The glycoprotein forms approximately 400 trimeric spikes which are tightly arranged on the viral surface. Matrix protein is associated both with the envelope and the ribonucleoprotein and may function as the central protein during the rhabdovirus assembly.

Rabies virus, a member of the genus Lyssavirus in the family Rhabdoviridae, is a neurotropic virus that causes fatal encephalitis in warm-blooded animals. This virus has a non-segmented, single-stranded, negative-sense RNA genome that is approximately 12 kb, comprising of the same five genes nucleoprotein (N), phosphoprotein (P), matrix protein (M), glycoprotein (G) and polymerase (L). encoding for five proteins (Fig. 1.6). Gene encoding for the G protein is known to play a predominant

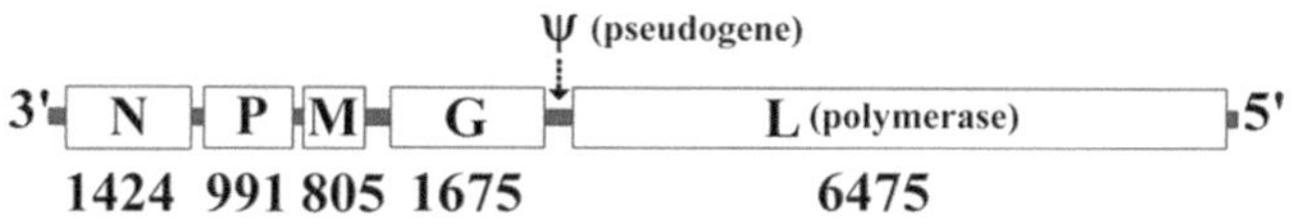

Fig. 1.6 Rabies virus genome

role in the pathogenesis of rabies virus. The genome of this virus is tightly encapsidated by a viral nucleoprotein. RNA-N complex will act as the template for the process of transcription and replication performed by the viral specific RNA-dependent RNA polymerase and its cofactor, phosphoprotein. The domestic dog is a primary reservoir and vector of rabies transmission along with other animal species, such as cat, ferret—Badger, fox, pig, cattle, donkey. An outbreak of pig rabies emerged in a rural pig farm in Sihui of southern China's Guangdong Province in March 2011, resulted in the death of 14 pigs. A virulent wild RABV strain, GD-SH-01, was isolated from the brain tissue of a rabid pig, and its complete genomic nucleotide sequence has been determined recently.

1.6 Segmentation in Viral Genomes

Segmented virus genomes are those that are divided into two or more physically separate molecules of nucleic acid, all of which are then packaged into a single virus particle. There are three viral families Arenaviridae, Bunyaviridae and Orthomyxoviridae containing a (−) sense RNA genome and does not exist as a continuous molecule, but is divided into several segments. These segmented genomes require the presence of RNA-dependent RNA polymerase enzyme to perform the synthesis and replication of mRNA, which reaches the cell along with other viral components during infection.

Segmentation of the viral genome consists of a wide number of advantages and disadvantages. There is an upper limit to the size of a non-segmented virus genome due to the physical properties of nucleic acids, specifically the shearing tendency of long molecules and the maximum length of the nucleic acid chain of each particular virus that can be packed into the capsid. The problem of shearing is particularly relevant for single-stranded RNA, which is much more fragile than the double-stranded DNA. The Physical breakage of the genome results in its biological inactivation, as it cannot be completely transcribed, translated even or replicated. With genome segmentation, there is a risk of shearing by an increase in the total potential coding capacity of the entire genome. Segmentation of the genome enables the virus to generate reassortants. During this process, RNA molecules of different viral strains get reshuffled in doubly infected cells during replication and morphogenesis. As a result, progeny viruses can obtain new combinations of RNA segments and thus gain novel properties. This mechanism, referred to as an antigenic shift, is more common and well-studied in influenza-A viruses. The main disadvantage of the segmentation strategy is that all the individual genome segments must be packed into each virus particle or the virus will be defective as a result of the loss of genetic information. In general, it is not understood how this control of packaging is achieved.

1.6.1 Influenza Virus

Influenza is the most commonly circulated viruses which cause flu in humans by infecting their respiratory tract. Compared with other respiratory infections such as common cold, flu often causes severe illness. World health organization in February 2014 has divided the influenza viruses into three sections; seasonal influenza, pandemic influenza and zoonotic or variant influenza. Seasonal influenza viruses, categorized into A, B and C type infect humans and cause flu every year, especially in the winter months, spreading from person to person through sneezing, coughing, or touching the contaminated surfaces. These seasonal viruses are capable of mild to severe infection, which might even lead to death, particularly in some high-risk individuals. Type-A influenza viruses are further divided into subtypes according to the specific variety and combinations of two proteins that occur on the surface of the virus, the hemagglutinin (H protein) and the neuraminidase (N protein). Currently, influenza A(H1N1) and A(H3N2) are the prominent seasonal influenza A virus subtypes. There are two type B viruses, which are named as Victoria lineage and Yamagata lineage after the areas where they were first identified. Type C influenza poses much less of a disease burden than influenza A and B causing mild infections, associated with sporadic cases and minor localized outbreaks.

Pandemic influenza occurs when this virus has not previously circulated among humans and to people not having proper immunity. These viruses may emerge, circulate and cause large outbreaks outside of the normal influenza season. As the majority of the population has no immunity to these viruses, a large proportion of a population is getting infected. In 2009, A(H1N1) virus which was not reported before, emerged and spread across the world causing the 2009 H1N1 pandemic. Since then, this virus is established in human populations as a seasonal influenza virus, as discussed above. Currently, there is no pandemic virus reported to be circulating in the world.

Zoonotic or variant influenza are the viruses that are circulating in animals, such as avian influenza virus subtypes A(H5N1) and A(H9N2) and swine influenza virus subtypes A(H1N1) and A(H3N2). Domestic animals such as horses and dogs also have their own influenza viral types. Even though these viruses may be named as the same subtype as viruses found in humans, all of these animal viruses are distinct from human influenza viruses and do not easily transmit to humans. Occasionally, some may infect humans, causing mild conjunctivitis to severe pneumonia and even leads to death. Usually, zoonotic influenza infections are acquired through direct contact with infected animals or contaminated environments, which do not spread rapidly among humans. If such a virus acquires the capacity to spread easily among people either through adaptation or acquisition of certain genes from human viruses, it would be considered an epidemic or pandemic, such as A(H3N2). When viruses of influenza A(H3N2) circulating in swine also began to infect people of the United States in 2011, they were isolated and labeled as variant (v) in order to distinguish them from human viruses of the same subtype and are named as swine flu, which refers to the influenza viruses infecting swine, and is never used when such viruses infect people. Other animal viruses such as A(H5N1), A(H7N7), A(H7N9), and A (H9N2) infecting humans are named as avian influenza viruses. When animal influenza viruses infect their natural animal host, they are named for that host, like avian influenza viruses, swine influenza viruses, equine influenza viruses, etc.

1.6.1.1 The Genome of Influenza Virus

The genomes of influenza-A and B virus are composed of eight segments and influenza-C is of seven. All segments are single stranded, (−) sense RNA. Proteins that are encoded by each segment of the genome were determined by their electrophoretic mobilities and mutant analysis. The size and the list of proteins encoded by each segment are listed in Table 1.3. All these eight segments have common nucleotide sequences at the 5′ and 3′ ends, necessary for the replication of the genome. These sequences are complementary to one another and the ends of the genome segments are held together by base-pairing to form a panhandle like structure that might be involved in the replication. The genome segments are not packed as naked nucleic acid but in association with the nucleoprotein which is visible in electron micrographs as helical structures, which shows some interaction between the genome segments and could be the reason for the ability of influenza virus particles to pack the genome segments within each viral particle with a low error rate. The exact mechanism of the interaction of genome sequences and proteins involved in this subtle mechanism is yet completely elucidated.

Influenza A virus is responsible for acute and highly contagious viral infection in the respiratory tracts of humans. Infection of this seasonal virus has been reported to cause approximately five million cases of severe illness and 250,000 to 300,000 deaths according to the World Health Organization (WHO). Reassortant influenza A strain arises when influenza virus variants are coincident in an animal population such as avian or swine reservoirs and genes from one variant become incorporated into the other. Zoonotic transmission of avian or swine influenza directly to humans as well as transmission of the reassortant virus has caused notable human pandemics over the last decade. The genome of the influenza A virus (family Orthomyxoviridae) consists of eight single stranded (−) sense RNA molecules spanning approximately 13.5 kilobases (kb). The length of the segments ranges from 890 to 2341 nucleotides and encodes up to 11 proteins (Table 1.3). Haemagglutinin and Neuraminidase genes encode the viral envelope proteins responsible for viral attachment and spreading the infection from cell to cell. The other six viral genes, PB2, PB1, PA, NS, M, and NP encode viral internal proteins required for viral replication and assembly. Among the Influenza A viral proteins, the replicative complex such as polymerases PB2, PB1, and PA together with nucleoprotein play a crucial role in viral infectivity in a different range of host environments. NS gene is found to be important in regulating the host cell response with an additional function of optimizing the viral replication inside the host. The

Table 1.3 Size of influenza virus segments and the proteins they encode

RNA segment	No. of nucleotides	Encoding protein	No. of amino acids
1	2341	Polymerase PB2	759
2	2341	Polymerase PB1	757
3	2233	Polymerase PA	716
4	1778	Haemagglutinin HA	566
5	1565	Nucleoprotein NP	498
6	1413	Neuraminidase NA	454
7	1027	Matrix protein M1	252
		Matrix protein M2	97
8	890	Non-structural protein NS1	230
		Non-structural protein NS2	121

C-terminus of the NS protein has been shown to play a pivotal role in the inhibition of interferon expression.

1.6.1.2 Reassortment of Influenza Viral Genomes

Viral reassortment is a type of genetic recombination, very specific to the segmented RNA viruses. In this process, co-infection of multiple viruses into the same host cell might leads to the shuffling of gene segments to generate a progeny of viruses with novel genome combinations. Reassortment is found to occur in all viral families whose genomes are segmented. But, this reassortment is most prominently studied in influenza viruses as a primary mechanism for interspecies transmission and the emergence of pandemic viral strains. The emergence of new influenza virus genes in humans and their subsequent establishment to cause pandemics have been consistently linked with the reassortment of novel and previously circulating viruses. Reassortment of the segmented viral genomes takes place by the entry of multiple virus particles into a single host cell, followed by the concomitant production of genome segments. Models related to two alternative mechanisms have been proposed for the reassortment.

1. Random packaging model posits the incorporation of the viral RNA into the virions without any discrimination, permitting the likelihood of forming viable reassortants with an entire genome set occurs by chance.
2. Selective packaging model states the packaging of a virus particle into eight unique viral RNA segments through specific packaging signals.

Experimental evaluation of RNA interactions during the assembly of viral segments has revealed an epistatic interaction of packaging signals among viral segments, which might play an imperative role in directing the reassortment. Through the experimental swapping of packaging signals between influenza viruses of different types, Essere et al. were able to overcome the bias observed towards specific genotypes. Baker et al. have shown that the swapping of packaging signals of two different influenza species has enabled the reassortment of viable particles that have not been naturally observed, indicating the central role of packaging signals in reassortment. Despite the lack of a mechanistic understanding of the function of packaging signals, these observational studies highlight some important implications for viral evolution through epistatic interaction between gene segments and the emergence of novel reassortants.

Influenza virus exhibits a very high level of mixed infections in its hosts, ranging up to 25% of all its infections in avian hosts involving multiple influenza subtypes. Large-scale genomic analysis has identified various levels of restrictions on random reassortment between co-circulating influenza viruses, which differ from the host, subtype and preferential genetic combinations. The highest frequency of influenza reassortment was observed in their natural reservoir, wild aquatic birds, where viruses of different subtypes will frequently exchange their gene segments leading to a tremendous genomic novelty. Reassortment leads to a significant increase in transient amino acid mutations, primarily on the surface glycoprotein hemagglutinin, leading to antigenic change. However, evolutionary change in the reassortment might also be due to the selection pressure induced by herd immunity. The emergence of drug-resistant mutations might have acquired the reassortment, as shown in the emergence of amantadine-resistant H3N2 viruses and oseltamivir-resistant seasonal H1N1 viruses, suggesting that reassortment con-founds the available methods of virus control.

1.7 Multipartite Viral Genomes

Genome segmentation is one of the most common traits in a broad variety of viruses, which is particularly striking in the case of multipartite viruses as their genome segments achieve complete independence at the apparent cost of reducing their infectivity. Multipartite viruses have their genomes fragmented up to eight segments; each of them packed into a separate capsid and contains at least one gene essential for the virus to complete its infection cycle. These viruses require complementation, in order to complete each genomic segment with the rest of segments for producing viral offsprings. The complementation requirements for multipartite viruses have a strong impact on the way they are transmitted. Multiple viral particles may have to enter into each cell so that at least one representative of each segment will be present. The multiplicity of infection (MOI) is a key component in the biology of multipartite viruses. Another significant feature of multipartite viruses is their asymmetry in host distribution. All multipartite viruses identified till date are found to infect only plants and none of these viruses infecting animals have been described. Reason for this asymmetry could be the complementation requirement of the segments, acts as a limiting factor leading to a bottleneck of viral extinction.

1.7.1 The Genome of Gemini Virus

Geminiviruses are characterized by the particles of size 20 × 30 nm, which are twinned, isometric containing circular, single stranded DNA genomes of approximately 2.7 kb. These viruses are classified into two major subgroups based on host plants, insect vector and number of DNA components. One subgroup contains viruses possessing a single DNA component, which infect monocotyledonous plants and are transmitted through leafhoppers. The second sub-group consists of viruses possessing a bipartite genome, which is designated as DNA A and B, infecting dicotyledonous plants and are transmitted by whiteflies. Beet curly top virus (BCTV) demonstrates the characteristics of both the groups. The genomes of Mastrevirus and Curtovirus genera consists of a single stranded (+) sense, DNA of 2.7 kb packaged into the virions. But both (+) and (−) sense strands are found in the infected cells containing protein-coding sequences. The genome of genus Begmovirusis is bipartite consists of two circular, single-stranded DNA molecules of 2.7 kb. Both the strands differ from each other completely in the DNA sequence, except for a common 200 nucleotides non-coding sequence involved in DNA replication. Each strand is packed into entirely separate capsids. The establishment of a productive infection requires both parts of the genome, it is necessary for a minimum of two viral particles bearing at least one copy each of the genome segments for infecting a new host cell. Both DNA strands of the virus found in infected cells contain coding information, present in overlapping open reading frames. Regulating the expression of such a high density of genetic information could be the possible reason for the multipartition of the genomes.

1.8 Evolution of Viral Genomes

Viruses are the most abundant biological entities on this planet. Virology studies show the occurrence of at least 10^{33} viruses on our planet, which is roughly 10 times higher than the bacterial number. They are found everywhere including the oceans, soil, abundantly in plants and in the human body. A healthy human being consists of 10^{13} cells, which harbors 10^{14}–10^{18} bacteria and an unknown number of viruses. These numbers pronounce that the genetic complexity of bacteria is 100 times greater than that of our own genome, making the bacteria as our second genome. With respect to the total genetic information of the human body, we are 99% bacteria, harboring more than 1.5 kg of 1500 different kinds of bacteria in our guts. Viruses form our third genome with over two hundred viruses have been identified in human gut samples based on similarities to known viruses. Archaea and fungi are also present in our guts, making humans a superorganism as well as a complicated ecosystem. Bacteriophages are viruses that can lyse bacteria, which gave them their name. It all started with RNA as the main building block of life on this planet, which is widely accepted today. It is unknown how the RNA has been formed, but it is hypothesized that hydrothermal vents with extreme temperatures of 400 °C to freezing temperatures at the bottom of the oceans might have lead to the synthesis of RNA with the catalytic support from the clay and energy supplied from the chemical reactions of the rocks composed of metal-rich granite helped life to evolve. RNA has evolved into catalytic oligonucleotides known as ribozymes. This catalytic RNA is capable of cleaving or joining the RNA molecules, which can replicate and mutate. Plant pathogens such as viroids reflect the properties of the early RNA world. Viroids are the ribozymes, which are widespread in plants proficient of damaging them. They look like remnants of a pre-protein world, consisting of non-protein-coding naked RNA without any protein coat. They are small with a few hundred nucleotides of single stranded RNA folded to form a hairpin-loop like structure, which might protect against environmental threats. It has been suggested that a viroid may have entered the human body and evolved into a human virus by acquiring genetic information for a protein from the host such as hepatitis delta virus (HDV), as HDV antigens are related to a human protein and HDV is the only known virus to be a catalytic ribozyme and pathogen in humans.

The next step in the viral evolution might be the plant viruses, such as tobacco mosaic virus (TMV). These viruses are extremely stable with rod-shaped structures. Many of the plant viruses do not synthesize their own replicase, instead use their host cellular RNA replicases, which are considered to be one of the first and oldest polymerases. Almost all plant viruses are small with immense genetic information. The limitation of a simple small RNA molecule to further increase in its length would have made it more unstable. This might have lead to the accumulation of several similar molecules in some kind of protective compartment, as reflected in the present day viruses with segmented genomes. RNA viruses developed their strategies to protect the ends or their RNA with some of them having tRNA like structures, which could fold back and bind to the target RNA for the transfer of individual amino acids. This might be the beginning of peptide synthesis.

The transition of DNA from RNA is a matter of speculation. DNA is much less multifunctional when compared with RNA but is having long-term memory and stabilizes the genetic information. In contrast, RNA is highly fragile and variable. The transition of RNA to DNA occurs in telomeres of chromosomes catalyzed by telomerase, which is a complex of reverse transcriptase, RNA-dependent DNA polymerase, and an endogenous RNA, consisting of sole repeats of a few ribonucleotides. Enzyme closely related to the telomerase is reverse transcriptase (RTase), which copies RNA into DNA, which is the hallmark of retroviruses. RTases are also been isolated from bacteria, which could have been the leftover from ancient retrovirus infections. Alternatively, RTase could also be a precursor of retroviruses. Sequences homologous to RTase have been identified in phage genomes, with the probable function of increasing the diversity of phages by allowing their transmission into the new hosts. Finally, it was the viruses, who invented the DNA, which preceded the formation of the three domains of life and the enzyme RTase appears to be an important evolutionary link between RNA and DNA.

1.9 Conclusions

Viral genomes are considered as one of the most rapidly evolving organisms in biology due to their short replication time and a large amount of the offspring's released from the infected host cell. Viral genomes play a significant role in our increased understanding of epidemiology, diagnosis, surveillance and their evolution. As the basic structure of the virus genomes are highly conserved, the determination of genomes in most of the viruses, understanding their pathogenicity and identification of new viruses is being conveniently done. These genome sequences have been deposited in the databases such as NCBI and Gene Bank and are being maintained. As the new viruses are being identified, the amount of the sequenced viral genomes is going to get increased enormously which might need the invention of next generation sequencing systems and accurate annotation and assemblies.

Archeal Genomics

2

Abstract

This chapter provides information about the unique features of archaea, shreds of evidence about their extraterrestrial life and archaeal viruses. Concepts related to the structure, shape, stiffening of the Archaeal genomes has been elaborated by including the ESCRT proteins, CRISPR repeats, MITES, etc. Information related to archaeal plasmids, horizontal gene transfer and integrase mediated insertion and deletions of archaeal DNA has also been discussed. The full details about the genomes of methane-producing archaeon *Methanococcus Jannaschii* and sulfate-reducing Archaea *archaeoglobus fulgidus* have been provided.

2.1 Introduction

Based on the ideas proposed and established by the eminent personalities like Lamarck, Darwin, Mendel and Huxley, the life forms have been categorized on their evolutionary origin. Traditionally, the differentiation of living things was done as plants and animals until the microscope was invented. During the middle of eighteenth Century, Haeckel has categorized the divisions of life, which were expanded into 5 kingdoms by the end of 1960, as Monera, Protista, and Fungi along with the Animalia and Plantae. Later, the two-empire system, comprising of prokaryotes (lacking a nucleus) and the eukaryotes (containing a membrane-bound nucleus), have gained maximum recognition as a swift model for classifying the life along with the 5 kingdom system. It was believed that these unicellular prokaryotic organisms lacking a true nucleus must be a less complex predecessor of the more complex eukaryotic cells. Improvement in the technology for the DNA and its sequencing has redefined the evolutionary divisions of life by providing a shift from phenotypic taxonomy to genotypic computationally aided phylogenetics. Earlier, both archaea and bacteria were combined as prokaryotes (ormonera), based on the similarities in their features. Sequencing of ribosomal RNA genes has eventually led to the separation and clustering of prokaryotes, ultimately giving rise to the main

K. V. Chaitanya, *Genome and Genomics*,
https://doi.org/10.1007/978-981-15-0702-1_2

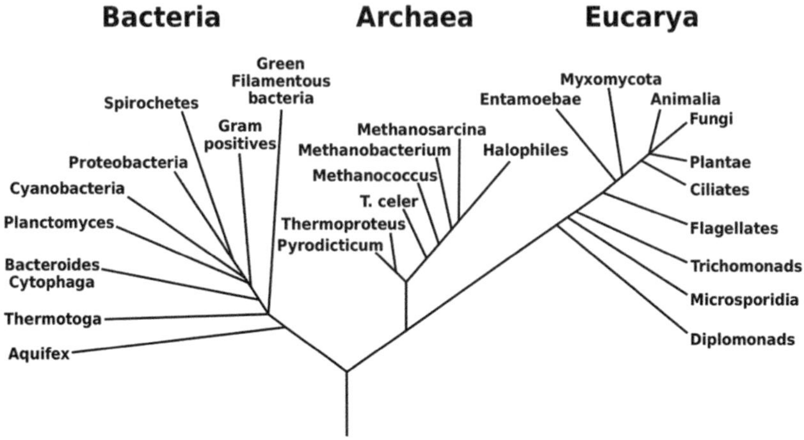

Fig. 2.1 Tree of life comprising of three distinct lineages; bacteria, archaea and eucarya

and top-level classification systems, defining 3 major domains of life, archaea, bacteria, and eukarya which are presently used (Fig. 2.1).

2.2 Archaea, the Third Main Domain of Life

Woese and Fox in 1977 proposed Archaea as a third main domain of life, based on ssrRNA sequence cataloging. There was a profound difference in the conserved primary sequence and in the secondary, tertiary structure of Archaea, Bacteria and Eukarya ssrRNA. This representation had a profound shift in the paradigm from the already existing dichotomy of eukaryotic and prokaryotic life forms. Sequences of 16S rRNA were used to design domain-specific oligonucleotides for their usage as probes and PCR primers for determining the presence of archaea in the environments. In addition to the ssrRNA sequences, a great deal of evidence for archaea was determined by Graham et al. (2000) by identifying over 350 genes (most of them with unknown function) specific to Archaea but lacking homologs either in bacteria or in eukaryotes. This unique bunch of genes with unknown function represents about 15% of the archaeal genomes and supports the belief that the archaea are an ancient lineage of major evolutionary importance. The salient information provided by Graham et al. was almost based on the sequence data of Euryarchaeota and the subsequent availability of more archaeal genomes has led to the identification of numerous signature insertion/deletions (indels) as well as signature proteins of Crenarchaeota and Thaumarchaeota, which have helped in further identification and classification of archaea. Further, the extremophilic properties have made many archaeal organisms a favorite starting point for the theories concerning with evolution of life in the hostile conditions of early Earth. Stetter (2006) has suggested that hyperthermophiles might have inhabited the early earth ~3.9 Gyr ago, when it was hot and anaerobic, which was consistent with the abilities of many archaea to thrive in extreme thermal environments, such as hydrothermal vents.

2.3 Unique Features of Archaea

2.3.1 Cell Envelopes and Cell Structure

Even though the cell wall composition of archaeal groups is highly diverse, the major striking feature of archaeal cell walls, which separates them from bacteria, was the lack of murein in their envelopes. Murein is ubiquitously found in the bacterial cell walls. A few methanogenic archaea were reported to possess an archaeal specific polymer pseudomurein. Another unique feature of archaeal cell walls is the cell wall component facing outside the cytoplasmic membrane is a two-dimensional array of protein or glycoprotein termed as S-layer. Even though the bacterial cell walls possess S-layers as their external envelope, they are separated from the cytoplasmic membrane by a layer of murein in Gram-Positive bacteria, murein along with an outer membrane in Gram Negative bacteria. The S-layer of *Halobacterium salinarum* was the first prokaryotic glycosylated protein to be shown. Another fundamental trait unique to the Archaea cell is the presence of ether-linked lipids in their cytoplasmic membrane. Archaeal cytoplasmic membranes are composed of diphytanylglycerol diethers which form a lipid bilayer. In the case of thermoacidophiles, membranes are composed of diphytanylglycerol tetraethers (phytanyl chains of 40 carbons), which span the cytoplasmic membrane in a very stable lipid monolayer. Phospholipids in the other two domains consist of linear fatty acids ester-linked to a glycerol backbone. Presence of unique lipids provides specific adaptation to extreme environments, with ether linkage is much more resistant than the ester linkage to hydrolysis upon exposure to the extremes of pH and high temperatures, commonly found in many archaeal habitats. Archaea are also characterized by unique features such as structure and composition of the ribosome, DNA-dependent RNA polymerase, unusual resistance to antibiotics and a variety of modifications to tRNAs. There are novel twists on lipoylation, histones lacking the N- and C-terminal extensions that are commonly found in eukaryotic histones, N-linked glycosylation system.

2.3.2 Unusual Appendages

Archaea have a variety of appendages extending from the cell surface with unique features. Among them grappling hook appendages called *hami* are found in large abundance on the surface of a euryarchaeon. Other unique appendages are the hollow tubes, called cannulae, which connect cells of the *hyperthermophile Pyrodictium abyssi*. Archaeal flagella are the appendages, which has resemblances with bacterial ones. They differ from their bacterial namesakes, with genetic and structural evidence, suggesting that archaeal flagella are related to bacterial type IV pili with a structure unlike that of any bacterial pilus.

2.3.3 Exosome

Both prokaryotic and eukaryotic cells harbor a highly conserved RNA processing and degrading protein complex called exosome, which plays a central role in the maturation of rRNA, small nucleolar RNA (snoRNA), small nuclear RNA (snRNA) and in the decay of mRNA. In Bacteria, an endoribonuclease RNase E forms the central component of a protein complex called the degradosome. In eukaryotes, exosome contains a central ring made up of six 30–50 exoribonuclease subunits. These subunits form two distinct paralogous groups as well as a number of associated protein factors such as RNA-binding proteins and RNA helicases. An archaeal exosome also consists of a central hexamer ring with six 30–50 exonuclease activity and a number of peripheral proteins similar to that of Eukaryotes. Archaeal exosomes contain two ribonuclease subunits Rrp41, which are homologous to eukaryotic Rrp41, Rrp46, Mtr3p, and Rrp42, homologous to eukaryotic Rrp42, Rrp43, and Rrp45p. The central ring is composed of three Rrp41–Rrp42 heterodimers. The activity of the complex resides within the active sites of the Rrp41 subunits, all three of which face the same side of the hexameric structure, whereas the Rrp42 subunits are inactive but contribute to the structuring of the Rrp41 active site.

2.3.4 Archaeal Virus Families

Viruses infecting Archaea have been fascinating virologists because of their unusual morphological, structural and unique genetic components. Viruses that specifically infect various archaeal species have been identified to possess unique structural characteristics. Examination of high temperature biomes, mesophilic and highly halophilic environments has led to the identification of a large variety of archaeal specific viruses that are unique both in their ultrastructure and genetic makeup of their genomes. Genome analysis of these viruses has revealed that 90% of their genes were lacking homologs. Many unusual morphotypes have been reported among archaeal viruses such as a bottle-shaped, fusiform, droplet, linear and spherical forms, leading to the classification of these viruses as novel families.

Unusual archaeal viruses are the large-tailed spindle viruses (LTSVs) and smaller spindle-shaped fuselloviridae archaeal viruses (SSVs) infecting only Archaeal species. Large-tailed spindle viruses consist of circular dsDNA genomes in a lemon-shaped or spindle-shaped virions that have tails protruding from one or both ends. Till date, five large spindle viruses; *Acidianus* two-tailed virus 1 (ATV1), *Sulfolobus tenchongensis* spindle shaped virus 1 and 2 (STSV1&2), *Sulfolobus monocaudavirus* 1 (SMV1), have been isolated from crenarchaeal hosts replicating in high-temperature hot springs around the world. The fifth one is Acidianus tailed spindle virus (ATSV) isolated from Crater Hills Alice Spring, at 80 °C, pH 2 in Yellowstone National Park (USA). The genome of ATSV is double stranded circular DNA of 71 kb, sharing 25% of its genes with STSV1 and 2. All members of SSV are small (60–90 nm) enveloped, spindle shaped possessing a double stranded circular DNA of size 14–17 kb as their genomes.

2.3.5 Archaea and Extra Terrestrial Life

As most of the archaea were found at the limits to life on this planet, they have often been proposed to resemble with the life found outside our planet. The nature of Earth-like organisms that could exist on other planets has varied, with methanogens often mentioned due to their adaptation to anaerobic niches with little or no organic carbon and especially with respect to the possible biogenic formation of the methane detected on Mars. Some experiments even suggest that terrestrial methanogens could survive under Mars-like conditions. It has recently been suggested that anaerobic methane-oxidizing archaea (ANME) may be able to use the methane on Mars.

2.4 Archaea and Eukaryotes

Ever since their recognition by Carl Woese and his co-workers in 1997, archaea have been prominently featured in hypotheses for the origin of eukaryotes, as eukaryotes and Archaea represented sister lineages in Woese's 'universal tree'. The evolutionary link between Archaea and eukaryotes was further reinforced through the studies made on the archaeal genomes, which revealed that most of their genes, their transcription machinery, and their genetic information-processing machinery are more similar to eukaryotes than that of bacteria. Early eukaryotic genomes were chimeric by nature, comprising genes of both archaeal and bacterial origin, in addition to the genes that are specific to eukaryotes. Yet, many of the bacterial genes could be traced back to the alphaproteobacterial progenitor of mitochondria, the nature of the lineage from which the eukaryotic host evolved remained obscure. This lineage might either descend from a common ancestor shared with archaea (following Woese's classical three-domains-of-life tree) or have emerged within the archaeal domain (so-called archaeal host or eocyte-like scenarios). Recent phylogenetic analyses of universal protein data sets have provided increasing support for models in which eukaryotes emerge as sister to or from within the archaeal 'TACK' superphylum, a clade originally comprising the archaeal phyla Thaumarchaeota, Aigarchaeota, Crenarchaeota, and Korarchaeota. In support of this relationship, comparative genomics analyses have revealed several eukaryotic signature proteins (ESPs) in TACK lineages, including distant archaeal homologs of actin and tubulin, archaeal cell division proteins related to the eukaryotic endosomal sorting complexes required for transport (ESCRT)-III complex and several information-processing proteins involved in transcription and translation. These findings suggest an archaeal ancestor of eukaryotes that might have been more complex than the archaeal lineages identified so far. Still, the absence of missing links in the prokaryote-to-eukaryote transition currently precludes detailed predictions about the nature and timing of events that have driven the process of eukaryogenesis.

2.5 Archaeal Genomes

Archaeal organisms are the only detectable life forms under the conditions that are simply unimaginable by any usual standards of biology such as in the vicinity of hydrothermal vents maintained above the boiling temperatures, saturated with hydrogen sulfide or in extreme salinity. In moderate environments, archaea coexist with bacteria and eukaryotes with increasing ecological importance. Early studies have revealed that archaea are highly unusual and are clearly distinct from bacteria particularly in the membrane structure with archaeal glycerolipids differing from bacterial and eukaryotic cells and archaea do not contain murein, which the predominant component of bacterial cell walls.

Major differences between archaea and bacteria can be found in the organization of their information-processing systems. The structures of ribosomes, chromatin and the presence of histones, sequence similarity between proteins involved in translation, transcription, replication and DNA repair pointing towards a closer relationship between archaea and eukaryotes than between either of these and bacteria. Polymerases, helicases and other key components involved in DNA replication are not homologous or orthologous in archaea and eukaryotes. Replication of double stranded DNA as the principal form of replication of the genetic material was invented twice, independently: once in bacteria and once in the ancestor of archaea and eukaryotes. Contrastingly, many of the archaeal metabolic pathways closely resemble bacteria rather than their eukaryotic counterparts. The above studies support the status of archaea as a distinct domain of life with specific connections to eukaryotes, emphasizing their unusual and unique nature of genomes.

Whole genome shotgun sequencing of *Methanococcus jannaschii* (recently renamed as *Methanocaldococcus jannaschii*) was the first archaeal genome sequenced. This genome was mysterious, producing only 38% of the known genes. Detailed computational analysis and the technology available at that time have made the general functional predictions for 70% of the unknown genes, showing the existence of a solid connection between the genomes of archaea and the other two life forms. Currently, 150 archaeal genomes were sequenced with 39 genomes belonging to Crenarchaeota, 105 belonging to the Euryarchaeota, 1 genome each belonging to Korarchaeota and Nanoarchaeota, 3 belonging to the Thaumarchaeota and 1 genome belonging to an unclassified Archaea, established by rRNA phylogenetic analysis.

Table 2.1 Size of archaeal genomes with G + C contents

S. No.	Name	Genome size (Da)	G + C content (%)
1	*Methanobacterium thermoautotropicum*	1.1×10^9	49.4
2	*Methanobrevibacter arboriphilus*	1.8×10^9	27
3	*Methanosarcina barkeri*	1.9×10^9	41
4	*Methanococcus voltae*	1.8×10^9	30
5	*M. Thermolithotrophicus*	1.1×10^9	35
6	*Halobacterium halobium*	2.3×10^9	65.5
7	*Halococcus morhuae*	2.3×10^9	64.5
8	*Thermoplasma acidophilum*	8.4×10^8	46

2.5.1 Structure and Organization of Archaeal Genomes

2.5.1.1 Size and G + C Contents

Genomes of archaea are structurally similar to that of Eubacteria ranging from 8.4×10^8 Da for *Thermoplasma acidophylum* to 2.3×10^9 Da for *Halobacterium halobium*. The G + C content of Archaea varies from 25.8% in *Methanosphaera stadtmaniae* to 68% in *Halobacterium sodomense* (Table 2.1).

2.5.1.2 Archaeal Genome Signature

An archaeal genome signature is defined as a set of genes that functions uniquely within a specific lineage. Genomic signatures describe particular taxa with intricate detail. They help in providing a better understanding of the structure and function of the genome, which can be compared across huge evolutionary distances. Genome signatures of all major lineages in aggregate will provide a fundamental understanding of the evolution of modern cell types. Archaeal members share many genes related to antibiotic resistance, characteristic modified nucleotides in tRNA, ether-linked isoprenoid lipids, proteinaceous cell walls, novel coenzymes and unique structures of the DNA-dependent RNA polymerase.

Prokaryotic genomes mainly tend to be optimized for compactness and the presence of long oligonucleotide repeats in their genomes would be evolutionarily unfavorable. Long oligonucleotide repeats over 25–30 bp have been identified in bacterial, archaeal and organellar genomes at a relatively high frequency, believed to participate in a variety of events relevant to prokaryotic genome plasticity such as amplification, deletion, inversion, translocation or transposition. In archaeal genomes, most of the repeats were found to be located in the non-coding and intergenic regions, with few of them are mobile. However, some repeats in the genome of *A. fulgidus* gets transcribed into sn mRNA and are presumed to play regulatory roles. It is likely that these repetitive elements are propagated by forces similar to those acting on other mobile elements such as insertion sequences and transposons.

Cox and Mirkin (1997) have shown that most of the enriched repeats have the potential of forming DNA secondary structures such as H-DNA, Z-DNA, cruciform, slipped structures and all of these repeats adopt a stem-loop conformation. This observation has shed some light on the evolution of large repeat sequences when Ogata and Miura (2000) found that long DNA sequences of more than 20 kb can be synthesized from a short DNA segments with palindromic or quasi-palindromic repetitive structure by hairpin elongation in the absence of a complementary DNA template in a few hyperthermophilic archaea, such as *Pyrococcus* spp. Genomic expansion by such methods along with homologous recombination and strand slippage mechanisms may be a primary feature of archaeal genomes, which are considered to be the primordial ancestors of higher life forms. Formation of such structures may lend greater resilience to the archaeal genomes under extreme conditions such as high temperature, salt, pH or pressure. Characteristics of forming secondary structure may also play a vital role in archaeal gene expression and regulation. Most of the oligonucleotide repeats arise due to tandem duplication, hyperploidization, strand slippage, transposition or double-strand break repair by insertion. A study on the long repeats of archaeal genomes has shown that the direct repeats are more common than inverted repeats in and interspersed repeats are mostly created as tandem repeats followed by the successive rounds of opposing processes such as recombination and deletion. These repetitive elements are interspersed throughout the genomes and may be under the same influences as mobile genetic elements. Non-autonomous MITE (Miniature Inverted Repeat Element)-like mobile elements that are propagated by transposases was identified in archaeal genomes contributing to the archaeal genome evolution by genomic rearrangements.

2.5.1.3 Shaping and Stiffening of Archaeal Genomic DNA

The genome present in the nucleus of the organisms belonging to three major domains of life is very long, compared with the respective sizes of the cells or the cellular compartments to which genomes are confined. Hence, it is important to keep the genomic DNA tightly compact and fit. In order to facilitate the compaction, cells utilize certain physical mechanisms such as molecular crowding and supercoiling, synthesizing architectural proteins that shape the structure of the genome. Even though no homology exists between the architectural proteins in the organisms from three different forms of life, their function in shaping the genome is highly conserved. Architectural proteins HU and H-NS, shape the bacterial genomes by inducing bends into DNA or by promoting the formation of a loop along the DNA into a structure, referred to as nucleoid. Eukaryotic genomes are organized by wrapping the DNA by histone proteins, yielding a nucleosomal fiber. Archaeal cells also lack a discrete nucleus, but their genome is organized into a nucleoid by applying thee strategies of both bacteria and eukaryotes. In the two main branches of archaea, Euryarchaea encodes a legitimate homolog of eukaryotic histone protein. Four or eight copies of the histone fold domain oligomerizes to form higher-order structures, such as the tetrasome and octamer, which act as the principal protein units of chromatin in Euryarchaea. Crenarchaea lack histones but encodes architectural proteins and mechanisms as observed in bacteria.

Histone like DNA associated proteins were isolated and characterized in archaea. Among them, HTa protein from *T. acidophilum* is very well characterized. This histone protein binds to DNA at a ratio of 4 protein molecules per 40 bp and condenses linear double stranded DNA into compact globular structures of 5–6 nm in diameter. HTa protein also prevents the diffusion of ssDNA strands during short exposures of *T. acidophilum* to high denaturing temperatures, facilitating the rapid renaturation of the genome upon returning to the optimum temperatures.

ALBA (Acetylation Lowers Binding Affinity) is one of the most abundant, highly conserved, non-specific, double-stranded DNA-binding proteins that are found throughout the archaeal domain, involved in the Archaeal chromatin organization. ALBA has been characterized in *Sulfolobus*, shown to coat DNA densely without compacting it. ALBA is acetylated on a lysine residue and deacetylated in a similar fashion of eukaryotic chromosomal proteins. Size of the protein is ~10 kDa per subunit that forms dimers in solution. ALBA shares a common fold with DNase I and the c-terminal domain of the translation initiation factor IF3, suggesting its origin within a group of ancient nucleic acid-binding domains. Homologs of ALBA protein was detected in eukaryotes suggesting that it was an ancient protein, which was originated prior to the divergence of the conserved genomic cores of the eukaryotes, crenarchaea, and euryarchaea from their common ancestor. While ALBA is the dominant chromosomal protein in crenarchaea, homologs of ALBA were not detected in the chromatin of eukaryotes and euryarchaea, suggesting that the highly-conserved ALBA homologs may have additional and undiscovered functions.

2.5.1.4 Endosomal Sorting Complexes Required for Transport Proteins (ESCRT)

Endosomal sorting complexes required for transport proteins (ESCRT) are essential for coupling the protein sorting to vesicle formation. ESCRT complex initially recognizes and sort ubiquitylated proteins called cargo into late endosomes (multivesicular bodies) for transporting to the lysosome for degradation. ESCRT complex consists of four protein complexes ESCRT-0, ESCRT-I, ESCRT-II, ESCRT-III along with the vacuolar sorting protein 4 (VSP4). ESCRT-0, ESCRT-I, ESCRT-II are involved in the recognition of the cargo proteins at the endosomal membrane, which will be followed by the association of the ESCRT-III. After association, ESCRT-III complex polymerizes on the outer endosomal membrane surface, giving rise to membrane deformation and budding away from the cytoplasm into the lumen of the endosome along with the ESCRT-bound cargo, resulting in the generation of intraluminal cargo containing vesicles. VSP4 is then recruited to the endosomal membrane by the interactions between the microtubule interaction and transport domain (MIT) of VSP4 and the MIT-interacting motif (MIM) domain of ESCRT-III proteins, which will disassemble the ESCRT-III complex during membrane invagination. Homologs of the eukaryotic ESCRT Vps4 and ESCRT-III were identified in the members of crenarchaea. Archaeal ESCRT homologs are also involved in the cell division. The structural basis of the interaction between Vsp4 and ESCRT-III proteins is highly similar in archaea and eukaryotes, indicating that this structural and functional similarity predates the divergence of the archaeal and eukaryotic lineages. The common ancestry of the archaeal and eukaryotic ESCRT proteins means that the proteins are likely to function by similar mechanisms.

2.5.1.5 Repeated Sequences in the Archaeal Genomes

Mobile genetic elements such as insertion sequence (IS) elements, transposons contribute to the plasticity and evolution of genomes by providing the basis for genome rearrangements, homologous recombination and facilitating the acquisition or deletion of genes during horizontal gene transfer and transposon activity.

2.5.1.5.1 Insertion Sequence (IS) Elements

Insertion sequence (IS) elements were identified in almost all archaeal organisms whose genomes were sequenced. Archaea shows a high amount of variation with respect to the diversity and the number of IS elements. Archaea with large genomes such as *Sulfolobus solfataricus* and *Halobacterium* NRC-1 are rich in these elements, whereas other archaea contain a very few with no element were detected in *Methanobacterium thermoautotrophicum. S. solfataricus* contains up to 200 IS elements with the size ranging from 0.7 to 2.0 kb, constituting ~10% of the genome. *Halobacterium* NRC-1 bearing 82 IS elements with 24 in its chromosome and 58 in two of its megaplasmids. *Archaeoglobus fulgidus* and *Thermoplasma volcanium* are carrying 18 and 27 IS elements respectively in their genomes. Both *Pyrococcus abyssi* and *Methanococcus jannaschii* contains a single copy of putative IS elements, identified on the basis of their sequence similarity with known transposases. These IS elements are similar to that of bacteria prior to their size and structure ranging from 520 bp to 1895 bp in length with terminal inverted repeats of 8–29 bp. Transposition of IS elements in archaeal genomes causes a duplication of the target DNA. Archaeal IS elements consist of at least one ORF.

2.5.1.5.2 Clustered, Regularly Interspaced Short Palindromic Repeats (CRISPR)

CRISPR/Cas (CRISPR-associated) system is one of the most elaborate defense systems identified to date, providing the acquired and heritable immunity to the prokaryotes against phages and other parasites. CRISPR contain arrays of repeated sequences interspersed short non-repetitive DNA segments called spacers (20–50 bp). A Spacer can be transcribed and processed into small CRISPR RNA (crRNA) molecules, which will base pair with nucleic acids of the invading organism and mark them for the degradation. CRISPR/Cas systems are capable of continuously acquiring the new spacers. CRISPR sequences are found in more than 90% of archaeal genomes and several archaea have been used as model organisms for elucidating the mechanism of CRISPR function. The single unit transcription of CRISPR and spacers before processing into small RNAs was first demonstrated in *Archaeoglobus fulgidus*. In *P. furiosus* the CRISPR/Cas system uses guide RNAs to specifically target foreign RNA for destruction. Even though the ultimate goal of CRISPR/Cas systems is to provide protection to the host genome from the invasion of viruses and harmful plasmids, many of the spacers acquired by archaea are capable of targeting random genes from diverse sources, providing a glimpse of the nucleic acids that an archeon has been exposed to.

2.5.1.5.3 *Miniature Inverted Repeat Transposable Elements* (MITES)

Miniature Inverted Repeat Transposable Elements (MITEs), are common components of eukaryotic genomes, which might have evolved either due to deletion of IS element denoted as type I MITE or by accruing the accumulation of terminal inverted repeats, similar to those of a known transposon considered as type II MITE. Type I MITEs were identified in the genomes of four archaea. In *M. jannaschii*, two type I MITES were identified showing sequence similarity with IS element 703 (ISE703). One is found to exist in a single copy on a plasmid, whereas the second one is occurring in eight chromosomal copies each with a 21-bp terminal inverted repeats. Nine copies of MITE AM1 (306 bp) is found to occur in *A. pernix* genome. Two copies of MITE TM1 (598 bp) were identified in *T. volcanium,* which were derived from ISE1247. All these elements possess terminal direct repeats of 8 bp.

Type II MITEs were first detected in the *S. solfataricus* genome, in which they are abundant and constituting up to 0.6% of the genome. One hundred and forty three copies of Type II MITEs have been detected till date, classified as SM1–4, based on their sequences terminal inverted repeats. Type II MITES are also having sequence similarity with IS elements, SM1 is 87% identical with ISC1048, SM2 is 87% identical with ISC1217, SM3 is 81% identical ISC1058 and SM4 is found to be 49% identical to ISC1173. Type II MITE-like elements have also been detected in bacteria such as *Streptococcus pneumonia* and *Neisseria meningitides.*

2.6 Plasmids of Archaea

Plasmids are either circular or linear extrachromosomal units found in microorganisms of Bacteria, Archaea, and Eukaryota and are transmitted by conjugation. They are one of the most important vehicles necessary for the communication of genetic information, facilitation of rapid evolution and adaptation abilities. Plasmids are key genetic tools used to manipulate and analyze microorganisms through introduction, modification or removal of target genes. Currently, there are 4602 completely sequenced plasmids in NCBI plasmid database. Among them, 137 are found in archaea, out which 112 were in Euryarcheoata and 23 were in Crenarchaeota. Plasmids of Euryarcheoata are ~119 kb in size with a GC content of 52.1%, while those of Crenarchaeota are 20 kb in size with 39.2% GC content respectively. The best characterized plasmids among Euryarcheoata are pME2001, pURB500, and pC2A and the Rep genes have been reported in pME2001 and pURB500 although multiple regions (or genes) were required for replication of pURB500. Plasmid pHH1 carries the genetic information for gas vacuole formation and harbors all IS elements, whose presence and mobility might provide extra stability to the pHH1. Plasmid pHV2 of *H. volcanii* is ~6354 bp consisting of 4 large ORFs, which would direct the synthesis of four polypeptides of sizes

90, 25, 23 and 21 KDa respectively. The majority of Crenarchaeota plasmids were identified in Sulfolobales, which were classified into pRN-type and pNOB-type groups. pRN-type, is a small plasmid of size <10 Kb is the hybrid of virus and plasmid that coexists intracellularly with the fusellovirus SSV1 and can also be packed into viral particles. The pNOB8-type plasmids are large (30 kb) and are conjugative. Shuttle vectors are available for a few genera within Euryarcheoata, such as halophiles, methanogens, and thermococcales. Vectors specific to Crenarcheota are available only for Sulfolobales. Archaeal plasmids are introduced by transformation and through conjugative transfer in a small number of plasmids belonging to Sulfolobus.

2.7 Horizontal Gene Transfer (HGT)

Horizontal gene transfer (HGT) plays a major role in the evolution of microbial genomes. It facilitates microbes to rapidly acquire new capabilities and adaptation to the changes in the environment. Most of the genes involved in HGT are also associated with pathogenesis, symbiosis, metabolism and prominently antibiotic resistance. HGT among bacteria is relatively well-studied (discussed in Chap. 3). Horizontal transfer of genes from bacteria to archaea is considered to be one of the important mechanisms involved in the origin of major archaeal clades. HGT between bacteria and archaea was shown to be possible almost for all genes, except for a small fraction of genes, which are toxic to the recipient organism. Genomes of the genus Methanosarcina are the largest among archaea and the large genome size in this genus is due to massive HGT from bacteria. Methanosarcinaceae is the family of archaea possessing the largest set of metabolic pathways and inhabited diverse environments among Archaea. Methanosarcinales consists of traits such as the presence of cytochromes, genes encoding for the A, K, and N subunits of reduced coenzyme F420 (F420H2) dehydrogenase, bacterial-type phosphoglycerate mutase, bacterial adenylate kinase, non-histone chromosomal protein MC1, involved in chromosome condensation and the long variant of condensin subunit ScpB. Horizontal transfer of a short operon from *Clostridia* have completely changed the metabolic capabilities of Methanosarcinales, which allowed these organisms to use methyl compounds as substrates for methanogenesis. All functional genes of *Methanosarcina mazei* were horizontally transferred from bacteria to *Methanosarcina* spp. and 30% of genes in *M. acetivorans* were predicted to be of bacterial origin. ~50% of *M. burtonii* genome is having oligonucleotide composition and high transposon content might be due to horizontal transfer. Analysis of specific patterns of gene gain in archaea using the arCOG database revealed that groups belonging to methanogenic archaea have acquired a number of genes (~1000 for Haloarchaea, 100 for Methanosarcinales from eubacteria) by horizontal gene transfer.

2.8 Integrase-Mediated Insertion and Deletion of Archaeal DNA

There is increasing evidence suggesting the entry and exit of DNA from archaeal chromosomes is facilitated by an archaeal-type integrase. This was first demonstrated in the *Sulfolobus* virus SSV1, encoding an integrase that facilitates the insertion and excision from the downstream half of a $tRNA^{Arg}$ gene in the chromosome of *Sulfolobus shibatae*. During the insertion, the integrase gene is partitioned into two ORFs which flank the linearized virus into the chromosome. Both the ORFs carry a 44 bp direct repeat required for both insertions as well as deletion events. Excision involves the recombination at these direct repeats, which regenerates the circular virus carrying an intact integrase gene. Equally, partitioned integrase genes are found in other archaeal chromosomes of *S. solfataricus*, consisting of an inserted plasmid pXQ1 in one part and border a large 67-kb integron-like segment in the second part. A similar type of partition was also identified in the genomes of *A. pernix* and *P. horikoshii*, implying that the integrase-based mechanism of insertion/deletion is a general event occurring in the archaeal genomes.

2.9 Genome of Methanogenic Archeon *Methanococcus Jannaschii*

Archaea, since their discovery in 1977 was believed to be the most significant evolutionary distinctions chosen between Prokaryotes and Eukaryotes. Archaea, although appears to be prokaryotic, are not specifically related. They resemble Eukaryotes at the biochemical, molecular levels and in many other related respects. The nature of the archaea and their relationships with Prokaryotes and Eukaryotes were extensively studied at the molecular level by sequencing the genome of an extremophile Methanococcus jannaschii, which provides the complete genetic complements and biochemical pathways of the three main domains of life.

Methanococcus is a genus of coccoid methanogens. Methanogens are microorganisms that produce methane as a metabolic byproduct under anoxic conditions. They are commonly found in wetlands, responsible for marsh gas and in the guts of animals such as ruminants and humans, responsible for the methane content of belching in ruminants and flatulence in humans. In marine sediments, biomethanation is generally confined to the regions, where sulfates are mostly depleted. Others are extremophiles, found in the extreme environments such as hot springs and submarine hydrothermal vents as well as in the "solid"rock of the Earth's crust, kilometers below the surface. They are all mesophiles, except the thermophilic *M. thermolithotrophicus* and the hyperthermophilic *M. jannaschii*, which was discovered by J.A. Leigh from a sediment sample collected from the seafloor at the base of a 2600 m-deep"white smoker" chimney located at 21 °N on the East Pacific Rise. *M. Jannaschii* grows at a pressure of more than 200 atm, over an optimum temperature of 85 °C. It is strictly anaerobic and produces methane. Described as 'raisin-like' in appearance under the microscope, *M. jannaschii* has a thin, tail-like flagellum on one end that gives the cell mobility. It grows in thick mats and shares its habitat near fissures in the Earth's crust with a few other hardy microbes and colonies of giant tube worms. *M. jannaschii* lives exclusively at high temperatures. Hence, its molecules are heat stable, which could be a useful property for many industrial and pharmaceutical processes. The genome of *M. jannaschii* offers an opportunity to develop heat-resistant products, such as new detergent additives or stable enzymes for the textile and chemical industries. Understanding how *M. jannaschii* produces methane could help researchers learn to genetically engineer organisms to produce quantities of methane for use as a source of renewable energy and synthetic chemicals. *M. jannaschii* also appears to produce metal-binding proteins that transport toxic compounds out of the cell, with potential applications for the concentration and clean-up of toxic wastes.

By comparing the genes of *M. jannaschii* to those of Prokaryotes and Eukaryotes, scientists hope eventually to better understand the evolution of all three major branches of life. After their discovery 20 years back, archaea were originally believed to live only at the extreme environmental conditions of temperature and pressure, but are now thought to be far more common and to make up a significant part of the world's biomass. They are suspected of playing important but still unknown roles in the earth's ecology, including its carbon and nitrogen cycles. *M. jannaschii* was the first archaeal genome to be completely sequenced and third microorganism ever to be completely sequenced and published, revealing many novel aspects.

2.9.1 The Methodology of Genome Sequencing

The complete genome sequence of *M. Jannaschii* was obtained by whole genome random sequencing method. A small insert plasmid library of average insert size 2.5 Kbp and a large insert λ library of average insert size 16 Kbp were used as substrates for sequencing. The λthe library was used to form a genome scaffold and also for verifying the orientation and integrity of the contigs formed from the assembly of sequences from the plasmid library. All the clones of the library were sequenced from both the ends, with the average length of the reads 481 bp. A total of 36,718 sequences were assembled using TIGR Assembler. The co-linearity of the sequenced genome to the in vivo genome was confirmed by comparing the restriction fragments generated by digestion with six rare cutter restriction endonucleases such as Aat II, Bam HI, Bgl II, Sma I, Kpn I and Sst II. Another co-linearity was provided by the genome scaffold comprising of 339 large insert λ clones covered up to 88% of the main chromosome.

2.9.2 Properties of the Genome

The genome of *M. jannaschii* consists of three main elements.

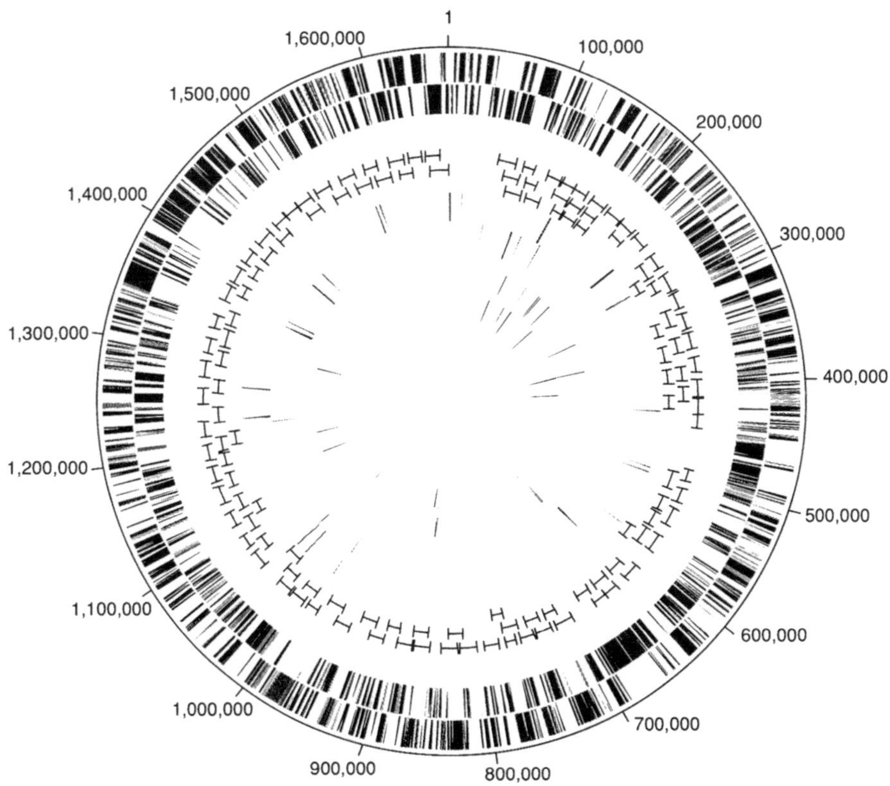

Fig. 2.2 Diagrammatic representation of the *M. jannaschii* large circular chromosome. First circle: predicted protein coding regions on the plus strand, Second circle: predicted protein coding regions on the minus strand, Third circle: coverage by λ clones, Fourth Fifth circles: Representing the Plus and Minus strands respectively (Courtesy Carol J. Bult, TIGR, Rockville, USA)

1. A large circular chromosome of 1,664,972 bp with a G + C content of 31.4%, predicted to contain 1682 protein coding regions (Fig. 2.2).
2. A large circular extra chromosome of 58,407 bp with a G + C content of 28.2%, consisting of 44 predicted protein coding regions.
3. A small circular extra chromosome of 16,550 bp with the G + C content of 28.8%, predicted to contain 12 protein coding regions. The sequences of *M. jannaschii* large circular chromosome and extra chromosomes have been deposited in the Genome Sequence Database with the accession numbers L 77117, L77118 and L77119.

Methanococcus jannaschii genome has 1738 genes. Among them, only 38% matches with the existing genes in the sequence databases, with an assigned function. 6% of the predicted protein coding regions had matches to hypothetical proteins. ~100 genes of the *M. jannaschii* genome had a marginal similarity with the genes from the public sequence database and could not be assigned with any function. *Methanococcus jannaschii* genome was like nothing scientists had ever seen before. More than half its genes were completely new. Only 44% of its genes matched with the DNA sequences of known ones. For instance, only 11% of the genes *in H. influenzae* and 17% of those in *M. genitalium* matched a sequence from *M. jannaschii*. By comparison, those two bacteria were highly similar with 83% of *M. genitalium* genes having a counterpart in *H. influenzae*. Majority of the identified genes are related to cell division, energy production, and metabolism.

2.9.2.1 Transcription and Translation Machinery

Among the vital multicomponent information processing systems such as replication, transcription, and translation, translation machinery of Methanococcus is having similarities with both Prokaryotes and Eukaryotes. *M. jannaschii* genome has two rRNA operons named as A and B along with a 5S RNA gene associated with tRNAs. Majority of the proteins associated with the ribosomal subunits of Prokaryotes, Eukaryotes are present in *M. jannaschii*. Translation elongation factors EF-1α and EF-2 are more similar to the Eukaryotes. Additionally, *M. jannaschii* genome consists of 11 genes oriented with the initiation of translation. 35 tRNA genes were identified in *M. jannaschii* genome with almost all amino acids encoded by two codons are having a single tRNA, except glutamic acid, which is having two tRNAs. A tRNA also exists for Selenocysteine (UGA codon). 4 genes were identified in Methanococcus genome with internal stop codons, such as Selenocysteine.

The transcription initiation system of *M. jannaschii* is much similar to that of Eukaryotes. The central molecule in the regulatory region is TATA Binding Protein (TBP) and Transcription factor B, which are the parts of vital complexes and additional factors such as TFIIA and TFIIF. These complexes and factors were not found in the *M. jannaschii* genome. 5 genes corresponding to the histone proteins were identified in the *M. jannaschii* genome, with three on the large circular chromosome and two on the large circular extra chromosome. These genes are homologs to the Eukaryotic H2a, H2b, H3, H4 and eukaryotic transcription related CAAT Binding factor CBF-A.

2.9.2.2 Inteins

Self splicing portions of a peptide sequence encoding a DNA endonuclease activity are called inteins. The sequences remained after the excision of the intein is called exein. Exeins are spliced together after the excision of one or more inteins to form functional proteins. 14 genes were identified in *M. jannaschii* genome containing 18 putative inteins. DNA polymerase gene of *M. jannaschii* has two inteins in the same locations as in *Pyrococcus* sp. strain KOD1.

2.9.2.3 Repeated Genetic Elements

At least three families of repeated genetic elements were identified in *M. jannaschii* genome. Members of the ISIMJ1 forms the first family, are repeated 10 times in the large circular chromosomes and once on the large circular extra chromosome. All members are possessing 16 bp terminal inverted repeats. The second family comprises 2 ORFs, identical across 928 bp. These ORFs are 23% identical at the amino acid level to the COOH terminus of a transposase from *Lactobacillus lactis*. Terminal inverted repeats were not found in this family. 18 copies of the repeat sequences belonging to the third family were found throughout the genome of *M. jannaschii*. None of these sequences are having coding potential and are composed of a long segment followed by the 1–25 tandem repetitions of a short segment, separated by a unique sequence. Similar short tandem repeat sequences were identified in bacteria and other archaea believed to be involved in chromosomal partitioning during cell division.

2.9.3 Identification of Proteins by Shotgun Proteomics

Even though the genome of *Methanococcus jannaschii* has been sequenced and the predicted protein functions have been assigned, proteome, which provides the products of the actually expressed genome, will eventually give greater insights in studying the importance of these macromolecules in the organism's lifestyle. Using shotgun proteome method called Multidimensional Protein Identification Technology (MudPIT), researchers were able to identify the actual amount of proteins, assigning them with specific functions.

2.9.3.1 Shotgun Proteome Analysis

Shotgun proteome analysis is a method of identifying the proteins in complex mixtures using a combination of HPLC and mass spectrometry. In shotgun proteomics, the proteins in the mixture are subjected to a protease digestion and the resulting peptide fragments are separated by Multidimensional chromatography, which allows the separation of complex protein mixtures by using multiple columns with different stationary phases. These columns are coupled orthogonally, which means the fractions emerging from the first column can be selectively transferred to other columns for additional separation. This enables separation of complex mixtures that cannot be separated using a single column. One of the most important applications for this technique is proteomics, where complex protein digests are separated by multi-dimensional liquid chromatography instead of using the two dimension gel electrophoresis. In the first dimension, fractions of the peptide mixture elute from an ion-exchange column by a salt step gradient. Each fraction is trapped on a small reversed-phase trapping column and then separated after the valve switches to a reversed column (the second dimension). The trapping column is first used to prevent salt from entering the mass spectrometer (ion suppression). Second, the column allows an enrichment step, which together with the low flow rate in the 2nd dimension provides high detection sensitivity. Tandem mass spectrometry is then used to identify the peptides. MudPIT is a method used to analyze the highly complex samples necessary for large-scale proteome analysis by electrospray ionization, tandem mass spectrometry (MS/MS) and database searching. As it is most frequently used, MudPIT couples a two-dimensional liquid chromatography

(2D-LC) separation of peptides on a microcapillary column with detection in a tandem mass spectrometer.

Protein mixtures were first digested in complicated protein mixtures by allowing *M. jannaschii* to grow on a complex medium enriched with H_2. This mixture was further separated and resolved by loading on a multidimensional chromatographic capillary column. Out of 1775 genes predicted from the sequenced genome, 963 accounting up to 54.2% were identified to be involved in the amino acid biosynthesis, cofactors biosynthesis, prosthetic groups and carriers, DNA metabolism, energy metabolism, synthesis of purines, pyrimidines, nucleosides, and nucleotides, fatty acid and phospholipids metabolism, transcription and other cellular processes. Due to the abundance of these proteins in the genome, they are considered as housekeeping proteins, crucial in allowing cell maintenance and growth. Out of the 963 genes identified, 137 (9%) were proposed to be involved in energy metabolism. This group of proteins is having a high abundance in the *M. jannaschii* proteome functions in the methanogenesis, and also many electron carrier proteins such as ferredoxin and subunits of proton transporting ATP synthase. The second major portion of the functions of the genes in the synthesis of ribosomal proteins. 67% of the predicted genes are involved in the synthesis of a cellular envelope. Majority of proteins that are expressed by these genes are S-layer structural proteins. Around 80% of the predicted genes are involved in fatty acids and phospholipids metabolism. Even though a large number of genes have unknown functions or are hypothetical genes, it was reported that these genes are most likely crucial to cellular growth.

The genome structure of *Methanococcus jannaschii* has provided a solid support for the hypothesis that archaea are indeed a separate domain of life, which is entirely different both from Bacteria and Eukaryota. It also hypothesizes that Archaea is more closely related to Eukarya than that of Prokarya. This hypothesis was made several years before genome sequencing even existed. Even though the genome of *M. jannaschii* is quite small compared with the genome of well-known bacterial species such as *E. coli*, it was believed that this genome is highly specific to the autotrophic lifestyle of an organism. Till date, 52% of the proteins in the genome have been identified and are assigned with appropriate functions. However, their hyperthermophilic properties and their special mechanisms provide the support for their adaptation to the extreme environments leading to the conclusion that these proteins are strictly involved in the energy production, specifically methanogenesis, cell division, metabolism and also in maintaining the organism's lifestyle. Additionally, this genome has also revealed a rich collection of mobile elements (both autonomous and non-autonomous), providing a shred of evidence for the occurrence of genome re-arrangements and complex integration events. Nonautonomous MITE-like elements appeared to be more spread in archaea than bacteria. This data provides the first detailed insight into the mechanisms related to the evolution of archaeal genomes.

2.10 Genome of *Archaeoglobus Fulgidus*

Sulfate reduction is part of the sulfur cycle essential to the biosphere. Biological sulfate reduction is restricted only to a few prokaryotes such as Eubacteria and archaea. Archeoglobales are the unique archaeal sulfate reducers, which grow at extremely high temperatures and are highly unrelated to other sulfate reducers. Archaeoglobales are strictly anaerobic, occurs in the hydrothermal environments and in subsurface oil fields. High-temperature sulfate reduction by Archaeoglobales produces iron sulfide contributing to the deep subsurface oil-well souring, leading to the corrosion of iron and steel in oil and gas processing systems. *Archaeoglobus fulgidus* VC-16 is the type strain of the Archaeoglobales, whose cells are irregular spheres with a glycoprotein envelope and monopolar flagella. Cell growth occurs at high temperatures of 83 °C, with a minimum division time of 4 h. This organism grows organoheterotrophically using a variety of carbon and energy sources but can grow lithoautotrophically on hydrogen, thiosulphate, and CO_2. The genome of *A. fulgidus* strain VC-16 was sequenced as an example of a sulfur metabolizing organism in order to gain further insight into the Archaea through genomic comparison with the genome of *M. jannaschii*.

2.10.1 Genome Sequencing and Assembly

A. fulgidus VC-16 genome has been sequenced by Whole-genome random sequencing. The culture was derived from a single cell isolated by optical tweezers, provided by Stetter (University of Regensburg). Cloning, sequencing, and assembly of the genome were performed by TIGR. One small-insert and one medium-insert plasmid libraries were generated by random mechanical shearing of genomic DNA. One large-insert λ library was generated by partial Tsp509I digestion of genomic DNA and its ligation to l-DASHII/EcoRI vector (1-clone). During the initial phase, 6.7-fold sequence coverage was achieved with 27,150 sequences from plasmid clones (average read length 500 bases) and 1850 sequences from l-clones. Both plasmid and l-sequences were assembled using TIGR assembler, resulted in 152 contigs separated by sequence gaps and five contigs separated by physical gaps. Sequences from both ends of 560 l-clones served as a genome scaffold, verifying the orientation, order, integrity, and contigs. Gaps were closed by editing the ends of sequence traces or by primer walking and physical gaps were closed by combinatorial PCR followed by its product sequencing. The final genome sequence is based on 29,642 sequences, with 6.8-fold sequence coverage. The linkage between the terminal sequences of 2101 clones from the small-insert plasmid library and 8726 clones from the medium-insert plasmid library supported the genome scaffold formed by the l-clones, with 96.9% of the genome was covered by l-clones. The assembled sequence was differing in 20 positions from the 14,389 bp of DNA in a total of 11 previously published *A. fulgidus* genes.

2.10.2 ORF Prediction and Gene Identification

Open Reading Frames (ORFs) were identified using a combination strategy based on two programs, Gene Smith (H.O.S., unpublished), for evaluating the ORF length, separation, overlap, and CRITICA (J.H.B. & G.J.O., unpublished), a coding region identification tool using comparative analysis. ORFs were searched against a non-redundant protein database using BLASTX10 and sequences shorter than 30 codons without a database match were eliminated. Appropriate frameshifts were recognized and were corrected. Remaining frameshifts, where the corresponding regions need to be authentically considered were annotated as authentic frameshift. 527 hidden Markov models were searched with HMMER based on conserved protein families (PFAM version 2.0), to determine ORF membership in families and superfamilies.

2.10.3 Features of the Genome

The genome of *A. fulgidus* consists of a single, circular chromosome of 2,178,400 bp comprising of 92.2% of protein coding regions with an average G + C content of 48.5%. Three regions with a low G + C content of 39% were identified in the genome, among which two regions are rich in the genes encoding for lipopolysaccharide biosynthesis. Two regions of 53% G + C content were identified, encoding the genes for large rRNAs, proteins involved in haemAB and several transporters (Fig. 2.3).

2.10.3.1 Open Reading Frames

Gene Smith, CRITICA and BLASTX10 searches predicted 2436 ORFs covering up to 92.2% of the genome. The average size of an ORF is 822 bp, which is similar to that of *M. jannaschii* (856 bp) and smaller than eubacteria (949 bp). All open reading frames in the genomes were searched against a non-redundant protein database, resulting in 1797 putative identifications that were assigned biological roles. Predicted start codons were 76% ATG, 22% GTG and 2% TTG. Unlike *M. jannaschii*, no inteins were identified. *A. fulgidus* contains a large number of gene duplications, contributing to its larger genome size. The average protein relative molecular mass (Mr) is 29,753, ranging from 1939 to 266,571, similar to that of prokaryotes.

2.10.3.2 Multigene Families

719 genes belong to 242 families were identified in *A. fulgidus* genome. Of these, 157 families contained genes with biological functions and most of these families contain genes oriented with the energy metabolism, transport and binding, fatty acid and phospholipid metabolism categories. ATP-binding subunits of ABC transporters is the largest superfamily, consisting of 40 members. Catabolic degradation and signal recognition systems in *A. fulgidus* are reflected by the presence of two large superfamilies, acylCoA ligases, and signal-transducing histidine kinases.

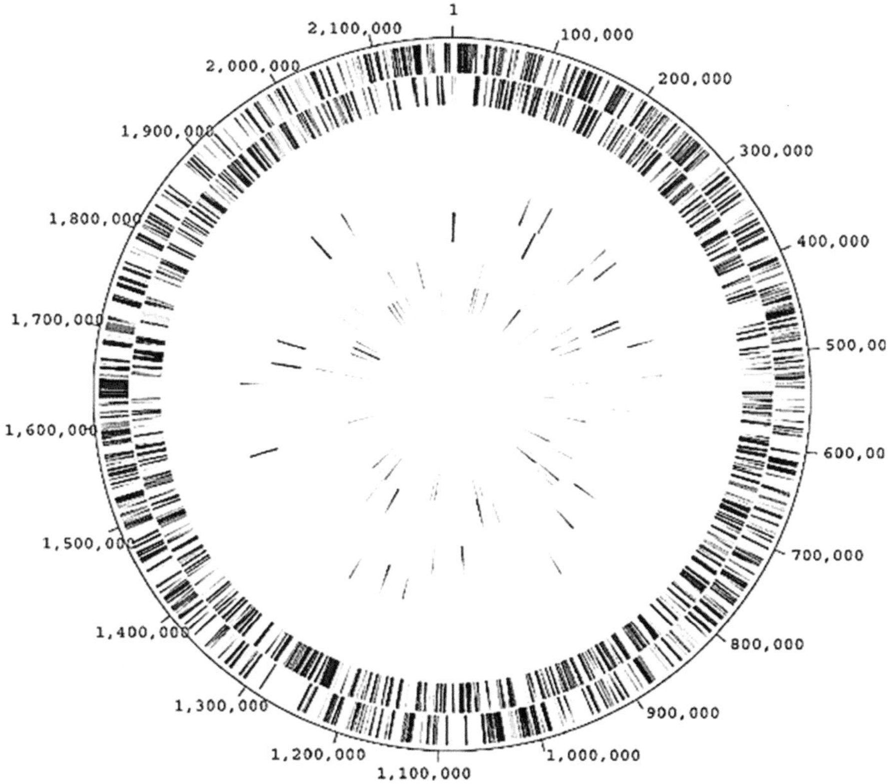

Fig. 2.3 Diagrammatic representation of *A. fulgidus* genome. The first circle: predicted protein-coding regions on the plus strand. The second circle: predicted protein-coding regions on the minus strand. Third and fourth circles: IS elements and other repeats on the plus and minus strands. Fifth and sixth circles: tRNAs, rRNAs, and sRNAs on the plus and minus strands, respectively (Courtesy Hans-PeterKlenk, TIGR, Maryland USA)

2.10.3.3 Repetitive Elements

The genome of *A. fulgidus* contains three regions of short direct repeats. Among them, two repeat regions SR-1A and SR-1B consisting of 48 and 60 copies of the 40 bp repetitive sequences, whereas the third repeat region SR-2 consists of 42 copies of a 37 bp repeat, interspersed with a unique sequence of 41 bp. These repeated sequences are similar to the short repeated sequences of *M. jannaschii*. 9 classes of long repeated sequences of 500 bp length (LR1-LR9) with >95% sequence identity were found in *A. fulgidus* genome. LR-3 is a novel element with 14-bp inverted repeats and consists of two genes. LR-4 and LR-6 encode putative transposases, which were not identified in *M. jannaschii*. The remaining LR elements are not similar to the known IS repeat elements.

2.10.3.4 Transporters

A. fulgidus have the provision of synthesizing several transporters for the importing carbon-containing compounds. This might probably contribute to increased ability in switching from autotrophic to heterotrophic growth. *A. fulgidus* have branched-chain amino-acid ABC transport system and a transporter for the uptake of arginine and lysine similar to that of *M. jannaschii*. *A. fulgidus* genome encodes proteins for dipeptide, spermidine/putrescine, proline/glycine-betaine, and glutamine uptake, as well as transporters for sugars and acids, which might provide necessary substrates for numerous biosynthetic and degradative pathways. Many *A. fulgidus* redox proteins were predicted to require iron because of the presence of both oxidized ($Fe_3{}^+$) and reduced ($Fe_2{}^+$) forms of iron transporters. Ten distinct transporters facilitating the flux of the physiological ions K^+, Na^+, $NH_4{}^+$, $Mg_2{}^+$, $Fe_2{}^+$, $Fe_3{}^+$, NO^{-3}, $SO_2{}^{-4}$ and inorganic phosphate (Pi) were also identified. Most of these transporters have homologs in *M. jannaschii* and are therefore likely to be critical for nutrient acquisition during autotrophic growth. *A. fulgidus* has additional ion transporters for the elimination of toxic compounds including copper, cyanate, and arsenite. *A. fulgidus* genome contains two paralogous operons of cobalamin biosynthesis-cobalt transporters, cbiMQO.

2.10.3.5 Sensory and Regulatory Networks

The genome of *A. fulgidus* has complex sensory and regulatory networks, which contain over 55 proteins with presumed regulatory functions. At least 15 signal-transducing histidine kinases were identified, but contain only nine response regulators, suggesting a high degree of cross-talk between kinases and regulators. Even though *A. fulgidus* is rich in regulatory proteins, it is lacking regulators for a response to amino-acid, carbon starvation and DNA damage. *A. fulgidus* also contains a homolog of the mammalian mitochondrial benzodiazepine receptor, which functions as a sensor in signal-transduction pathways.

2.10.3.6 Replication, DNA Repair, Cell Division Transcription, and Translation

The genome of *A. fulgidus* possesses two family BDNA polymerases related to the catalytic subunit of the eukarya -l delta polymerase, similar to that of Sulfolobales. It also has a homolog of the proofreading $\in$ subunit of *E. coli* Pol III, which was not previously observed in the Archaea. The DNA repair system of *A. fulgidus* is more extensive, when compared with that of *M. jannaschii*, consisting of a eukaryal Rad25 homolog, a 3-methyladenine DNA glycosylase and exodeoxynuclease III, reverse gyrase, topoisomerase I and VI. *A. fulgidus* consists of a recognizable type I restriction-modification system but does not contain type II system. The cell-division machinery is highly similar to that of *M. jannaschii*, with orthologues of eubacterial *fts* and eukaryal *cdc* genes. However, several *cdc* genes of *M. jannaschii*, such as *cdc23*, *cdc27*, *cdc47*, and *cdc54* were found to be absent in *A. fulgidus*. The transcriptional and translational systems of *A. fulgidus* is quite distinct from their eubacterial and eukaryaotic counterparts, whose RNA polymerase contains large universal subunits and five smaller subunits. *A. fulgidus* has a homolog of eukaryotic

TBP-interacting protein. Translation process in *A. fulgidus* parallels *M. jannaschii* with a few exceptions such as the presence of only one rRNA operon with an Ala tRNA gene in the spacer and lacks contiguous 5S rRNA gene. Genes for tRNAs were identified in *A. fulgidus,* with five of them containing introns in the anticodon region, might have got removed by the intron excision enzyme EndA. The RNA component of the tRNA maturation enzyme RNase P is present in *A. fulgidus* but lack glutamine synthetase and asparagine synthetase. The relevant tRNAs might have aminoacylated with glutamic and aspartic acids respectively. *A. fulgidus* possesses tRNA synthetase for both Cys and Ser unlike *M. jannaschii,* in which the former was not identifiable and the latter was unusual. *M. jannaschii* has a single gene belonging to the TCP-1 chaperonin family, whereas *A. fulgidus* has two genes that encode for two subunits a and b of the thermosome.

2.10.3.7 Synthesis of Essential Compounds

A. fulgidus genome has the capability of synthesizing many essential compounds, amino acids, cofactors, carriers, purines and pyrimidines, whose biosynthetic pathways show a high degree of conservation with *M. jannaschii. A. fulgidus* shows high rate of similarity with *M. jannaschii* in their biosynthetic pathways for siroheme, cobalamin, molybdopterin, riboflavin, thiamin, and nictotinate. Biosynthesis of amino acids has the highest conservation between these two organisms. Among 78 *A. fulgidus* genes assigned for amino-acid biosynthetic pathways, 73 (94%) have homologs in *M. jannaschii.* In both archaeal species, amino-acid biosynthetic pathways resemble those of *Bacillus subtilis* than those of *E. coli.* None of the biotin, pyridoxine, and ferrochelatase (the terminal enzyme in haem biosynthesis) biosynthetic genes were identified in *A. fulgidus.* Still, biotin, pyridoxine, and cytochromes can be detected in the cell extracts of *A. fulgidus.* These cofactors may be obtained by mechanisms, which are yet to be recognized.

The complete genome of *A. fulgidus* clearly demonstrates that this organism has a great diversity of electron transport systems with some unknown specificity. Sequencing of this genome also provides a new wealth of information regarding its exploitation of the environment by reducing the sulfur oxides has been well characterized. *A. fulgidus* has also been characterized as a scavenger with numerous potential carbon sources and its gene complement reveals the extent of its capability. *A. fulgidus* appears to obtain carbon from fatty acids through β-oxidation, by the degradation of amino acids, aldehydes, and organic acids. *A. fulgidus* has extensive gene duplication when compared with other fully sequenced prokaryotes. Proteins encoding these duplicated genes are not identical and the presence of duplicated proteins suggests metabolic differentiation, especially during carbon recycling. These observations suggest that gene duplication is an important evolutionary mechanism for increasing the physiological diversity in Archaea.

2.11 Comparative Genomics of *A. Fulgidus* and *M. Jannaschii*

Comparison of *A. fulgidus* and *M. jannaschii* genomes provides the basis for some preliminary conclusions regarding the nature of Archeal genomes. A comparison of the gene content in these two organisms reveals that the gene conservation varies significantly and the genes involved in transcription, translation, and replication are highly conserved. Around 80% of the genes in *A. fulgidus* have homologues in *M. jannaschii* in these categories. Biosynthetic pathways are also highly conserved in these two Archaea, with ~80% of the *A. fulgidus* biosynthetic genes are homologous to *M. jannaschii* and 35% of the central intermediary metabolism genes are having homologs, which reflects their minimal metabolic overlap. More than half of the ORFs in *A. fulgidus* genome (1290) has not been assigned with any biological function and 639 of them have no database match. Remaining 651 ORFs are conserved and their hypothetical proteins are having sequence similarity with the two-thirds hypothetical protein homologs of *M. jannaschii.* These hypothetical proteins will increase our understanding of the genetic repertoire of the archaea and also on prokaryotic genetic diversity.

2.12 Conclusions

Study of Archaeal genomics is important in increasing our understanding regarding two critical transitions in the evolution of life. The first one is the split between bacteria, archaea and eukaryotic lineages and the second one is the origin of eukaryotes. Archaea are a valuable source providing a great value of information regarding the origin and evolution of eukaryotes, as they appear to have retained primitive traits while eukaryotes have undergone major changes. Indubitably, archaea resemble the common ancestor of the archaeo-eukaryotic line of descent more closely than eukaryotes. Archaeal genomics is the best way to reconstruct this critical intermediate in the evolution of life. Lack of established model systems for archaeal experimental biology and difficulty with large-scale experimentation, archaeal genomes are even more crucial for understanding their functions. So far over 150 archaeal genomes have been sequenced resulting in a sound wealth of knowledge relevant not only to archaea but also to studies in bacterial and eukaryotic cells, significantly contributing to answering large questions in biology. Still, a large number of inter-domain and archaeal-specific genes are left to be characterized and it is likely that working archaeal genomes will continue to shed significant light on the aspects of biology.

Structure, Function, and Evolution of Bacterial Genomes

3

Abstract

Organization of bacterial genomes, differences in the primary and secondary chromosomes among the bacteria were well discussed in this chapter. Details of the bacterial genome rearrangements have been discussed with the genomes of K-12 and pathogenic O157: H7 *E. coli* strains, *Mycoplasma genitalium* normal and synthetic genomes, which made the first synthetic life possible on the planet. Key concepts related to the bacterial genome rearrangements, the evolution of bacterial genomes and the genetic diversity of the infectious bacteria were also discussed in detail.

3.1 Introduction

Bacterial genome is concerned with studying the entire hereditary information of bacteria, which can be applied for analysing the outbreaks of bacterial infections and even bacterial evolution. Two decades after the first bacterial genome of *H. influenza* has been completely sequenced, till date, more than 30,000 bacterial genomes have been sequenced, and their genome information is publically available in different databases (Table 3.1). Projects such as Genomic Encyclopedia of Bacteria and Archaea (GEBA) are adding more information not only their genomes but also on their genetic diversity. The most significant factor leading to an increase in the genomes is the vast reduction in the cost of sequencing due to technical advancements. In this chapter, the organization of bacterial genomes was discussed along with their sequencing strategies using *E. coli* and *Mycobacterium* genomes as the main type studies.

K. V. Chaitanya, *Genome and Genomics*,
https://doi.org/10.1007/978-981-15-0702-1_3

Table 3.1 Number of genomes sequenced in six bacterial phyla

S. No	Phyla	Number of genomes sequenced
1	Actinobacteria	4059
2	Bacteroidetes/Chlorobi group	932
3	Cyanobacteria	340
4	Firmicutes	9628
5	Proteobacteria	14,268
6	Spirochaetes	525
7	Other	1500

Source: GenBank

3.2 Structure and Organization of Bacterial Genomes

3.2.1 Bacterial Chromosomes and Plasmids

The genomes of bacteria are very simple structures with a remarkable diversity due to the presence of dynamic chromosomes. A bacterial chromosome can be defined as a DNA molecule that contains the necessary information for replication and continuing the life of a bacterial cell under normal growth conditions. For a long time, it was believed that all eubacterial genomes are composed of a single circular chromosome. Later, it was revealed that bacterial genomes also contain linear chromosomes as found in *Borrelia*, *Streptomyces*, *Rhodococcus fascians,* and *Coxiella burnetii*. The number of chromosomes also vary in bacteria with two chromosomes were found in *Rhodobacter spheroides*, *Brucella melitensis*, *Leptospira interrogans* and *Agrobacterium tumefaciens*. Genus *Borrelia* is unique among bacteria with linear chromosomes of the size ranges from 900 to 920 kbp in length (the known range of bacterial chromosome sizes is 580–9300 kbp). The linear chromosomes have covalently closed hairpin-like telomeres. In *Agrobacterium* genome, there is one circular chromosome and one non-homologous linear chromosome. Unlike eukaryotic chromosomes, bacterial chromosomes lack centromeres and there may be a partitioning system based on membrane adherence.

Linear Plasmids were identified in all members of *Borrelia* and at least 10 *Streptomyces* species as well as a number of bacteria with circular chromosomes. *Borrelia* has a very complex plasmid with 12 linear molecules and nine circular molecules, which are present in approximately the same number of molecules per cell as the chromosome. A linear plasmid was also reported in the Gram-negative *Thiobacillus versutus*. Duplication of the genomes is initiated at the origin of replication site, which may proceed unidirectionally or bidirectionally. The structure of the origin of replication, the *oriC* locus, has been extensively studied in a range of bacteria, found to consist essentially of the same group of genes in nearly identical order. The *oriC* locus is well defined in bacteria.

3.2.2 Bacterial Genomes with Primary and Secondary Chromosomes

Just as various elements of DNA enter into the chromosome, part of a chromosome might also get integrated into another replicon within the nucleus of a cell. During this integration, an essential gene can be the part of an extrachromosomal DNA, as in the case of *R. sphaeroides*. In *B. cereus*, the size of the chromosome varies between 2.4 and 6.3 Mb. Three strains of *B. cereus* with a chromosome size of 2.4 Mb have another replicon of more than 2 Mb, comprising the locus for the housekeeping gene pyruvate. This structure can be considered as a giant plasmid and also as a secondary chromosome. In *Leptospira interrogans*, housekeeping gene *asd* is localized on a 350 kb replicon, rather than on the chromosome of 4.4–4.6 Mb indicating the existence of the second chromosome in this bacterium. The above identifications indicate that complex genomes in bacteria are more common with the presence of two or more chromosomes. They can be termed as a primary chromosome, comprising of the genes necessary for life and a secondary chromosome, containing one or a few housekeeping genes.

3.2.3 Insertion Sequence (IS) Elements

Insertion sequences (IS) are short DNA segments of typically 1–2 kb in length, with an ability to translocate within and among replicons. These mobile genetic elements are reported to encode the genes related to the transposition and the regulation of transposition leading to a numerous molecular and genetic phenomena such as shifting, replication, gene activation, repression, deletion, rearrangement, recombination and transfer, which might be responsible for the induction of mutations that can change the fate of these bacteria. Several hypotheses have been proposed to explain the persistence of IS elements in the bacterial population. The selfish DNA hypothesis asserts that IS elements could persist due to their ability to self replicate, with no contribution to the genome. Horizontal gene transfer (HGT) is another factor responsible for the spread of IS elements among bacteria. Adaptive mutations are also believed to play a pivotal role in the persistence of IS elements. Both IS elements and their host's genomes can encode mechanisms, leading to the suppression of transposition activities, called transposition burst. IS transposition burst promotes the adaptation of bacteria to a high osmolarity environment by increasing the rate of adaptive mutations. IS elements have been considered as a major source of bacterial genetic diversity, as their movements mediate changes, which are beneficial to their host genomes.

An IS element can increase the fitness of an organism by expression and mobilization of antibiotic resistance genes, thus promoting adaptive evolution. There are several examples of IS-mediated gene expression leading to an increase in the antimicrobial resistance. Insertion of an IS element in the upstream of a protein coding sequence often result in the enhanced expression, leading to different phenotypes depending on the function of the over-expressed gene. Insertions of the same IS elements in the nearby locations can generate a composite transposon, capable of mobilizing the intervening sequence and transferring it to new genomic locations. IS elements are also of epidemiological importance, owing to their ability to form composite transposons and mobilize antibiotic resistance determinants.

The IS finder database currently contains over 500 distinct IS elements. During transposition, some IS elements generate direct repeats or target site duplications in the sequences into which they get integrated. Rates of transposition vary between IS elements and host species, but are frequently in the order of the rate of nucleotide substitutions, making IS activity as one of the most dynamic evolutionary forces that play in many bacterial genomes. IS elements also have been implicated in large changes to genome structure by expanding the copy number in microbial genomes. The subsequent loss of IS elements results in the inactivation of genes, formation of pseudogenes and rearrangements of the genome.

3.2.4 Conjugative Transposons

Conjugative transposons are genetic elements that are capable of integrating into the bacterial chromosome, and also have the ability to mobilize the non-transmissible plasmids from cell to cell when they fuse with such plasmids. Normally, the conjugative transposons remain integrated within the bacterial chromosome. Occasionally, they are excised from the chromosome to form a circular intermediate, a copy of which is transferred to a recipient cell, presumably through a mating bridge, just like a self-transmissible plasmid. The copy transferred to the recipient cell is eventually integrated into the chromosome. A conjugative transposon possess an attachment site, with which it is inserted into the chromosome. Replication of the circularized free transposon occurs by rolling circle model originating from a site, OriT, as in the case of a self-transmissible plasmid. Conjugative transposons have been reported in several bacteria, where they act as potential carriers of bacterial genes. Conjugative transposons excise and integrate into DNA, through a different method by having a covalently closed circular transposition intermediate and do not duplicate their target site while integrating into the DNA. They act like plasmid by having a covalently closed circular transfer intermediate and are transferred during conjugation. But the circular intermediate of conjugative transposon does not replicate, as in the case of plasmids.

3.2.5 Invertrons

The ends of linear chromosomes called telomeres poses two problems that are not applicable to circular chromosomes. 1. The free double-stranded DNA ends are very sensitive to degradation by intracellular nucleases, necessitating a mechanism to protect them. 2. The ends of linear DNA molecules must have a special mechanism for DNA replication. These problems are solved by the properties of the telomeres. Bacterial chromosomes contain two different types of telomeres hairpin telomeres and invertron telomeres. Invertrons are the genetic elements composed of DNA with inverted terminal repeats at both 5′ and 3′ ends. Inverted terminal repeats at the 5′ end are covalently bonded to terminal proteins at 5'end, involved in the initiation of DNA replication. DNA polymerase interacts with the 5′ terminal protein at the telomere and catalyzes the formation of a covalent bond between the terminal protein and a dNTP. The dNTP bound to the terminal protein has a free 3'-OH group, which acts as the primer for the chain elongation. Invertrons also function as viruses, linear DNA plasmids, transposable elements, and sometimes combinations of two of these properties. They differ from retroviruses and related retro-type transposons, by having direct repeats on both their genomic ends and exploit RNA intermediates for replication of their DNA.

3.2.6 Integrons

Integrons are highly versatile genetic elements that are commonly found in the bacterial genomes, involved in the acquisition and expression of exogenous genes. They are ancient genetic elements that are considered as hot spots for the genomic complexity, generating phenotypic diversity and shaping adaptive responses to the bacteria. They play a major role in the acquisition, expression, and dissemination of antibiotic resistance genes, particularly in Gram-negative bacteria. They occur in all environments and are able to move between species and lineages over evolutionary time frames and have access to a vast pool of novel genes whose functions are largely unknown. All integrons have three essential activities whose combined action will express exogenous genes. 1. They possess an *intI* gene, which encodes for an integrase (IntI), a member of the tyrosine recombinase family, characterized by the presence of RHRY (with Y being the catalytic tyrosine) amino-acids in the box 1 and box 2 conserved motifs. IntI catalyzed recombination between *attI* and/or *attC* sites resulting in insertion or excision of gene cassettes. 2. An integron-associated recombination site, *attI*. 3. An integron-associated promoter Pc, which facilitates the expression after the recombination of the cassette. Integrons acquire new gene cassettes consisting of a single open reading frame bounded by a 59-bases cassette-associated recombination site, called as *attC*. Circular cassettes get integrated by the site-specific recombination between *attI* and *attC*, mediated by the Intl. This process can be reversible, leading to the excision of the cassettes as free circular DNA elements. Insertion at *attI* site allows the expression of an incoming cassette, driven by the adjacent Pc promoter.

3.2.6.1 Classes of Integrons

Based on the differences and divergence in the sequences of *intI*, integrons are classified into four classes, from class1 to class 4. Most of the studies had been conducted on class 1 integrons of Gram-negative microorganisms. Class 1 integrons are reported in ~9% of the sequenced bacterial genomes and class 1 integron platform is one of the most ubiquitous platforms, which has been reported among bacteria with clinical significance. This class of integrons is more similar to Tn402-

like transposons of Tn3 transposon family. Class 1 integrons are not capable of self moving, but moves within the or between species with the aid of mobile genetic elements, such as conjugative plasmids, and serves as vehicles for the intraspecies and interspecies. Site-specific recombination in Class 1 Integrons are mediated by IntI1, recognizing three types of recombination sites; *attI1*, *attC* and secondary sites with recombination between *attI1* and *attC* sites, showing more efficiency than the previous two sites. This increased efficiency has enabled the class 1 integrons with increased expression levels through a promoter located in the 5′-conserved segment. Role of class 1 integrons in providing the antimicrobial resistance has been well studied in Gram-negative bacteria, associated with a wide variety of resistance gene cassettes that encodes streptomycin, spectinomycin and Trimethoprim resistance. Class 1 integrons have also been reported in Gram-positive bacteria such as *Corynebacterium*, *Streptococcus*, *Enterococcus*, *Staphylococcus*, *Aerococcus,* and *Brevibacterium.*

Class 2 integrons are associated with the Tn7 transposon family members such as Tn1825, Tn1826 and Tn4132, carrying their recombination site *attI2* and promoter Pc. Its 3′ conserved region contains 5 tns genes tnsA, tnsB, tnsC, tnsD and tnsE, which functions in the movements of a transposon, mediating the mobility of class 2 integrons *via* a preferential insertion into a unique site within bacterial chromosomes. The homology of intI2 gene amino-acid sequences are found to be less than 50% with that of intI1 and remains unfunctional due to the replacement of the termination codon with glutamic acid at 179th position of the amino acid, producing a shorter and inactive polypeptide, unable to catalyse the recombination reaction resulting in the formation of a pseudogene. Class 2 integrons are considered to be a major contributor to the distribution and widespread antibiotic resistance among bacteria. This class of integrons is reported to be present only in the Gram-negative bacteria such as *Acinetobacter*, *Enterobacteriaceae*, *Salmonella,* and *Psuedomonas*, with a low occurrence and prevalence than class 1 integrons.

The occurrence and identification rate of class 3 integrons have been ranged from 0 to 10%. Class 3 integrons are structurally similar to Class II integrons, with *IntI1* and *IntI3* found in soil and freshwater proteobacteria, *IntI2* found in the marine γ-proteobacteria, sharing a similar function. Integration of cassettes into the *attI3* site occurs with significantly lower recombination frequencies. Class 3 integron was first identified from *Serratia marcescens* isolates in Japan in 1993. Its presence has been reported in bacteria such as *Acinetobacter* spp., *Alcaligenes*, *Citrobacter freundii*, *Escherichia coli*, *Klebsiella pneumoniae*, *Pseudomonas aeruginosa*, *Pseudomonas putida*, *Salmonella* spp. and *Serratia marcescens*. Surveillance of 587 Gram-negative bacteria has demonstrated a high-level resistance to both ceftazidime and sulbactam-cefoperazone, with 0.7% of the isolates harboring class 3 integrons.

Class 4 integrons were first reported in *Vibrio* isolates, with its existence pre-dating the antibiotic era. This class of integrons was distinguished from other classes by two unique features;

1. Incorporating hundreds of gene cassettes in *V. cholera* at a rate of 216 unidentified genes in 179 cassettes, occupying approximate 3% of the genome.

2. High homology between the *attC* sites of these gathered cassettes.

Identification of class 4 integrons has been limited to bacterial families such as *Vibrionaceae* and genera of *Shewanella*, *Xanthomonas*, *Pseudomonad,* and *proteobacteria*. This class of integrons has been found to carry gene cassettes imparting resistance to the antibiotics chloramphenicol and fosfomycin.

3.2.6.2 Integrons in Foodborne Bacteria

Food-borne infections and diseases are one of the leading concerns in public health and food safety, which have been reported to be caused by a large variety of pathogens that contaminate food and food-related products. Major foodborne pathogens such as *Staphylococcus aureus, E. coli* O157, *Vibrio parahaemolyticus*, *Salmonella* species, and *Listeria monocytogenes* are identified from food poisoning, contamination of various foods such as milk, pork, chicken, veal, beef, turkey and lamb meat, in food production animals such as cattle, chickens, pigs and cows, responsible for 14 million illnesses, 60,000 hospitalizations and 1800 deaths annually. Antibiotic-resistant foodborne pathogens have been considered to be a major contributor to both health-care and associated food-borne illnesses. Antimicrobial resistance conducted on 96 foodborne strains of *Salmonella*, *E. coli* and *S. aureus* reveals the role of class 1 integrons and the abilities of biofilm formation, which may offer significant guidance in effective control on the dissemination of antibiotic resistance of foodborne bacteria.

3.2.7 Integratable Plasmids and Phages

Addition of DNA into the bacteria will confer new or altered properties to their genomes. tRNA genes are one of the important sites for integration and chromosome expansion, as observed in *Mycobacterium smegmatis* and BCG. Phage L5 integrates into a 43 bp *attB* core site, overlapping a putative tRNA gene in these bacteria through site-specific recombinations, without altering the tRNA sequence. Integration of bacteriophages usually occurs in tRNA genes and several virulence genes, such as pathogenicity islands of various sizes. Since the discovery of transposons by Mc Clintock in 1950, questions regarding their evolution were frequently asked. Integration of transposons into the bacterial chromosome might provide the adaptation and advantage by promoting rapid evolution through increased plasticity to bacterial genomes by deletions and rearrangements. A bacterial rearrangement in response to environmental stress factors involving IS elements have been documented in *E. coli* cultures stored for long durations. Developmentally regulated rearrangements are reported to occur in *Anabaena* during heterocyst formation and in *B. subtilis* during sporulation, where a non-replicative DNA circle is excised from the chromosome catalyzed by a site-specific recombinase, *SpoIV CA*, leading to the formation of a gene fusion encoding for sigma K transcription.

3.3 Genome Rearrangements in Bacteria

Bacterial genomes are highly dynamic with remarkable stability from one generation to the other. At the same time, they are also highly plastic on the evolutionary time scale, mostly due to the genome rearrangements and the specific activities of transposon elements. Bacterial chromosomes are highly complex and at the same time show certain dynamic characteristics that give more flexibility to their genomes. Genome rearrangements in bacteria that include deletions, duplications, amplifications, insertions, inversions, translocations that occur in the bacterial genomes, might cause mutations and genetic instability affecting the phenotype. Few of the mutations can be silent not affecting the phenotype, while other mutations can lead to a phenotypic variation, evolution, and even speciation. Genome rearrangements can change the amount of genetic information present in the chromosome and even can disrupt the genes. Most of the genomic rearrangements result in the formation of new sequences at the rearrangement sites, with the capability of changing the protein function and sometimes even the expression of proteins. Some of the genomic rearrangements have an impact on the structure and integrity of the bacterial chromosome (s).

3.3.1 Rearrangements Due to Mobile Genetic Elements

Several types of genetic elements such as miniature inverted-repeat transposable elements (MITEs), repetitive extragenic palindromic (REP) sequences, bacterial interspersed mosaic elements [BIMEs], transposons, transposable bacteriophages, genomic islands, inteins, introns, retroelements, and integrons generate a wide variety of genomic rearrangements in the bacterial genomes. Transposable elements are highly abundant in bacteria, which lives under extreme habitats. These elements are the nucleotide sequences with specific ends, capable of moving within the locus and some times between genomes through excision and integration, independent of homologous recombinations. Most of the bacterial transposons contain short terminal inverted repeats and use transposases for recognizing and processing the element ends. They tend to duplicate their target sequence in which they integrate, thus creating a short direct-repeat called target site duplication sequence. The main function of transposase is to select the target site, which differs at the level of sequence specificity and stringency. The target sites of some elements such as Tn7 potentially display a high target specificity, whereas the elements such as Tn5 display low target specificity (bacterial transposons and their classification are discussed in detail in the section). Along with the transposable elements, other mobile bacterial elements such as MITEs, REP sequences, BIMEs, etc., also regulate their mobility for avoiding excessive mutagenesis, which would be detrimental for the bacterial cell.

3.3.1.1 Rearrangements Due to Miniature Inverted-Repeat Transposable Elements (MITEs)

MITEs are a short stretch of AT-rich DNA sequences with a length 0.1–0.5 kb, consisting of terminal inverted repeats. These elements often display a TA dinucleotide motif at their ends, flanked by target-site duplications. They are the DNA elements possessing a recognition sequence necessary for their mobility without encoding for a transposase. Several MITEs have been reported in bacteria including CE/NEMIS/CREE/SRE in *Neisseria*, RUP, BOX, and SPRITE in *Streptococcus*. Insertion of a MITE will lead to the addition of genetic material, inactivation of a gene modulates the transcription levels of neighboring genes by introducing a regulatory binding site or by changing the DNA topology at an insertion site. There is a possibility that the recombination of two different MITE sequences can lead to the deletion of large chromosomal regions or intrachromosomal rearrangements. Frequently, a MITE that encodes the insertion of one or several ORFs into a host gene will lead to an *in-frame* gene fusion and the formation of a novel protein. An inserted ORF encodes for a specific motif that is responsible for the change in its function or sometimes changes the localization of the protein. MITEs provide a scheme of actions on their host genome, which are slightly detrimental to highly beneficial. Further studies on bacterial MITEs are in the process to reveal their origins and intrinsic cellular functions.

3.3.1.2 REP and BMI Elements

Repetitive Extragenic Palindromic (REP) sequences are highly repeated imperfect palindromes with 20–40 nucleotides, mostly extragenic but transcribed in genomic regions. They were first discovered in the genomes of enteric bacteria, later found in other bacteria. *E. coli* genome consists of more than 600 REP sequences, corresponding to 1% of the bacterial genome. Most of the REP sequences are found as single occurrences but are more often organized in pairs or in clusters. REPsequences regulates the differential gene expression in operons, by folding into stable RNA structures, protecting the upstream mRNAs from exonuclease degradation. Bacterial Mosaic Interspersed Elements (BMIE) is a pair of REP sequences in an inverse orientation separated by a short linker sequence containing conserved sequence motifs. The genome of *E. coli* reported containing 250 BIMEs, located in the GC-rich regions of the genome. BIMEs specifically interact with a number of proteins that are involved in the bacterial nucleoid organization. Together, BIMEs and REP sequences have an important effect on the genome rearrangement, bacterial evolution, and speciation.

3.3.2 Transposons

Transposons are the mobile genetic elements with a variable size ranging from ~2.5 to 60 kb, consisting of long terminal inverted repeats that are intermittently arranged between one or multiple accessory genes that codes for antibiotic, heavy metal, phage resistance, catabolic, vitamin, antimicrobial compound synthesis, nitrogen fixation pathways. Based on structure and properties, transposons have been classified into three major types;

1. Composite (also called compound) transposons are flanked on both sides by the indistinguishable Insertion Sequences. At least one of the IS sequences encodes for a functional transposase, permitting their transposition together with the sequence that separates them.
2. Conjugative transposons are capable of transposing either intracellularly or excise to transfer intercellularly through conjugation. These transposable elements have the characteristic features of phage, plasmid, transposon and are transmissible among bacteria.
3. Mobilizable transposons are not capable of self transmission. They can be mobilized by conjugative elements.

Conjugative transposons, such as Tn5397 have a very strong preference for the insertion sites. Upon insertion, a conjugative transposon either disrupts a gene function or can modify the regulation of neighboring genes. Insertion of these transposons can also induce certain genomic rearrangements, such as deletions, duplications, inversions, induce the formation of co-integrates.

3.3.2.1 Transposable Bacteriophages

Transposable bacteriophages are the viruses that transpose their genomic DNA into a bacterial chromosome or plasmid and duplicates the inserted sequence at the insertion site during this process. These transposable phages stay in their bacterial host genomes as latent prophages or can replicate actively. Insertion of these phages into a gene or into its regulatory region might lead in its inactivation. Mutations created by these elements might inactivate the downstream genes of the same operon due to polar effect. Transposable bacteriophages can induce a variety of genomic rearrangements in bacteria through deletions, inversions or by the formation of cointegrates. Transposable phages can stimulate the mobility of other bacteriophages or even induce the recombination between transposable elements.

3.3.3 Genomic Islands in Bacteria

Genomic Islands (GIs) are the large DNA sequences found in the genomes of certain bacteria, plasmids and also in the bacteriophages. These islands encode for a number of accessory genes, which enhances the chances of survival or colonization of bacterium, offering a selective advantage. The size of a genomic island is usually between 10 and 200 kb in length. Small regions of the island show similar characteristic features are known as genomic islets. Insertion of a new genomic island can lead to a total change in the phenotype, its behavior, or even the lifestyle of the receiving organism. A genomic island can be a pathogenicity island (*Salmonella* SPI1), a fitness island (acid fitness island AFI of *E. coli*), a metabolic island (production island of *Xanthomonas xanthangum*), a resistance island IAbaR7 of

Acinetobacter baumannii, symbiosis island of *Mesorhizobium loti* strain R7A, a saprophytic island that encodes adhesins in some *E. coli* strains), an ecological island, which permits phenol degradation in *Pseudomonas putida*, defense island (strain ANA-3 of *Shewanella* sp.). Similar genomic islands in different bacteria may show distinct functions, under specific ecological conditions. A bacterium may contain several genomic islands in its genome responsible for different functions. GC percentage, frequency of repetitive sequences and the codon usage by the genomic islands are mostly different from the rest of the chromosome, which indicates that these islands might have got imported through horizontal gene transfer.

3.3.4 Inteins

An intein is a mobile DNA element that encodes for a peptide and undergoes self-splicing from the host protein, immediately after its translation. There are inteins that encode peptides of size from 134 to 608 amino acids within the frame of host proteins. A single host protein might contain several inteins. Through posttranslational processing, an intein self catalyzes its precise excision. This process is followed by the concomitant ligation of its flanking regions, resulting in the formation of a mature functional host protein. Inteins do not require any co-factors or other accessory proteins for protein splicing. They often encode for a homing endonuclease domain that is essential for their mobility. This domain is also encoded in the frame intein and host protein. Interestingly, a protein that contains an intein can be encoded by two partial genes localized at different places of a genome. Each partial gene encodes for a part of the protein fused to a part of the intein so that the original gene is disrupted within the intein. After translation, the two halves of the corresponding protein will be fused by the intein through trans-splicing, resulting in the formation of an active whole protein. Inteins generally interrupt the highly conserved regions of the essential proteins. Inteins are found to be located in the proteins involved in the metabolism and biosynthesis of nucleic acids, cell division, transcription, DNA replication, repair, and recombination. Inteins have become highly useful molecular tools as self-cleavable affinity tags for protein purification or to ligate expressed protein by trans splicing. Expression of inteins as DNA repair and recombination proteins during the same time would help the cell in recovering from an unwanted DNA cleavage.

3.3.5 Introns

An intron is an intragenic DNA element that encodes an RNA and splices out after its transcription. Two major groups of introns are reported in bacteria: group I introns and group II introns. Both of them are transposable elements. The size of bacterial introns ranges from 0.2-1 kb to 0.7-3 kb, respectively. These introns can be transcribed into self-splicing RNAs. Introns are widespread among the bacterial kingdom but are not so abundant.

A group I intron is transcribed into a self-splicing RNA with 10 helices capped by loops and are joined at the junctions. A site-specific homing endonuclease is used by the intron to invade another DNA molecule that is generally encoded within the terminal loop. In bacterial chromosomes, the group I introns are found in tRNAs, rRNAs, and essential genes. Some group I introns are not capable of splicing out after transcription, showing detrimental effects on the cell growth. Functionally, group II introns behave more as retroelements than as introns. They get transcribed into a highly structured RNAs with six distinct double-helical domains, which provides the catalytic activity for splicing. Group II introns also encode for a protein with an endonuclease, reverse transcriptase and RNA maturase activities in their fourth domain, responsible for their mobility. Bacterial group II introns are often fragmented. However, mostly they get integrated into mobile elements, which enhances their proliferation chances. Rare integration of group II intron into a host gene will inactivate the targeted gene only when the intron could not splice out. In bacteria, group II introns can also alter the sequence of the host protein by undergoing alternative splicing. They can induce deletions, inversions, or other chromosomal rearrangements in their host genome through this process.

3.3.6 Homing Endonucleases

Homing endonucleases are the functional mobile elements that are encoded within the introns and inteins of bacteria. They can also be freestanding in intergenic regions. Homing endonucleases bring mobility to the splicing elements in exchange for the capacity to target conserved genes without being detrimental to the host bacterium. A splicing element avoids being counter selected by not disturbing the function of its host protein. Deletion of an element may be counter-selected, as imprecise excision is likely to damage the host gene. This association increases the chances of homing endonuclease and the splicing element being maintained in a population and invading other bacteria through horizontal gene transfer. During the homing process, an endonuclease cleaves the target DNA followed by the gene conversion that occurs during the DNA repair, when the splicing element is copied into the previously empty allele. Insertion of a splicing element disrupts the recognition site of the homing endonuclease, preventing new cleavage. Homing endonucleases are encoded by short genes of size less than 1 kb. They recognize the specific DNA sequence of 12–44 residues, allowing a few single-base-pair changes within this target sequence, which is often present only once by a genome. This rare cutting characteristic feature helps to maximize their mobility and to minimize non-specific cleavage, making them excellent biological tools. Some homing endonucleases need associated proteins to regulate their activity and certain homing endonucleases also encode a maturase activity helping intron RNA splicing and a reverse transcriptase activity for the retrohoming of group II introns.

3.3.7 Retro Elements

A retroelement is a transposon that encodes a reverse transcriptase that functions through an RNA intermediate. Three important retroelements in are found in bacterial genomes; groupII introns, retrons and Diversity-Generating Retroelements (DGRs). Retrons are a little rare type retro elements of the size 2 kb, that gets inserted into prophages and chromosomes of a wide variety of bacteria. These elements mainly consist of three ORFs that encode for reverse transcriptase and a peculiar DNA/RNA hybrid molecule, described as multicopy single stranded DNA (msDNA). msDNA is a small, single-stranded cDNA molecule, covalently bound to an RNA molecule, which folds together into a stable secondary structure. Accumulation of msDNA is abundant in its host, at a rate of 1000 molecules per cell. Till date, no experimental evidence is available to show that the retrons are mobile elements. However, truncated copies of msDNAs are found inserted into some bacterial genomes. Upon integration, retrons replace various sizes of sequences in their host genomic DNA, leading to an increase in the mutagenic level. This increase might be due to the binding of cellular mismatch repair proteins to the mismatches on msDNA molecules. The exact function of msDNA molecules in bacterial cells is still unknown. It is hypothesized to be involved in helping the host bacteria in increasing their mutation rates when needed for their survival.

Diversity-generating retroelements (DGRs) are found in a wide range of bacteria. These DGRs are composed of two repeats with the length 150 bp and two ORFs. The first repeat is the variable repeat (VR), which forms 3′ end of a gene, as its sequence can undergo nucleotide substitutions at variable hotspots. Downstream to the VR is the second repeat, called template repeat (TR) with an invariant sequence and the ORFs. One ORF encodes for reverse transcriptase. Mutations in the sequence of VR exchange a dATP for a random deoxyribonucleotide. The mechanism creating this directed mutagenesis involves the reverse transcriptase and a unidirectional transfer of information from the TR to the VR. DGR induces a very high variability in the sequence of the protein encoded by the targeted gene. A number of DGRs are found in the surface proteins, assuming that DGRs may play an important role in the interaction of a bacterium with its environment. Future studies might reveal some unknown effects of these elements on genome rearrangements.

3.4 Evolution of Bacterial Genomes

Bacteria are considered as the most ancient, highly abundant and genetically diverse organisms on this planet. Information provided by the fully sequenced genomes states that the diversity among bacterial genomes is the result of genetic variations occurring due to minor changes in the DNA and intragenomic re-shuffling of DNA sequences in the genome due to transformation, conjugation, transposition, transduction. Changes in the DNA might occur because of the mutations, natural selection, and genetic drift. The role of mutations and natural selection in influencing the evolutionary trajectory of genome complexity are relatively well understood, compared with genetic drift.

It has been estimated that bacteria might have evolved ~3.8 billion years back on this planet. The currently existing bacterial life forms show enormous diversity, particularly their capabilities to adapt extreme environmental conditions, such as chemical composition, pH, temperature and pressure. Many bacteria have become symbionts or pathogens and have adapted to living in close community with other organisms and this adaptation might have occurred recently as evidenced by many numbers of evolutionary genes existing in today's bacteria. Many of the evolutionary genes and transposable elements have undergone a long evolution and are now fine-tuned for their present functions. Theory of molecular evolution postulates that the products of evolution and genes work hand in hand with other factors such as structural and stability features of biological macromolecules, the action of chemical and physical mutagens, as well as the chance of random encounters of interactive components.

3.4.1 Role of Mutations in Bacterial Genome Evolution

Point mutations are quite often being thought as a raw material of evolution, which brings small changes in the genome by substituting one nucleotide with another, referred as a single-nucleotide polymorphism (SNP) or insertion or deletion of a single nucleotide. Substitution of a single nucleotide in a DNA sequence encoding a protein may generate either a synonymous mutation, without any change in the encoded amino acid, or a non-synonymous mutation, leading to a change in the amino acid with an impact on that protein function. Single nucleotide insertion or deletion in protein coding sequences will result in a frameshift such that the codons located in the downstream to the mutated codon along with the stop codons will be translated from a different reading frame, leading to significant alteration of the encoded protein. Point mutations in non-protein-coding DNA sequences might also have consequences, particularly if they affect a regulatory element. Whatever may be the consequence, a point mutation in the organism might imply a selective advantage, used for comparing the genomes of related individuals for the construction of phylogenetic trees, rate, type and distribution of point mutations, which gives an insight of the evolutionary pressures on bacteria.

Increasing volume of whole genome sequence data available for bacteria reveals that the most of the point mutations are biased towards C/G → T/A transitions, particularly in the bacterial genomes with high GC contents, suggesting that in the absence of selection pressure, bacterial genomes would approach an equilibrium GC content of 20–30%. This observation suggests that mutational pressures alone are not responsible for the nucleotide composition in the majority of bacterial genomes and selection probably plays an important role. GC content of bacteria is also dependent on the habitat they were isolated from, suggesting that the environment may also be an important contributor in the genome composition. Understanding the dynamic flux of mutations and selection in the nucleotide composition of bacteria will give us a greater resolution of understanding of the evolution of bacterial genomes.

3.4.2 Role of Recombinations in Bacterial Genome Evolution

Homologous recombination is one of the vital mechanisms, allowing the maintenance of genetic diversity among bacterial populations. These recombinations also provide a mechanism for the bacteria to make large evolutionary leaps, such as the acquisition of drug resistance and antibiotic resistance. Homologous recombination involves the interaction of two sequences with a high nucleotide identity. However, more distantly related sequences are exchanged or inserted into the bacterial genomes through horizontal gene transfer. Site-specific recombination can also mediate the integration of phage genomes or conjugative elements by short stretches of homology between specific sequences of foreign and chromosomal DNA of bacteria.

Homologous recombination is also shown to be involved in the process of speciation as illustrated in whole genome shotgun study of two *Vibrio cyclitrophicus* populations that live in different habitats, which carry a few discrete regions in their genomes containing habitat-specific genes. These regions were introduced into the genome of one population by homologous recombination. There was also a possibility of genetic exchange within or between habitats, which might have accelerated the speciation process.

3.4.3 Evolution of Bacterial Pathogens

Bacteria are considered as one of the first inhabitants of the earth. Evolution of bacteria was started much before then the emergence of animals which involved competition, genetic exchange, and selection. Eukaryotes and multicellular organisms have evolved during the past one billion years and the proliferation of mammals specifically occurred during the past 65 million years. This evolution clock shows that bacteria became pathogenic in one million years. Understanding the similarities and differences among the bacteria and linking these aspects with evolution, geographical locations and disease associations of the bacterial pathogens will provide vital information for identifying the sources responsible for the infection, which might be useful in designing the preventive measures for controlling and treatment of a disease.

Whole genome shotgun sequencing of bacterial genomes and sequencing of multiple species, strains facilitated to list the complete set of genes responsible for pathogenicity, interaction with specific host, identifying the unknown systems such as horizontal gene transfer, mobile elements such as phages, plasmids and pathogenic islands responsible for the generation of pathogenicity and drug resistance among bacteria. The pairwise comparison of laboratory strain *E. coli* K-12 and pathogenic *E. coli* O157: H7 genomes have revealed that 26% of the genes are unique to the pathogenic strain. Increasing number of comparisons within the species led to the observation of enormous genetic diversity within the bacterial species in terms of unique gene number, which has formalized the concept of 'pangenome' referring to the total number of genes present in a species including the number of genes common to the genus and the number of accessory genes specific to the adaptation of that strain, most of them are often pathogenicity related.

Some important human bacterial pathogens like *Yersinia pestis*, *Bordetella pertussis* and *Salmonella typhi* are highly monomorphic with little mobile DNA and low nucleotide genetic diversity. Genome analyses of such pathogens have revealed many surprising and interesting features regarding their evolution. The genomes of these monomorphic bacterial pathogens are having a large amount of repetitive, selfish DNA in the form of insertion sequences and large numbers of pseudogenes, which were involved in the pathogenicity or host interactions. Many of the monomorphic pathogens were the host-restricted descendants of a broad range of ancestors, indicating that these organisms invaded a new niche as exemplified by the *Salmonella typhi* and *Yarsenia pestis*, which migrated from gastrointestinal to systemic disease by acquiring novel pathogenicity genes on plasmids (*Yarsenia pestis*), on mobile chromosomal elements (*Salmonella typhi*) accompanied by the accelerated mutations due to the evolutionary bottleneck leading to loss genes required for the previous niche.

Bordetella pertussis, animal pathogens (*Burkholderia mallei*, causes glanders in horses) and plant pathogens (*Xanthomonas oryzae* of rice) manifested another resourceful example of the niche change and increase in pathogenicity has occurred purely through genome degradation. The common factor linking all of these evolutionary transitions is the Neolithic revolution, during which the human population was dense and human settlements were more permanent, creating a new niche for human pathogens. Human civilizations have generated large monocultures of crop plants and domestic animals, creating similar novel niches for these pathogens.

3.5 Genetic Diversity of Pathogenic Bacteria

Even though the pathogenic bacteria constitute a very small proportion of the bacterial species, they are characterized by high genetic diversity and genotypic variations, which poses a major barrier for their disease control. Among the pathogenic bacteria, diversity of bacterial pathogens infecting humans and other animal hosts is highly studied compared with that of plant pathogens.

3.5.1 Mechanisms of Genetic Diversity

Bacteria possess several mechanisms of gene shuffling including point mutations, gene duplication, gene silencing, gene rearrangement and gene loss, which might have lead to high genetic diversification. Random genetic mutations (also called pathogenicity adaptive mutations) in the pre-existing genes result in the functional modification and elimination of original genes, contributing to bacterial virulence. These mutations result in the emergence of highly pathogenic bacteria by host-specific adaptations to that pathogen as in the case of uropathogenic *E. coli* strains carrying mutations in *fim H* gene for the adhesive subunit of type I fimbriae, which increases the ability to recognize the mono-mannose receptors, exhibiting higher tropism for their survival in uroepithelium. These mono-mannose specific *E. coli* strains are possessing a better ability to colonize bladder than that of wild types, providing a selective advantage to their virulence. Point mutations also generate antigenic diversity, by evasion from the immune response of the host, leading to stabilization. Deletions or loss of resident genes in an organism also plays an important role in the generation of pathogen diversity. Deletions in *the muc A* gene of *Pseudomonas aeruginosa*, leads to the overproduction of alginate, providing an additional advantage for their survival in the human lungs. Transposable elements such as transposons and IS elements promotes the transfer of genes between phylogenetically diverse populations. IS elements are responsible for DNA rearrangements such as deletions, inversions, gene amplifications and fusion of two DNA molecules by co-integrate formation. Transposons such as Tn5 propagates antibiotic-resistant genes and confers resistance to antibiotics such as kanamycin, bleomycin, and streptomycin. Integrons are also involved in the acquisition of several promoterless virulence genes by incorporating them next to its promoter, leading to enhanced resistance and pathogenicity of an organism.

3.5.2 Horizontal Gene Transfer

Horizontal gene transfer also called lateral gene transfer refers to the exchange of genes between different species. It is an evolutionary phenomenon, which challenges the traditional evolution of life and the core neo-Darwinist belief in the role of reproductive isolation between species in evolution. The role of horizontal gene transfer in bacteria was debatable until the development of high-throughput sequencing and comparative genomics, which has generated unprecedented pieces of evidence regarding the presence of this mechanism in bacteria. The extent of horizontal gene transfer in bacteria varies from 0% in *Borrelia burgdorferi* to 12.8% in *E. coli*. Horizontal gene transfer plays a vital role in the emergence of new and highly virulent pathogens like enterohaemorrhagic *E. coli* O157: H7, conferring novel capabilities to the recipients, which enable them to conquer new virulence niches, ultimately leading to diversification of natural populations as illustrated by *Streptococcus pyogenes* and penicillin-resistant *Streptococcus pneumoniae*. The presence of scarlet fever toxin-encoding *spe* A1 allele among the diverse lineages of *S. pyogenes* refers to the horizontal distribution of this allele among diversified clones of this organism. In penicillin-resistant *S. pneumoniae*, horizontal gene transfer followed by the recombination has resulted in the generation of molecular remodeled penicillin-binding proteins with reduced affinity to penicillin. Multiple horizontal gene transfers among *Staphylococcus aureus* have made them resistant to methicillin globally. Although transformation, transduction, and conjugation are the processes involved in the horizontal gene transfer of pathogenic bacteria, bacteriophage-mediated transduction of genetic material between organisms is the most common phenomenon that occurs in pathogenic bacteria. Phages carry virulence genes within their genomes, which can convert a normal strain to pathogenic upon lysogenization, as in the case of filamentous phage genome encoding the cholera toxin genes of *Vibrio cholerae*. Bacteriophages are also capable of mediating the transfer of large DNA fragments through transduction, generating genome variability.

3.5.3 Pathogenicity Islands

Pathogenicity islands (PAIs) acquired by the horizontal gene transfers are one of the major contributors to the pathogenesis of certain bacteria. Pathogenic islands are one of the subsets of genomic islands first described in *E. coli* but now have been identified in the genomes of pathogenic bacteria infecting plants, animals, and humans. Pathogenicity islands are characterized by large clusters of virulence traits, occupying large portions of the chromosome ranging from 10 to 100 Kb, residing at t-RNA or t-RNA like loci and are flanked by direct repeats, which mediates their mobility. Some bacterial strains also harbor pathogenicity islets, which are small DNA fragments of 1–10 Kb size. Comparison of the ORFs, G + C content and codon usage of the pathogenicity islands with the genome they reside reveals that they are acquired through horizontal gene transfer from either phages or integrases. Genomic comparison of virulent, less virulent and avirulent bacterial strains by subtractive hybridization suggests that pathogenicity islands play an important role in conferring virulence to pathogenic strains. Two pathogenic islands, PAI I (encoding *a*-hemolysin) and PAI II (encoding *a*-hemolysin and *P*-related fimbriae) are involved in the pathogenicity of uropathogenic *E. coli*. Acquisition of pathogenicity islands can also transform an organism into a potential pathogen, which plays an important role in the evolution and establishment of novel pathogens as studied in *Yersinia pestis*, which causes plague. The population genetic structure analysis of *Y. pestis* and two other species, *Y. pseudotuberculosis* and *Y. enterocolitica* have shown that highly virulent *Y. pestis* has emerged from *Y. pseudotuberculosis* through the horizontal gene transfer of its high pathogenicity islands (HPI).

3.5.4 Genetic Diversity and Origin of New Bacterial Pathogens

Genetic diversity studies have provided new insights for identifying the origin of several bacterial pathogens such as enterohaemorrhagic *E. coli* or *E. coli* O157: H7, an important food-borne pathogen identified two decades back, which caused large outbreaks of the fatal hemolytic uremic syndrome (HUS) and ulcerative colitis. 1300 isolates of *E. coli* O157: H7 were compared with 16 *E. coli* serotypes by Multi Locus Enzyme electrophoresis (MLEE) and probing for Shiga like toxin genes revealed that *E. coli* O157: H7 is closely related to *E. coli* O55: H7. Large-scale DNA sequencing data indicated that *Mycobacterium tuberculosis* was evolved ~15,000–20,000 years ago from *M. bovis* from a common ancestor. Similar studies also show that *Yersinia pestis*, originated from *Y. pseudotuberculosis* ~1500–20,000 years ago.

3.5.5 Techniques Used for Studying the Genetic Diversity of Pathogenic Bacteria

Genetic diversity of infectious microorganisms has been studied using various techniques such as restriction analysis of chromosomal DNA (REAC), restriction fragment length polymorphism (RFLP), randomly amplified polymorphic DNA (RAPD), pulsed-field gel electrophoresis (PFGE), amplified fragment length polymorphism (AFLP), variable number of tandem repeats (VNTR), REP repetitive extragenic palindrome-PCR, (REP-PCR), enterobacterial repetitive intergenic consensus-PCR (ERIC-PCR), BOX-PCR based fingerprinting, multilocus enzyme electrophoresis (MLEE) and single locus and multilocus sequence typing (MLST) including multi-virulence locus sequence typing (MVLST). Using these techniques, enormous data has been generated on the genetic diversity of bacterial pathogens. However, many of these techniques suffer from the drawbacks like high cost, lack of reproducibility, tedious methodology, difficulty in interpretation of results and non-applicability. In spite of these drawbacks, only MLST seems to be the technique for undertaking genetic diversity of pathogenic bacteria in the present situation. With high throughput DNA sequencing and high-density re-sequencing microarrays becoming economical and within the reach of most of the laboratories, studies on genetic diversity using high throughput sequencing is going to change rapidly.

3.6 Genome of *Escherichia coli* K-12 Strain

Escherichia coli is a Gram-negative rod-shaped bacterium that is commonly found in the lower intestine of warm-blooded organisms (endotherms) and is a major component of the biosphere. Forms of *E. coli* even exist as pathogens, responsible for enteric, urinary, pulmonary, nervous infections. Harmless strains are part of the normal gut flora, which can benefit their hosts by producing vitamin K2 and by preventing the establishment of pathogenic bacteria within the intestine. *E. coli* are not always confined to the intestine and their ability to survive for brief periods outside the human body makes them an ideal indicator organism to test environmental samples for fecal contamination. These bacteria can also be grown easily and its genetics are comparatively simple and can be manipulated or duplicated through the process of Metagenics, making it one of the best-studied prokaryotic model organism and an important species in biotechnology and microbiology. *E. coli* was discovered by a German pediatrician and bacteriologist Theodor Escherich in 1885. In 1946, Joshua Lederberg and Edward Tatum were the first to describe the phenomenon of bacterial conjugation using *E. coli* as a model and still, it remains as the primary model for studying the conjugation. *E. coli* was an integral part of the first experiments for understanding the phage genetics and early researchers such as Seymour Benzer, used *E. coli* and phage T4 to understand the topography of the gene structure. Prior to Benzer's research, it was not known whether the gene was a linear structure, or if it had a branching pattern. The long-term evolution experiments using *E. coli*, begun by Richard Lenski in 1988, have allowed direct observation of major evolutionary shifts in the laboratory. In this experiment, one population of *E. coli* unexpectedly evolved the ability to aerobically metabolize citrate, which is extremely rare. As the inability to grow aerobically is normally used as a diagnostic criterion to differentiate *E. coli* from other closely related bacteria such as *Salmonella*, this innovation may mark a speciation event observed in the lab. By combining nanotechnologies with landscape ecology, complex habitat landscapes can be generated with details at the nanoscale. On such synthetic ecosystems, evolutionary experiments with *E. coli* have been performed in order to study the spatial biophysics of adaptation in an island biogeography on-chip. *E. coli* K12 is a cultivated strain, well-adapted to the laboratory environment. K-12 strain has lost its ability to thrive in the intestine and to form biofilms, unlike wild type strains. These properties protect the wild type strains from antibodies and other chemical attacks but require a large expenditure of energy and other material resources.

In September 1997, Frederick Blattner, Guy Plunkett III, Craig Bloc et al. have published the details of the 4,639,221 base pairs of *E. coli* K-12 genome by choosing MG1655 strain as representative because of its minimal gene manipulations. All K-12 derivatives, including MG1655, carry the *rfb-50* mutation, where an IS5 insertion results in the absence of O-antigen synthesis in the lipopolysaccharide.

3.6.1 Genome Sequencing

Sequencing of *E. coli* K-12 was carried out in sections using the M13 Janus shotgun strategy, which has proved to be one of the most efficient strategies for data collection. Janus shotgun involved 1. Initial random sequencing with a four-five folds redundancy in the Janus vector, 2. Computerized selection of the templates to be re-sequenced from the opposite end, 3. Limited primer walking. The first 1.92 Mb region of the genome positioned from 2,686,777 to 4,639,221 bp was sequenced from the overlapping set of 15–20 kb MG1655 λ clones by means of radioactive chemistry, which was deposited in GenBank during 1992 and 1995. The genome was sequenced using dye terminator fluorescence sequencing (Applied Biosystems), which provided greater speed with low cost and avoided electrophoretic compression artifacts, owing to its 50.8% G + C content, occurred practically in every gene of *E. coli*.

For the next segment of the genome positioned from 2,475,719 to 2,690,160 bp, the DNA was obtained for sequencing through pop out plasmid approach, by directly excising the non-overlapping segments from the chromosome, gel-purified and shotgunned for sequencing. The largest portion of the genome positioned from 22,551 to 2,497,976 was sequenced from M13 Janus shotguns prepared from 11 I–

Sce I fragments of 250 kb. Fragments prepared from I–Sce I method has the ability to select the size of fragment to be shotgunned, eliminating the redundant sequence at the borders between segments and DNA sequencing without intermediate cloning steps. As the isolated DNA was never subjected to any amplification, deleterious genes when present in the multicopy form are not subjected to any re-arrangements. Each I–Sce I fragment shotgun contained 15–30% random clones from all locations of the genome, arising from randomly sheared genomic fragments comigrated in the pulsed-field gel. The region between positions 0 and 22,551 did not yield a suitable I–Sce I fragment, so three lambda clones were used for the final completion of the genome, with one of them containing a deletion, to be finished by shotgun sequencing of a long-range PCR fragment. Other gaps of 36.9 kb of the genome were also re-sequenced by long-range PCR, with amplimers used directly as sequencing templates or as source material for shotguns. The completed genome sequence was deposited in GenBank on 16 January 1997.

3.6.2 Annotation of the Genome

Annotation of the sequenced *E. coli* K-12 genome was performed for (i) identifying genes, operons, regulatory sites, mobile genetic elements and other repetitive sequences in the genome; (ii) assign or suggest functions to the genome to the possible extent and (iii) relate the *E. coli* genome sequence to other organisms whose genomes are completely sequenced. The annotation included 4288 original and proposed protein-coding genes with one-third of these genes are well characterized. The postulation of genes in uncharacterized base sequences was done by selection from the ORFs on the basis of codon usage statistics, link's database of NH_2-terminal peptide sequences from *E. coli*, computer prediction of signal peptides, upstream matches to the Shine-Dalgarno ribosome binding site and other related information. Assignment of NH_2 termini posed problems as most ORFs contain multiple in-frame starting codons. This method preserves the most coding information for analysis, but it may not reflect the situation in vivo. Functions of previously known *E. coli* proteins were collected from the GenProtEC and EcoCyc databases. The functions of newly translated sequences were imparted by sequence similarity. Each gene in the sequence was assigned a unique numeric identifier such as "b". When no name has been assigned to the identified gene, it is referred to by this number. A specific physiological role was assigned if most of the hits were for a specific function such as alcohol dehydrogenase. If a less specificity was found among the hits, a general function was assigned to an ORF when a majority of the hits for one type of function, such as a class of enzymes. When the functions of the hit sequences were varied and there was no solid agreement even for the type of function, no function was assigned to the query ORF and it was counted as unknown.

The average distance between *E. coli* genes is 118 bp. Around 70 intergenic regions that are larger than 600 bp were re-evaluated for the presence of ORFs using Geneplot, DNASTAR Inc. and were searched against the entire GenBank database for DNA sequence (BLASTN) and protein coding (BLASTX) features. It was revealed that 15 of these regions contain previously unannotated ORFs which were overlooked because of their small size. Among the 55 intergenic regions, 11 regions contain sequence features such as long UTR leader sequences.

The remaining 44 are large intergenic regions falling into three classes:

1. 29 putative gene regulatory regions separated by more than 600 bp, 13 regions between transcribed ORFs among which 11 have at least one predicted promoter for each ORF. There are 16 regions between ORFs transcribed in the same direction containing one predicted promoter for the downstream ORF and several containing a terminator for the upstream ORF.
2. 7 large repetitive sequences including a huge intergenic region of 1730 bp consisting of repeat sequences such as REP or LDR.
3. 7 intergenic regions larger than 600 bp with no predicted regulatory or coding functions. 5 of these regions contain sequences which encode proteins of at least 50 amino acids, which are not expressed. These regions might contain additional, yet undiscovered functions.

Promoter and protein binding site sequences located in the upstream regions of 2436 genes were searched along with the potential regulatory sites located at the 400-bp upstream of each and every gene. The codon adaptation index (CAI), which measures the extent to which codon usage agrees with an *E. coli* reference set from highly expressed genes was calculated for each ORF according to Sharp and Li (1987). Clusters of four or more adjacent genes with CAI values as low as 0.25 were identified in this genome. Genes with exceptionally low CAI values may be with recent horizontal transfers reflecting the optimal codon usage or mutational spectrum of their previous host. The annotated sequence with accession number U00096 is available at the National Center for Biotechnology Information (NCBI) through the Entrez Genomes division, GenBank, and the BLAST databases. The version discussed here is M49.

3.6.3 Overview

E. coli genome consists of 4,639,221 bp of circular duplex DNA, with genes that encode Proteins accounting for 87.8% of the genome, stable RNAs encodes 0.8% and noncoding repeats contain 0.7% and 11% of the genome for regulatory and other functions. The origin and terminus of replication divide the genome into two oppositely replicated halves termed as replichores (Fig. 3.1). Replichore 1 replicates clockwise, presenting the leading strand of *E. coli* and the complementary strand is the leading in replichore 2. Many features of *E. coli* genome such as 7 ribosomal RNA (rRNA) operons, 53 of 86 tRNA genes are oriented with respect to replication and are expressed in the direction of replichore 1.55% of protein coding genes are also aligned with the direction of replication.

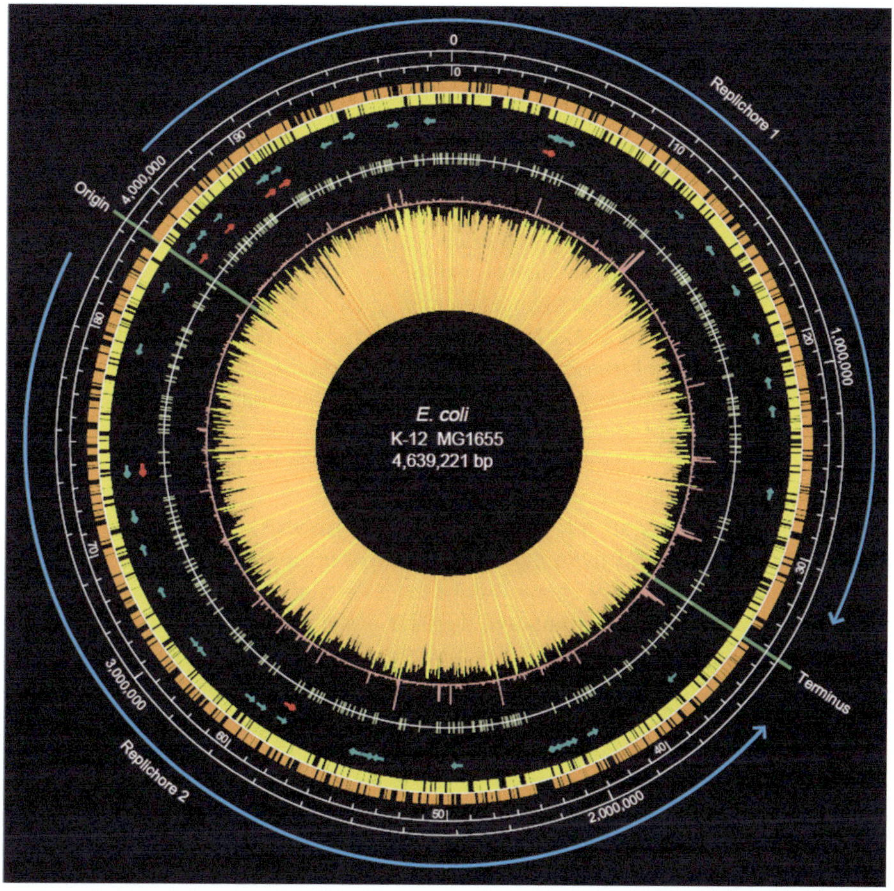

Fig. 3.1 The overall structure of the *E. coli* genome. The origin and terminus of the genome are sown in green lines, blue arrows indicate the replicores 1 and 2. Two outer rings represent the distribution of the genes with orange boxes representing the presented strand and yellow boxes represent genes in the opposite strand. Red arrows indicate the location of rRNA genes and their direction of transcription. tRNA genes are represented in green arrows. The next circle describes the positions of REP sequences around the genome as radial tick marks. Central burst represents the histogram of Codon Adaptation Index (CAI), comprising of long yellow rays representing the clusters of low CAI (<0.25) and orange burst representing the inverse CAI (Courtsey Frederick R. Blattner et al. 1997)

3.6.4 Compositional Organization of the Genome

Leading strands of both replichores have a greater abundance of G (26.22%) than C (24.58%), A (24.52%), T (24.69%). 6 new tRNA genes were discovered in this study with 4 genes *valZ*, *lysY*, *lysZ* and *lysQ* at positions 780, 291–780, 875 are part of the *lysT* operon, consisting of a duplicate of *valT* and three duplicates of *lysW*. The other two genes form a single-gene transcriptional unit, *asnW*. At position

2,056,049–2,056,124 is a duplicate copy of *asnT* and *ileY*, at position 2,783,782–2,783,857 is a copy of *ileX*, differing in a single compensating base pair change in the aminoacyl stem of the tRNA. 6 *E. coli* enzymes constituted a pathway for the degradation of aromatic compounds such as phenylpropionate, but only two genes *mhpB* and *mhpE* have been previously identified from it. Based on the similarity searches, an operon that starts with the monooxygenase gene mhpA, followed by the known dioxygenase gene *mhpB*, the hydrolase gene *mhpC*, the hydratase gene *mhpD*, the dehydrogenase gene *mhpF* and the known gene *mhpE* coding for 4-hydroxy2-oxovalerate aldolase were identified at positions 367,835–373,095; b0347–b0352. The next gene upstream at the positions 366,811–367,758; b0346 could be the regulator for this pathway, as its sequence is similar to a number of transcriptional regulators. Another unrecognized operon of *E. coli* genes at positions 2,667,052–2,671,269 Kb resembling *Pseudomonas* genes for the degradation of the aromatic compounds such as toluene, benzene, and biphenyl were also identified. Further research is needed to study the physiological and biochemical significance of this pathway.

E. coli genome has an array of 14 flagellar synthesis genes (b1070–b1083), among which only two of them *flgM* and *flgL* have been previously reported in the region between *flgM* and *flgL*, homologs of the *Salmonella typhimurium flgA* (basal-body P-ring formation), flgB (putative flagellar basal-body formation protein), *flgC* (putative flagellar basal-body formation protein), *flgD* (basal-body rod modification protein), *flgE* (flagellar hook protein), *flgF* (putative flagellar basal-body formation protein), *flgG* (flagellar basal-body formation protein), *flgH* (flagellar L-ring protein precursor), *flgI* (flagellar P-ring protein precursor), *flgJ* (flagellar protein), and *flgK* (flagellar hook-associated protein 1) genes were identified in *E. coli* genome at position 1,128,637–1,140,209 bp. The arrangement of genes in this cluster is almost identical to that of the cluster at 26.5 centisomes on *Salmonella* chromosome, making the flagellar systems of *E. coli* and *S. typhimurium* are essentially identical.

3.6.5 Open Reading Frames and Gene Function

Among the 4288 ORFs annotated in the genome sequence, 1853 are previously described genes. Complete list of *E. coli* ORFs is available at www.genetics.wisc.edu/. The distribution of start codons corresponds to ATG 3542, GTG 612, TTG130, ATT 1, CTG 1 and the distribution of translation termination codons corresponds to TAA 2705, TGA 1257 and TAG 326.4288 ORFs were searched for matches to the link database of peptides excised from two-dimensional gels, which confirmed the expression of 30 hypothetical ORFs. The longest ORF encodes a 2383 amino acid protein of unknown function, resembling several bacterial attaching and effacing proteins. The size of an average ORF is 317 amino acids, with 4 ORFs ranging from 1500 to 1700 amino acids, 51 ORFs between 1000 and 1500 amino acids and 381 ORFs are smaller than 100 amino acids. 40% of the ORFs are completely uncharacterized. *E. coli* consists of 281 transport and binding proteins, in contrast to 123 transport proteins in *Haemophilus* and 34 in *Mycoplasma genitalium*. The

Table 3.2 Distribution of *E. coli* proteins among functional groups

S. No.	Functional class	Number	Percentage (%)
1	Regulatory function	45	1.05
2	Regulatory proteins	133	3.10
3	Cell structure	182	4.24
4	Putative membrane proteins	13	0.30
5	Putative structural proteins	42	0.98
6	Phage, transposons, plasmids	87	2.03
7	Transport and binding proteins	281	6.55
8	Putative transport proteins	146	3.40
9	Energy metabolism	243	5.67
10	DNA replication, recombination modification, and repair	115	2.68
11	Transcription, RNA synthesis, metabolism, and modification	55	1.28
12	Translation, posttranslational protein modification	182	4.24
13	Cell processes (adaptation, protection)	188	4.38
14	Biosynthesis of cofactors, prosthetic	103	2.40
15	Groups, and carriers		
16	Putative chaperones	9	0.21
17	Nucleotide biosynthesis and metabolism	58	1.35
18	Amino acid biosynthesis and metabolism	131	3.06
19	Fatty acid and phospholipid metabolism	48	1.12
20	Carbon compound catabolism	130	3.03
21	Central intermediary metabolism	188	4.38
22	Putative enzymes	251	5.85
23	Other known genes	26	0.61
24	Hypothetical, unclassified, unknown	1632	38.06

number of proteins involved in the *E. coli* translation mechanism is 182 against 141 proteins of *Haemophilus* and 101 proteins of *Mycoplasma*. Among the 1827 characterized *E. coli* proteins, 75 pairs of isozymes, or multiple enzymes with identical or nearly identical function were described. An additional 11 groups of potentially redundant enzymes have been identified. 1345 proteins of *E. coli* genome have at least one paralogous sequence. The largest number of significant hits to a single protein was 37 belonging to the ABC transporter proteins. There are 54 ABC transporter proteins and an additional 26 members related to this family in the *E. coli* genome (Table 3.2).

3.6.6 Operons, Promoters and Protein Binding Sites

E. coli genome comprises of a total of 2584 predicted and known operons. Among them 73% are with only a single gene, 16.6% are with two genes, 4.6% are possessing three genes and only 6% are having four or more genes. All the operons have at least one promoter, either known or predicted. Among the 2405 operon regions with predicted promoters, 68% contain a single promoter, 20% contain two promoters, and 12% contain three or more promoters. 603 regions in the *E. coli* genome corresponding to 16% of operon regions and 10% of inter-operonic regions were identified to be regulatory. Within the regions of predicted sites, 89.2% are regulated by a single protein, 8.4% by two proteins and 2.4% by three or more proteins. 81.2% of these regions contain only a single site, 12.2% of the regions have two sites and 6.6% have three or more sites. These numbers are more or less consistent with the distribution of regulatory sites among a set of promoters, where transcriptional regulation has been well studied. In this collection of 132 promoters, 73% are regulated by one protein and 43% of the promoters contain only a single site for the regulatory binding.

3.6.7 Repeated Sequences and Insertion Sequences

A number of repeated and insertion sequences have been characterized in the *E. coli* genome, with *Rhs* elements of 5.7–9.6 kb forming the largest repeated sequences comprising 0.8% of the genome. Even though the strain comparisons suggest that these sequences might be mobile elements, they have no known function. Palindromic sequences of 40 bp length are referred to as REP, BIME, or PU constituting the largest class of repeated sequences. There are around 581 repeated sequences in 314 REP elements containing up to 12 tandem copies accounting for 0.54% of the genome with unknown origin and function. The genome of *E. coli* K-12 contains autonomous transposable elements, which play a significant role in the generation of many spontaneous mutations such as insertions, deletions, duplications, and inversions.

3.6.8 Cryptic Prophage and Phage Remnants

E. coli K-12 genome carries λ bacteriophage along with the defective or cryptic lambdoid prophages *DLP12*, *Rac* and *Qin*. These cryptic, prophages have lost their functions essential for lytic growth and infection but still retained a few functional genes. In addition to the prophages, numerous isolated genes were also identified in the *E. coli* genome, showing similarity with the bacteriophage genes, named as phage remnants. Presence of cryptic prophages and bacteriophages in the *E. coli* genome implies a phage origin, with the last vestiges of a cryptic prophage ravaged by deletions. These genes may actually be homologs encoded both by a bacteriophage and its host, with no ready indication as to which genome was the original carrier.

3.7 Genome of Enterohaemorrhagic *E. coli* O157: H7

Enterohemorrhagic *E. coli* O157: H7 strain was first recognized as a gastrointestinal pathogen in 1982 causing hemorrhagic colitis and hemolytic uremic syndrome (HUS) in humans, sometimes leading to their death. Since then, its occurrence has become a worldwide public health problem, causing major out brakes in Japan, Unites States of America, etc., infecting thousands of human beings especially young children. Enterohemorrhagic *E. coli* O157: H7 bacterial strain (RIMD 0509952) was isolated from one of the Sakai outbreak patients. This strain is characterized by the production of two Shiga toxins, Stx1 and Stx2 and contains two plasmids, pO157 and pOSAK1. Virulence factors contributing to the infection of Enterohemorrhagic *E. coli* O157: H7 have been partially characterized, but the mechanism of hemorrhagic colitis and HUS are not completely understood. Genome sequence of the Enterohemorrhagic *E. coli* O157: H7 strain isolated from the outbreak of Sakai, Japan has been determined and compared with the genome of *E. coli* K-12.

3.7.1 Genome Sequencing and Annotation

The initial stages of sequencing were done by the whole genome random shotgun method. A pUC18-based library containing 1–2 Kb inserts was constructed and sequenced. Around 50,156 clones corresponding to the library were sequenced using a forward sequencing primer. The sequence data were assembled using Phred/phrap/consed. After the preliminary assembly, two types of clones were selected; clones with inserts within 1.5 Kb from the contig ends which are oriented outside, and the clones whose opposite ends of the inserts covering the regions having ambiguity in their sequence. A total of 19,969 clones belonging to these two criteria were sequenced using a reverse primer. A λ-phage based library has also been constructed with the inserts of 20-Kb size. 86 clones containing the sequences, which are non-homologous to *E. coli* K-12 at both the ends of inserts were selected, assembled into 111 contigs, with the length larger than 1 Kb. 286 low-quality sequences identified by the visual inspection were PCR amplified and re-analyzed. Gap closing was also performed by using the PCR amplified fragments and physical map of the Enterohemorrhagic *E. coli* O157: H7 chromosome.

For annotation, all O157 Sakai-specific sequences larger than 19 bp were defined by comparing the whole chromosomal sequence with that of K-12 MG1655 (Accession no. U00096) using the MUMmer program. The ORFs in the strain-specific regions and in the conserved regions of the two strains were identified and annotated separately using Genome Gambler version 1.41, GLIMMER 2.01, and BLAST. ORFs larger than 150 bp were searched, and the conserved ORFs were annotated according to K-12 MG16555 and the *E. coli* Index (http://web.bham.ac.uk/bcm4ght6/res.html). Eight small conserved ORFs were newly identified in this study. Genes corresponding to tRNA were identified by tRNA scan, other small RNAs and paralogous gene families were identified by BLAST search.

Table 3.3 Characteristic features of *E. coli* O157 genome

Components	Chromosome	pO157	pOSAK1	Total
Length of the sequence (bp)	5,498,450	92,721	3306	5,594,477
G + C content (%)	50.5	47.6	43.4	50.4
ORFs	5361	83	3	5447
Protein coding region (%)	88.1	82.5	52.8	88
Average ORF length (bp)	904	922	579	904
rRNA (16S, 23S, 5S)	7	0	0	7
tRNA and tmRNA	103	0	0	103
Non-classical RNA	13	0	0	13

3.7.2 Outline

The complete sequence of the *E. coli* O157 genome is 5,498,450 bp, distributed on a single circular chromosome. This strain consists of a large virulence pO157 plasmid (92,721 bp) and a cryptic pOSAK1 plasmid (3306 bp), making the size of whole genome to 5,594,477 bp, the second largest bacterial genome sequenced. The genome encodes 5361 protein-coding regions, 7 rRNA sets (16S, 23S, and 5S RNAs), 102 tRNAs, 1 tmRNA and 13 small RNAs including RNase P, 6S RNA, and 4.5S RNA. 88.1% of the genome forms protein-coding regions with the average ORF length 904 bp. The G + C content of the entire chromosome is 50.5 mol%. Characteristic features of the O157 genome were displayed in Table 3.3.

3.7.3 Comparison Between *E. coli* O157 and *E. coli* K-12 Genomes

E. coli O157 chromosome is 859 Kb larger than that of *E. coli* K-12 MG1655. Approximately 4.1-Mb of the sequence is conserved among these two strains, with no large re-arrangements such as translocation or inversion in the conserved regions. Both these *E. coli* strains have shown a remarkable level of nucleotide sequence conservation of 4.1 Mb, which is 98.31% with 2027 gaps. The 4.1-Mb of the conserved sequence might be representing the chromosome backbone, which remains conserved in most of the *E. coli* strains. However, the conserved backbone region of *E. coli* O157 is interrupted by large strain-specific loops, especially in the regions surrounding the replication termination site (*ter*). Even though the lengths of the replichores 1 and 2 are almost equal in K-12, replichore 1of O157 is 290 Kb longer than replichore 2. There is 296 strain specific loops in O157 (S loops) and 325 in K-12 (K loops). Among these, 203 loops are located at analogous sites on the two chromosomes at different sequences, which are considered as “hot spots” for the integration of foreign DNAs or for recombination. Most of the large loops (larger than 10 Kb) are prophages or prophage-like elements. In the O157 genome, 21 loops corresponding to prophages or prophage-like elements have been identified in the genome. One large loop (numbered 108) of 91.8 Kb size is composed of two λ-like phages integrated in tandem. The total length of loops in the O157 genome is

1,393,071 bp, making up to 25.3% of the chromosome. The length of the loops in the O157 genome corresponds to the whole genome size of *Borrelia burgdorferi* (1.44 Mb), the causative agent of Lyme disease. Comparative analysis of codon usage between the genes on the backbone and those on the S-loops suggests the abundance of foreign genes on S-loops. Codons that are used more frequently in the backbone genes are less frequently used in the S-loop genes and codons that are less frequently used in the backbone genes are more frequently used in the S-loop genes, indicating atypical codon usage in the S-loop genes of *E. coli* O157 genome.

3.8 Genome of *Mycoplasma genitalium*

Genus Mycoplasmas belongs to the class Mollicutes, comprising a large group of bacteria lacking cell wall with a characteristically low G + C content. These highly diverse organisms are parasitic on a wide range of hosts including humans, animals, insects, plants and even cells grown in tissue culture. Apart from their role as potential pathogens, Mycoplasmas are of critical interest because of their small, reduced genome size and its contents than any other prokaryote. *Mycoplasma genitalium* is a small parasitic bacterium which lives on the ciliated epithelial cells of the primate's genital and respiratory tracts. *M. genitalium* is the smallest known free-living bacterium and the second-smallest bacterium after the recently-discovered endosymbiont *Carsonella ruddii*. Until the discovery of Nanoarchaeum in 2002, *M. genitalium* was also considered to be the organism with the smallest genome. *Mycoplasma genitalium* was originally isolated in 1980 from urethral specimens of two male patients with non-gonococcal urethritis. Infection by *M. genitalium* can be transmitted between partners during unprotected sexual intercourse. *Mycoplasma genitalium* is a sexually transmitted bacterium causing female reproductive morbidities including pelvic inflammatory disease, which adversely affects the pregnancy.

3.8.1 Genome Sequencing, Assembly and Annotation

The genome was sequenced by whole genome random ("shotgun") sequencing. For maintaining the order of contigs, each template was sequenced from both the ends. A total of 9846 sequencing reactions were performed by five different individuals at an average of 8 AB 373 DNA sequencers per day for 8 weeks. The assembly of 8472 high-quality *M. genitalium* sequence fragments was performed using TIGR ASSEMBLER, along with 299 random genomic sequences, generating 39 contigs of size ranging from 606 to 73,351 bp, consisting of 38,06,280 bp primary DNA sequence data. Contigs were arranged in order by ASM ALIGN program, ensuring that there were no physical gaps in the sequence assembly. The order of the contigs was confirmed by comparing the order of the random genomic sequences that were incorporated into the assembly with their known position on the physical map of the *M. genitalium* chromosome. For detecting the overlaps, these 39 contigs were searched against each other with GRASTA (a modified FASTA). 11 overlaps were detected during this process, reducing the gaps to 29. Templates for each sequence gap were identified. All gaps of the size >300 bp were closed by primer walking. All electropherograms were visually inspected with TIGR EDITOR for initial sequence editing. The locations where a discrepancy could not be resolved or a clear assignment was made, the automatic base calls were left unchanged. For each of the 53 ambiguities remained and the 25 potential frameshifts detected after sequence-similarity searching, the appropriate template was re-sequenced with alternative sequencing chemistry such as dye terminator versus dye primer for resolving the ambiguities. 99% of the *M. genitalium* genome was sequenced with better than single-sequence coverage and the mean sequence redundancy was 6.5-fold. The error rate was assessed to be less than 1 in 10,000 bases on the basis of frequency of shifts in ORFs, overall quality of raw data and fold coverage. The genome sequence of *M. genitalium*, version 1.0 has been deposited in the Genome Sequence DataBase (GSDB) with the accession number L43967.

3.8.2 Overview of the Genome

The genome of *M. genitalium* is 580,070 bp, embedded in a single circular chromosome. The entire genome consists of 521 genes, among them, 482 are protein encoding, including genes required for DNA replication, transcription and translation, DNA repair, cellular transport, and energy metabolism. Overall G + C content of the genome is 32% with the content of A is 34%, C is 16%, G is 16% and T is 34%. The G + C content varies across the genome ranging between 27 and 37%, with the regions of lowest G + C content flanking the presumed origin of replication for this organism. The ribosomal RNA (rRNA) operon (44%) and the transfer RNA (tRNA) genes (52%) contains higher G + C content than rest of the genome, which may reflect the necessity of retaining essential G + C base pairing for secondary structure in rRNAs and tRNAs. The genome contains 74 *EcoRI* fragments, as predicted by both cosmid mapping data and sequence analysis.

3.8.2.1 Coding Regions

The coding regions of *M. genitalium* were predicted by searching for the ORFs, which are greater than 100 amino acids length. All open reading frames in the genome were searched with BLAZE against a non-redundant bacterial protein database (NRBP) developed at TIGR. Protein matches were aligned with PRAZE, a modified Smith-Waterman algorithm. 53 ORFs with low coding potentials and lengths smaller than 100 nucleotides were removed from the final set of putative coding regions. These processes resulted in the identification of 470 predicted coding regions, of which 374 were putatively identified and 96 had no matches to protein sequences from any other organism. The 374 predicted coding regions with putative identifications were assigned biological roles (Table 3.4).

Table 3.4 Gene content and functions in *M. genitalium*

Functions	No. of genes
Amino acid biosynthesis	1
Biosynthesis of cofactors	5
Cell envelope	17
Cellular processes	21
Central intermediary metabolism	6
Energy metabolism	31
Fatty acid and phospholipid metabolism	6
Purines, pyrimidines, nucleosides, and nucleotides	19
Regulatory functions	7
Replication	32
Transcription	12
Translation	101
Transport and binding proteins	34
Other categories	27
Unassigned role	152

3.9 Synthetic Genome of *Mycoplasma genitalium*

The genome of *Mycoplasma genitalium* has been identified as the smallest genome of any independently replicating cell that has been grown in the pure culture and over 100 of its 485 protein-coding genes are proved to be nonessential under optimal laboratory conditions when they are individually disrupted. Still, it is not clear which of these 100 genes are simultaneously dispensable. One approach for this question is to produce reduced genomes *via* chemical synthesis and introduce the synthetic minimum genome into the cells for testing their capacity to provide the essential genetic functions of life. In October 2007, a team of scientists headed by Dr. Craig Venter and Nobel laureate Hamilton Smith announced that they plan to create the first artificial life form in history by creating a synthetic chromosome, which they planned to inject into the *M. genitalium* bacterium, potentially resulting in an artificial species dubbed *Mycoplasma laboratorium* or *Mycoplasma* JCVI-1.0, after the research center in which it was created, the J. Craig Venter Institute in the United States of America.

3.9.1 Strategy for Synthesis and Assembly

The native 580,076-bp genome sequence of *M. genitalium* (accession no. L43967) was partitioned into 101 cassettes of 5–7 kb in length that were individually synthesized, verified by sequencing and then joined together in stages. The cassette boundaries were placed between genes so that each cassette contained the full length of the genes. This approach will simplify the future deletion or manipulation of genes in individual cassettes. Cassettes were overlapped with their adjacent neighbors by 80–360 bp. Cassette 101 was overlapped by cassette 1, completing the circle. Short "watermark" sequences were inserted at the intergenic sites of the cassettes numbered 14, 29, 39, 55 and 61. Watermarking has been done for differentiating the synthetic genome from the native genome. These watermarks were inserted at the sites known to tolerate transposon insertions. Additionally, a 2514-bp insertion in gene MG408 (*msrA*), encodes for an aminoglycoside resistance gene, was placed in cassette 89, as the strain with this specific deficit in its virulence cannot adhere into mammalian cells. Thus, the pathogenicity has been eliminated in the best available model systems. The synthetic genome of *Mycoplasma* JCVI-1.0 with all above insertions is 582,970 bp in length.

3.9.1.1 Synthesis of DNA

The synthesis and production of the cassettes were outsourced, to Blue Heron Technology, DNA2.0 and GENEART. The main challenges faced in this project were the assembly and cloning of synthetic DNA molecules, which were quite larger than the previously reported ones. A five-stage assembly strategy has been proposed for it.

1. In stage 1, sets of four neighboring cassettes were assembled by in vitro recombination and joined to a bacterial artificial chromosome (BAC) vector DNA to form circularized recombinant plasmids, with 24 Kb inserts. For example, cassettes 1–4 were joined together to form the A1–4 assembly, cassettes 5–8 were assembled to form A5–8, and so forth.
2. In stage 2, 25 A-series assemblies were taken three at a time to form B-series assemblies. For example, B1 was constructed from A1–4, A5–8, and A9–12 respectively, which reduced the 25 A-assemblies to only 8 B assemblies.
3. In stage 3, two B assemblies were taken at a time to make C-assemblies comprising one-fourth of the genome. These first three stages of the assembly were done by in vitro recombination and cloned into *E. coli*. Certain difficulties were encountered in carrying out the planned assembly and cloning of the partial and complete synthetic genomes in *E. coli*. For this reason, the final assemblies were carried out in *S. cerevisiae* by transformation-associated recombination (TAR) cloning.

3.9.1.2 Assembly by in vivo Recombination in Yeast

Half of the clones corresponding to the synthetic genome could not be obtained using BAC might be due to the instability of larger assemblies in *E. coli*. For this, *S. cerevisiae* yeast artificial chromosome (YAC) was used as a cloning host, which is believed to support at least 2 Mb of DNA in its centromere. Linear YAC clones were constructed by the ligation of an insert into a restriction enzyme cloning site. For the assembly, pTARBAC3 vector, which contains both YAC and BAC sequences was used. The sequence exactly matched their designed genome, can be accessed at GenBank with an accession number CP000925. On 24 January 2008, the same team reported having synthesized the complete 582,970 base pair genome of

M. genitalium by knocking out a key gene that enables the wild organism to cause disease. The final stage of synthesis was completed inside *M. capricolum,* which had its DNA removed, with the help of yeast cells. On 20 May 2010, they reported the success with a similar process, using the genome of *M. mycoides*, creating the first artificial life.

3.9.2 Minimal Number of Genes

After comparing the first two sequenced bacterial genomes of *Haemophilus influenzae,* and *Mycoplasma genitalium*, a version has been identified in these two genomes that consist of 250 genes for the minimal gene set. After the addition and comparison of multiple and complete genome sequences of other bacteria, the number of the minimal genes has reduced and it became clear that only 80 of the 250 minimal sets of genes are represented by orthologs of all the life forms. Viable knockouts were obtained for 15% of the genes from the minimal genome in *M. genitalium* including some of the universal genes, included in the first version of the minimal gene set. Construction of a minimal genome might enhance our understanding of the basic concepts of cell functioning by systematically detecting nonorthologous gene displacement and deciphering the roles of essential but functionally uncharacterized genes.

3.10 Conclusions

Genomics has allowed a significant advance in our understanding regarding the evolution of microbial genomes and the spread of pathogenicity among bacterial populations. Earlier, single-genome approaches have provided basic information on the bacterial genome diversity and identifying the determinants of pathogenicity. Large-scale whole-genome sequencing and multi-genome typing have elucidated the transmission pathways of pathogens at a global and local level in great detail. The increasing availability of genomes from large phenotypic collections and techniques such as MLST allowed the identification of genetic determinants for many pathogenicity-associated phenotypes using the genome-wide association approaches. Application of these new technologies will reduce the time and sample preparation needed for sequencing, as well as increasing the accuracy with minimum errors, allowing real-time sequencing for rapid outbreak detection and analysis of evolution as it occurs in nature. The continuing drop in the cost for genome sequencing will permit whole-genome approaches to the study of experimental evolution, radically expanding our understanding of evolutionary mechanisms. The future will definitely involve whole-genome sequencing in microbiology for tracking the transmission and spread of pathogens, as well as in the prediction of drug-resistance profiles. Genomic technology is improving in a great speed with reduced cost and increased sophistication, and is likely to be a fundamental tool for identification of novel bacteria as well as pathogens in the future.

Orgenellar Genome Analysis

4

Abstract

This chapter describes the resemblances of chloroplasts and mitochondria with bacteria along with the architecture of their genomes. Sequencing technologies for chloroplast genomes, phylogeny, and chloroplast genome engineering were discussed in detail with the genome description of protestant *Euglena gracilis*, which gives more information about the evolution of organelle genomes. Diseases due to the occurrence of mutations in the human mitochondrial genome have been emphasized along with the gene therapy for mitochondrial disorders in Humans. Mitochondrial genome sequencing and analysis of the Neandertal fossils will also provide a broader spectrum of information regarding the ancient DNA and their traits.

4.1 Introduction

Bacteria are the oldest single cellular organisms on earth. Fossil records indicate that the young Earth was covered by the mounds of bacteria. Some of them began making their own food using carbon dioxide in the atmosphere and energy they harvested from the sun by a process called photosynthesis, which produced sufficient oxygen and made the life on this planet. There is compelling evidence that mitochondria and chloroplasts were once primitive bacterial cells. As they have more resemblances with the Prokaryotic cells than that of Eukaryotes. Endosymbiotic theory describes the dependence of a host cell and ingested bacteria on one another for survival, resulting in a permanent relationship. Over the course of evolution, mitochondria and chloroplasts have become more adapted to stay within the cells and remained as organelle. Whole genome sequencing of mitochondria lineages revealed that they are very close to SAR 11 group of alpha proto bacteria which are free living. Similarly, sequencing of chloroplast genomes has grouped them very close to the cyanobacteria, supporting the endosymbiosis theory.

K. V. Chaitanya, *Genome and Genomics*,
https://doi.org/10.1007/978-981-15-0702-1_4

4.2 Resemblances of Chloroplast and Mitochondria with Bacteria

Mitochondria and chloroplasts have striking similarities with the bacterial cells. They have their own DNA, which is different from the nuclear DNA of the cell. Both these organelles use their own DNA for producing proteins and enzymes required for their function. Both of them are surrounded by a double membrane and reproduce by binary fission. Further evidence suggests that each of these organelles were ingested by a primitive host. Similar to bacteria, they replicate their own DNA and directs their own division. The starting codon in bacteria, chloroplasts, and mitochondria is N-formylmethionine, which is a different form of the amino acid methionine that starts the protein synthesis in Eukaryotes. Additionally, mitochondria use AUA as an alternate start codon. Ribosomes of chloroplast resemble bacterial ribosomes in their sedimentation coefficient and sizes of their RNA components. The function of chloroplast and mitochondria will be affected by the activity of similar antibiotics such as chloramphenicol, streptomycin, which inhibits the bacterial protein synthesis by binding to the ribosomes. Besides chloramphenicol, several other antibiotics also found to inhibit the chloroplast protein synthesis by binding to various sites on their ribosomes but not with the host cell ribosomal units, leading to a confirmation of the hypothesis that chloroplasts and mitochondria have close resemblances with bacteria and have evolved from symbiotic prokaryotes. Mitochondrial DNA has a unique pattern of maternal inheritance, that is inherited from mother to child and accumulates changes much slower than other types of DNA. Because of these unique characteristics, Mitochondrial DNA provides many important clues about the evolutionary history of the organisms.

4.3 Architecture of Organelle Genomes

Mitochondria and chloroplast are vital energy-producing eukaryotic organelles, which have originated independently by the integration of a free-living bacterium into a host cell more than 1.4 billion years ago and settled as endosymbionts. Mitochondria have evolved first, from an alphaproteobacterial endosymbiont and chloroplasts evolved later through the endosymbiosis of a cyanobacterium by a eukaryotic ancestor of Archaeplastida and laterally spread into different groups through endosymbioses. Mitochondria and chloroplast contain their own genomes and their DNAs have many traits in common. Both of their genomes are highly reduced when compared to the genomes of their ancestors. This might be due to the transfer of their genes to the host nuclear genome and are depending on nuclear-encoded, organelle-targeted proteins for attending most of their functions.

Human and mouse mitochondrial genomes were the first organelle DNAs and the first non-viral chromosomes which were completely sequenced in 1981 and *Marchantia polymorpha*, tobacco chloroplasts were the first complete plastid genomes sequenced in 1986. Till date, more than 7000 complete mitochondrial DNA (mtDNA) and plastid DNA (ptDNA) genome sequences are available in the databases, making the organelle genomes one of the highly sequenced types and the rate is steadily increasing. In addition to the genome sequence, a wide amount of information on the organellar genomes such as inheritance patterns, replication, transcriptional and translational processes, mutations and horizontal gene transfer, helps in enhancing our understanding of eukaryotic evolution, archaeology, forensics, and medicine. Mitochondria and chloroplasts are endless reservoirs of interesting and unconventional genomes. Most of the organelle genomes sequenced were of mitochondria (82%) compared with chloroplast (11%). But, in terms of genetic diversity, ptDNA is higher than that of mtDNA, contributing to more meaningful analysis towards increased understanding of cellular and genomic complexity.

4.3.1 Genome Size and Structure

Organelle genomes are much smaller in size compared with nuclear genomes. They are composed of Ds DNA, arranged either in linear, truly circular, or circular topologies. Size of the organelle genome varies by orders of magnitude with mtDNA exhibiting a wide range of sizes varying from giant mtDNAs of seed plants such as cucumber (1.6 Mb), *Silene conica* (11 Mb), and diplonemids (more than 500 kb) to apicomplexans (6 kb). Fragmented mitochondrial genomes are found in *C. velia*, with even smaller size lesser than 6 kb and diminutive mtDNAs were observed in fungi. Size of the chloroplast genome is bigger than the mitochondrial genome, with chlorophycean algae *Volvox carteri* (525 kb), *Floydiella terrestris* (520 kb) and the ulvophyte *Acetabularia acetabulum* (1 Mb). Small fragmented chloroplast genomes were also found in *peridinin dinoflagellates* (30 kb), non-photosynthetic algae, such as *Plasmodium* sp. (35 kb) and *Helicosporidium* sp. (37.5 kb).

Although the topology of organelle chromosome appears to be circular, they usually have more multifarious structures. Mitochondrial and plastid genomes of land plants and yeast mtDNAs, appears to be circular when assembled but are thought to exist inside their organelles as multigenomic and linear-branched structures. Linear mitochondrial genomes with well-defined telomeres are found in various green algae, protists, animals, and fungi. These telomeres have ornate conformations, such as hairpin loops, single-stranded overhangs, and/or covalently attached proteins, which might help in preserving the chromosome ends independent of telomerase.

4.3.2 Nucleotide Composition

One of the prominent features of organelle DNA, relatively constant across lineages is their nucleotide composition. All completely sequenced mt and pt. DNAs have a high range of biasedness towards adenine and thymine (AT) content than that guanine and cytosine (GC). 90% of AT is reported in mitochondria of yeast and insects. Chloroplast genomes are also rich in AT, reaching up to 87% in *Plasmodium falciparum*, but are having less nucleotide biasedness than mtDNA. GC rich organelle DNA is found in euglenozoans, green plants, animals and fungi. The highest documented mitochondrial GC content is 68% found in the lycophyte *Selaginella moellendorffii*. This is significantly higher than the highest GC chloroplast content of 58% reported in trebouxiophyte green alga. Interestingly, all known species with the high GC-rich ptDNA also have GC-rich mtDNA. Processes such as mutations, recombination, random genetic drift, and selection will affect the equilibrium nucleotide composition of a genome, leading to AT biasedness. AT richness of mt and pt. DNA reflects the endosymbiotic history of these genomes, their location within the cell, unique population-genetic features that define organelles and selection for metabolic and translational efficiency.

4.3.3 Chromosome Number

A shift from a single to the multipartite chromosome architecture has been identified in the mitochondrial genomes. Around 12 mitochondrial lineages have displayed fragmented mitochondrial genomes and even splintering of mtDNA was also found in cnidarians, Chlamydomonas, and some vascular plants. The level of fragmentation in mtDNA varied from 75 small circular DNAs (6–7 kb) of single gene modules in green alga *Polytomella piriformis* to thousands of giant intertwined networks and few dozens of large circular chromosomes in *Trypanosoma*. mtDNA of *Amoebidium parasiticum* is composed of hundreds of linear pieces varying from 0.3–8.3 kb. Fragmentation in chloroplast genomes is little rare, currently restricted to dinoflagellates, in which the ptDNA is fragmented into mini circles of 2–3 kb with one to a few genes per fragment.

Many copies of organellar chromosomes are present per cell, which may be in hundreds or thousands. The regulation of cellular copy number varies in different cells of the same organism. The leaf cells of the garden beet have about 40 chloroplasts per cell and each chloroplast contains 4–8 nucleoids (The regions of chloroplasts contain specific areas that are densly stained for DNA) and each nucleoid contains from 4 to 18 chloroplast DNA (cpDNA) molecules. Thus, a single cell of a beet leaf may contain $40 \times 8 \times 18 = 5760$ copies of the chloroplast DNA. The number of cpDNA varies from organism to organism. Photosynthetic protist *Chlamydomonas* contains only a single chloroplast per cell, but the single chloroplast contains up to 1500 cpDNA molecules, commonly observed to be packed in the nucleoids.

Haploid yeast cells contain up to 45 mitochondria, each having up to 30 nucleoids with 4–5 mitochondreal DNA (mtDNA) molecules in each nucleoid. Human cells contain 2–10 mtDNA molecules per mitochondria. The number of mitochondria per cell varies in different cell types along with mtDNA number per mitochondria. Several hundreds of mtDNA molecules are present in human fibroblast cells and ~100,000 mt DNA molecules are reported in human oocytes.

4.3.4 Non-coding DNA

Majority of the variations in the organelle genome size is due to differences in the contents of non-coding DNA, varying from 1–99% among mtDNAs and 5–80% among cpDNAs. Small organelle genomes, such as vertebrate mtDNA and *apicomplexan* cpDNA are densely coded without introns, whereas large genomes such as *Shorea conica* mtDNA and *Vlvox carteri* cpDNA are possessing high amounts of noncoding nucleotides, often riddled with repeats, introns, mobile genetic elements and foreign DNA in the case of mitochondrial genomes.

Introns are the non-coding regions existing between two exons within a single gene, which are much more complex, containing their own intron-encoded genes, involved in the intron migration and other processes. Among the existing four classes of introns, the only group I and group II are found to occur in the organellar genomes. These two groups belong to autocatalytic introns, capable of splicing themselves from a sequence without the aid of any proteins. The number of introns in organelle genomes varies from 0 to 150 and mitochondria genomes display more intron variability than plastid genomes. Group I introns were found in the *cob* and *cox I* genes of *Volvox carteri* with tandem arrays of short palindromic sequences that are related to each other. Other regions in the mtDNA have revealed similar palindromic repetitive sequences that are dispersed in the non-protein coding regions of the genome. The cox1 gene of *Agaricus bisporus* (29.9 kb) mt DNA contains 19 introns, which makes it the largest and highest intron-rich gene of eukaryotes organelle. Such repetitive sequences are thought to be one of the main reasons for the mitochondrial rearrangements during evolution. 149 introns are reported in the chloroplast genome of Euglena. Introns are considered as mobile genetic elements that have been added to the ancestral intronless genes during their evolution. All of the known Euglena chloroplast genes encode for ancient proteins involved in RNA synthesis, protein synthesis, ATP synthesis, and photosynthesis. All these ancient genes might have originated before the evolutionary divergence between eubacteria and eukaryotes. It is believed that all 149 introns of Euglena chloroplast are the descendants of mobile genetic elements that have invaded this genome. The evidence supports this hypothesis comes from the chloroplast genome containing these 149 introns in unique locations that are not found in other chloroplast DNAs. The chloroplast genome of *Euglena gracilis* contains 15 **twintrons** (discussed in human genome section), which must be removed sequentially for accurate splicing. Euglena chloroplast DNA also containing 46 individual group II introns and 18 more group III introns that are components of twintrons. The mtDNA of *S. moellendorffii*, nad4L is located within an intron of nad1.

4.3.5 Coding Regions

The contents of both nuclear and organellar genomes are similar with respect to the coding and non-coding regions. Organelle genomes also contain coding as well as noncoding regions and the ratio of these regions varies from one organism to another. Coding regions in organelle genomes are the DNA sequences give rise to mRNA, rRNA, and tRNA. Similar to the size and structure, organelle genomes show great variation in the gene number and organization. The mitochondrial genome of *Andalucia godoyi* has the largest and least-derived gene content, comprising of 100 genes with 66 of them encodes proteins. The mitochondrial genomes of dinoflagellates, apicomplexans have the most reduced coding regions with the genes coding for three or even fewer proteins, no tRNAs and in even lack complete rRNAs (*C. velia*). Chlamydomonas mtDNA also shows diminished coding contents (10–13 genes) and some land plants (*S. moellendorffii*), animals (the winged box jellyfish) and trypanosomes have lost all or most of their mitochondrial tRNA-coding regions. Plastid genomes are more gene-rich than that of mitochondria, coding up to 250 genes in red algae. Plants and algae that have lost photosynthetic capabilities have reduced plastid gene content to <75, but lowest chloroplast gene number was found in photosynthetic, peridinin plastids of dinoflagellates, which encodes <20 genes.

4.3.6 Genome Loss

During the course of evolution, the organelle of some organisms has lost their genomes partially and few organisms have lost it entirely. However, the genome loss is more in mitochondria than in chloroplast, which could be due to the disappearance of the genes coding for oxidative phosphorylation. Mitochondria-derived organelles such as mitosomes and hydrogenosomes without genomes were identified in several anaerobes and microaerophiles, such as *Giardia*, *Trichomonas*, *Mikrocytos,* and *Entamoeba*. Loss of the chloroplast genome is mainly due to the disappearance of the genes responsible for the photosynthesis. Non-photosynthetic plastids were found in the parasitic plant *Epifagus*, parasitic algae *Plasmodium* and *Helicosporidium*. Chloroplasts lacking genomes have been identified only in the green algal genus *Polytomella*.

4.3.7 Gene Fragmentation

Mitochondria encoded genes in a number of Eukaryotes are found to be fragmented and scrambled. This feature is particularly specific to most of the mitochondrial genomes sequenced till date. Well-studied examples of scrambled and fragmented mitochondrial genomes include the discontinuous and jumbled mitochondrial rRNA, protein-coding mitochondrial genes of various green algae, alveolates, and euglenozoans. Large genes corresponding to the large and small-subunit rRNA genes of *Chlamydomonas reinhardtii* have split into eight and four unordered coding segments, which come together after transcription through secondary-pairing interactions. RNA splintering was further noticed in *P. falciparum* mitochondrial rRNA, with at least 27 distinct modules encoding the LSU and SSU rRNAs. Fragmented genes are highly uncommon in chloroplast genomes, but have been documented in Chlamydomonas and dinoflagellates with peridinin as the main carotenoid.

4.3.8 Non-Canonical Genetic Codes and RNA Editing

At least 12 unique changes were identified in the universal genetic code of animal mitochondrial DNA, with five codon re-assignments. One or two mtDNA codon alterations as well as the occasional loss of start and stop codons were noticed in yeast. Overall, mitochondria that retain the universal genetic code have relatively rare exceptions. In contrast, no non-canonical genetic codes have been observed in the chloroplast genomes. Mitochondrial RNA editing is highly diversified in Eukaryotes such as fungi, land plants and dinoflagellates. Most elaborative display of RNA editing occurs in the kinetoplast of mitochondria, with ~90% of codons experience either uracil insertion or deletion. Chloroplast RNA editing has been reported to be confined to land plants and dinoflagellates. The extent of plastid editing in each of these groups is not as widespread and elaborate as that of mitochondria.

4.3.9 Horizontal Gene Transfer and Acquisition of Foreign DNA

Many mitochondrial genomes are bounded with horizontally acquired DNA, whereas chloroplasts are conspicuously lacking the foreign DNA in their genomes. In angiosperms such as Plantago and ginger, the mitochondrial genome contains gene mosaics formed by the conversion of native and foreign DNA sequences. mtDNA of the shrub *Amborella trichopoda* harbors the equivalent of six horizontally acquired mitochondrial genomes. Chloroplast DNA was thought to be impenetrable to foreign DNA, till the data obtained from various angiosperms uncovered mitochondrion-to-plastid and nucleus-to-plastid DNA migration events. There are also examples of chloroplasts such as diatoms, red alga *Gracilaria tenuistipitata*, and acquired genes from plasmid and also from bacterial genomes. Plasmids have been isolated and characterized from the mitochondria of land plants, fungi, and protists, but are mostly absent in chloroplasts with a few exceptions.

4.4 Evolution of Organelle Genomes

Genomes of mitochondria and chloroplast have evolved convergently as illustrated in *Selaginella moellendorffii*. Both the organellar genomes have high GC contents (68% and 68%), highly reduced gene density and gene number especially tRNA (0 and 13), inflated number of introns (37 and 11), huge C-to-U RNA editing (in thousands and hundreds) and very high levels of structural rearrangements. mtDNA has many peculiarities that are not found in the cpDNA, such as standard genes situated within introns, repetitive elements and a recombinant structure. Most of the land plants also follow the same trend in possessing higher levels of RNA editing, high intron densities and large, complex genomic structures in mtDNA. Even though, both the organelle DNAs has followed similar evolutionary paths with respect to their genome organizations, the mtDNA has undergone severe gene losses, widespread and elaborate forms of posttranscriptional editing and processing, number of gene isoforms and extensive gene fragmentations. mtDNA of dinoflagellates contain non-canonical start and stop codons, trans-spliced genes and oligonucleotide caps at both the 5′ and 3′ ends of mRNA, which were not noticed in the cpDNA. Mt. and cp DNAs of *Dunaliella salina* and *Volvox carteri* have undergone massive expansions, resulting in uncharacteristically long intergenic regions and large amounts of repetitive DNA, whereas both these organelle genomes have undergone contraction in prasinophyte *Ostreococcus tauri* and have a very little non-coding DNA. The degree of genome reduction is attributed to ~92% of the coding region within the mitochondrial genome and 80% of the coding region and one intron in the plastid genome. The co-expansion and co-contraction of mitochondrial and plastid genomes have been identified in a huge number of organisms with cpDNA number higher than mtDNA.

4.4.1 Evolution of Traits and Characteristics in Organelle Genomes

Similar peculiarities in the genomes of chloroplast and mitochondria of the same species or within the species were observed, which might be attributed to remarkable parallelisms or evolution of one system might from the other. The evolution of specific host-to-organelle protein targeting systems means that organelle genetic information is partitioned between the two genomes. Factors that govern the organelle processes can be considered intrinsic if they reside within the organelle genome or extrinsic if they reside within the nuclear genome. These processes may be dominated by one or the other type of factors, such as mitochondrial RNA editing governed by the intrinsic factors in trypanosomes and the same in the case of land plants is governed by extrinsic factors. Similarly, Non-canonical genetic codes can be intrinsic if the tRNAs are organelle-encoded or extrinsic if the tRNAs are nuclear. Predominant extrinsic processes evolved in one compartment could spread if key proteins find their way into another compartment. The appearance of rare and complex genomic anomalies in both plastid and mitochondrion genomes of the same species is not surprising if they are mediated by extrinsic factors. There is also increasing evidence for the existence of molecular crosstalk between mitochondria and plastids and the correlations of mtDNA and cpDNA might be due to this cross talk. Some mitochondrial mutations can have major effects on chloroplast functions. The complexity of the organelle genome might also due to the non-adaptive processes. A hypothesis called the mutational hazard hypothesis (MHH) proposed that genomic embellishments such as introns, fragmented genes, and RNA editing sites are mutationally burdened to the organism. So, they have a great tendency to accumulate these mutations at a low rate. This theory was supported by a very low mutation rate in the mitochondrial genomes of land plants.

Contradicting this theory, enormous mitochondrial genomes of angiosperms with a high mutation rate and no reduction in its population size, relative to the species with smaller mtDNAs was observed.

Whenever mitochondrial and chloroplast genomes acquire similar traits in a group or a species, their similarities and differences are more pronounced in mtDNAs than in cpDNAs, instead of reflecting on both organelles. The reason could be the endosymbiotic event that led to the origin of mitochondria hundreds of millions of years before the chloroplasts have originated. However, many genomic eccentricities observed in the mt and cp DNAs are forcing us to believe that both these organelles have arisen at the same time in both the compartments as studied in RNA editing of these organelles. Mitochondrial and plastid RNA editing in plants was thought to have evolved concurrently in the common ancestor of land plants. Still, the RNA editing is more elaborate in mitochondria than in chloroplasts, probably due to the small gene number. Changes in the genetic code are expected to occur more frequently in genomes with small protein coding sequences. Similarly, a small gene might also be responsible for high mutation rates as the selection against a mutator allele would be proportional to its effect on the rate of functionally deleterious mutations per genome.

Structural and physiological differences between the organelles have also lead to extreme architectures. Unlike plastids, mitochondria regularly fuse themselves in vivo with domestic and foreign mitochondrial DNA sequences and this fusion could be responsible for the acquisition of mtDNA sequences from mosses, green algae, angiosperms into *Amborella*. Another possible reason for the presence of exogenous sequences in mtDNAs might be due to the existence of an active DNA import system, whereas no such system has been identified in chloroplasts. These mechanisms explain only a few aspects of mitochondrial and chloroplast genome complexity. A broader understanding of the organellar genome evolution requires in depth knowledge of the processes involved in the modification of the DNA in the chromosomes such as mutations, recombinations, natural selection, and genetic drift.

4.5 Chloroplast Genomes

Chloroplasts are one of the most important reaction centers of plant and protistant cells that convert the solar energy into carbohydrates through the process of photosynthesis and are the main sources that sustain the life on our planet through oxygen re-release. Apart from photosynthesis, chloroplasts also play vital roles in the physiology and development of plant by involving in the synthesis of amino acids, nucleotides, fatty acids, phytohormones, vitamins and metabolites, important for the plant interactions with the surrounding environment and its response to heat, drought, salinity, etc. They also act as active centers for the assimilation of nitrogen and sulfur. Study of chloroplast genome will enhance our understating of plant biology and evolution. Chloroplast genomics has many vital applications including the development of crop plants resistant to abiotic and biotic stress, development of edible vaccines and biopharmaceuticals. Since the full-length sequencing of the tobacco chloroplast genome in 1986, over 800 chloroplast genomes were completely sequenced (NCBI, organelle genome database), which have enhanced our understanding of plant biology and its diversity, contributed significantly in the phylogenetic analysis of several plant species for resolving their evolutionary relationships.

4.5.1 Sequencing Technologies for Chloroplast Genomes

Traditionally isolated chloroplasts were sequenced by the amplification of entire genomes through rolling circle amplification or by screening the BAC or fosmid libraries using chloroplast genome sequences as probes. However, these methods are subjected to numerous challenges such as difficulty in the construction of quality libraries, increased probability of nuclear and mitochondrial DNA contamination, a large number of PCR reactions. Applying PCR reactions were also difficult with the chloroplast genome samples without any close relatives, whose genomes are sequenced or with highly rearranged chloroplast genomes. Advancement of technology has facilitated the sequencing of chloroplast genomes. Development of high throughput sequencing methods and next-generation sequencing systems have provided fast and cheaper methods for sequencing of chloroplast genomes with increasing accuracy. Moore et al. (2006) were the first to report the utilization of next-generation sequencing systems for sequencing the chloroplast genomes of *Nandina* and *Platanus*. Even though multiple next-generation platforms are available for sequencing the chloroplast genomes, Illumina is currently a major next-generation platform used for chloroplast genomes, which utilizes rolling circle amplification products and bioinformatics for performing the de novo assembly without any need for the reference genome sequences. A third-generation sequencer of PacBio systems comprising of single-molecule real-time sequencing is another widely used next-generation platform for chloroplast genome sequencing. This system provides long read lengths, which facilitates the de novo genome assembly particularly at the junctions between inverted repeats and single copy regions.

4.5.2 Chloroplast Genome Sequencing

Advanced sequencing technologies such as next-generation sequencing allowed the rapid sequencing of about 800 chloroplast genomes belonging to flowering plants, bryophytes, lycophytes, gymnosperms, green and red algae, photosynthetic dinoflagellate chromalveolates (NCBI organelle genomes, http://www.ncbi.nlm.nih.gov/genomes/genlist.cgi?taxid=2759&type=4&name=Eukaryotae%20Organelles). Availability of the genomic information has facilitated the increased understanding of extensive changes in chloroplast genome related to the evolution of this organelle. Features of these genomes and their sequence information are also being exploited to resolve the long-standing phylogenetic quest 'tree of life'. First two complete chloroplast genome sequences of tobacco and liverwort were sequenced through Sanger's di-deoxy method and the latest chloroplast genome sequence available was of Pigeonpea (*Cajanus cajan* (L.) Millspaugh), sequenced using Roche 454 GS FLX platform, available in public databases since 2016.

4.5.3 Structure of Chloroplast Genome

Chloroplast genomes of land plants are highly conserved and organized, comprising of either single circular or linear DNA with a quadripartite structure composed of large single copy (LSC) and small single copy (SSC) regions separated by two copies of IR repeats (Fig. 4.1). These IR repeats are densely populated with 16S, 23S and 5S rRNA genes. The genome consists of 120–130 genes involved primarily in photosynthesis and also in transcription, translation. These genes are divided into three broad categories (i) involved in photosynthesis (photo-system I and II—psaA, psaB, psbA, psbB, cytb6f, ATP synthases, rbcL and NAD(P)H genes, etc.), (ii) regulatory genes (tRNA genes-trna H, trnK; rRNA genes-rrn16, rrn5; RNA polymerase rpoA, rpoB; ribosomal subunit genes-rps2, rps3, rpl2, rpl16 etc.) and (iii) conserved ORFs such as *ycfs* and protein-coding genes like *matK*. Differences in the size of chloroplast genomes of plants are mainly due to the number of genes in the IR repeats that are duplicated. Copy number of chloroplast genome per plant cell is very high with each chloroplast consisting of 50–100 copies of its genome and each cell consisting of more than hundreds of chloroplasts making their copy number to 10,000 per cell. This feature of high copy number of genomes is exploited in genetic engineering of plastids for overexpression of transgenes, thereby allowing the recombinant proteins to accumulate at high concentrations of over 10% of the total soluble proteins.

Intergenic spacer regions in the genome consist of certain vital regulatory sequences. Size of the chloroplast genome shows enormous diversity between species, varying from 107 kb (*Cathaya argyrophylla*) to 218 kb (*Pelargonium*), which is completely independent of nuclear genome size. Few chloroplast genomes show structural rearrangements, with loss of IR regions, entire gene families, even the highly conserved introns. Majority of the intron loss has been observed in the chloroplast genomes of barley (*Hordeum vulgare*), bamboo (*Bambusa* sp.), cassava (*Manihot esculenta*), chickpea (*Cicer arietinum*), monocots such as members of Poaceae, eudicots such as Onagraceae, Oleaceae, and gymnosperms such as pinus. Figure 4.1 illustrates the chloroplast genome of *Glycine syndetica* one of the close lineage of soybean (*Glycine max*).

4.5.4 Phylogeny of Chloroplast Genomes

Plant cells comprise of three distinct genomes; nuclear, chloroplast and mitochondrial. Chloroplasts are believed to be evolved from the endosymbiosis of a cyanobacterium by the massive gene transfer to its nucleus, which can be evaluated by their gene content and number. Generally, gene content, number, and structure are highly conserved in the chloroplast genomes with a few genes absent in chloroplast

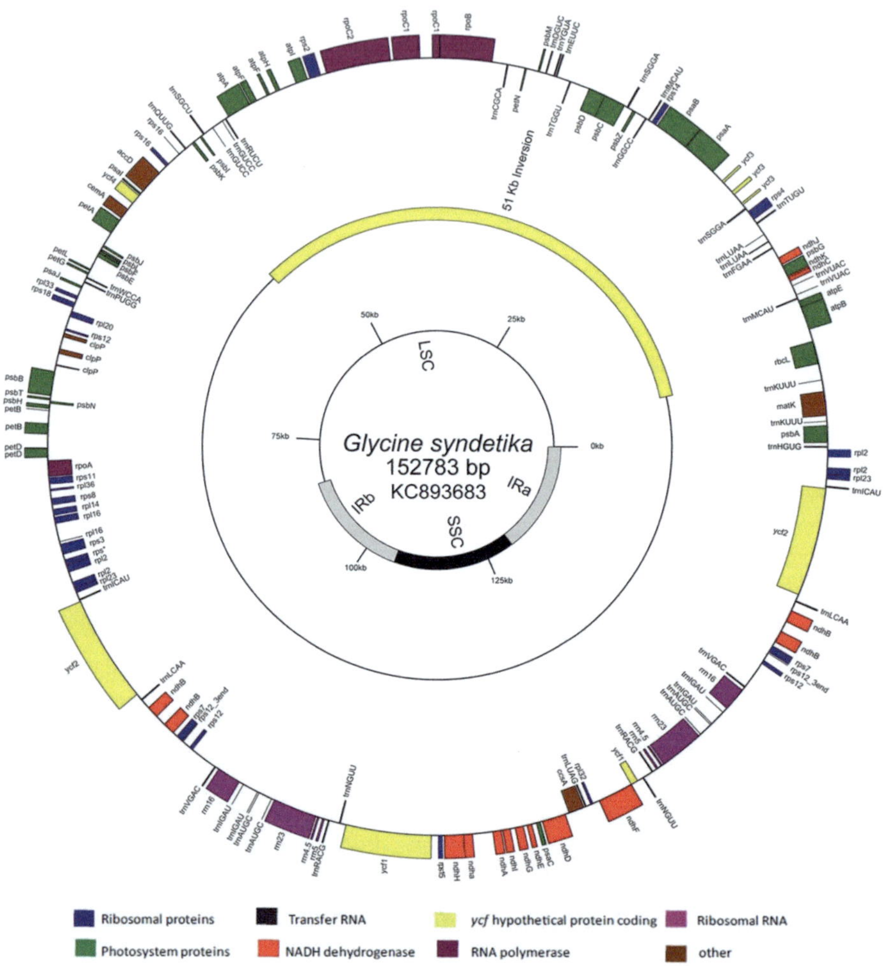

Fig. 4.1 Chloroplast genome map of *Glycine syndetika*. Gray bars on the inner circle represent the inverted repeats (IRa and IRb). The yellow bar on the middle circle represents 51-kb inversion region. Genes on the outside of the circle are transcribed in a clockwise direction and genes on the inside of the circle are transcribed in an anti-clockwise direction (Courtesy Henry Daniel et al. 2016)

genomes but are found either in nuclear or mitochondrial genomes of specific species. Loss of Translation Initiation Factor 1(*infA*), nuclear-encoded *rpl22* and *ndh* genes in the chloroplast genome and their intracellular transfer to the nuclear or mitochondrial genomes provide valuable information for phylogenetic analyses and evolutionary studies. Chloroplast *infA* is an essential homolog of *E. coli*, *infA* gene, which initiates the translation in collaboration with two nuclear-encoded initiation factors and mediating the interactions between mRNA, ribosomes and initiator tRNA-Met. Nuclear-encoded *infA* genes have been identified in *Arabidopsis thaliana*, soybean, tomato and ice plant, whose protein sequences contain chloroplast transit peptides. Studies using soybean and *A.thaliana* infA-GFP proteins have shown that nuclear-encoded *infA* genes are first translated in the cytosol and are then transported into chloroplasts. It was also noticed that chloroplast-encoded *infA* genes were subjected to deletions during angiosperm evolution.

57 chloroplast genomes belonging to 26 genera have reported the deletion of an essential *rpl22* gene from the chloroplast genome and its transfer into the nuclear genome. Nuclear-encoded *rpl22* contain a transit peptide, which is predicted to deliver this protein from the cytosol to chloroplasts. Eleven chloroplast genes encoding *ndh* subunits are involved in the photosynthesis. These *ndh* proteins assemble into the photosystem I complex and mediate the cyclic electron transport in chloroplasts, facilitating the chloroplast respiration. The entire family of *ndh* genes was deleted in gymnosperms, cactus and in the members of the family Geraniaceae. Partial deletions of these genes were also found in orchid chloroplast genomes and were identified in their mitochondrial genomes. Non-functional chloroplast *ndh* gene fragments are reported in the nuclear genome of Norway spruce. The massive amounts of deduced information on chloroplast genomes generated from different plant families and green algae would be useful in developing more efficient transplastomic technology for crop improvement.

4.5.5 Chloroplast Genome Engineering

Sequencing of the chloroplast genome is highly useful in the identification of closely related and breeding-compatible plant species. With the advent of novel genetic engineering techniques, it is possible to transfer the genes corresponding to the desirable traits from unrelated species readily into the commercial and economically important cultivars. Such genetically modified crops have revolutionized agriculture in the past two decades, drastically reducing the usage of chemical pesticides and herbicides with enhancing yields. However, there are a few limitations for transgenic plants generated by transferring the genes into the nuclear genome such as low levels of gene expression and the possibility of transgenes escape *via* pollen. One of the solutions for the above said problems is the introduction of foreign genes into the chloroplast genome. As the chloroplast genome copy number is very high, up to 10,000 transgene copies can be introduced into the chloroplast genome of each plant cell, resulting in extremely high levels of foreign gene expression. Another significant feature is, chloroplast genomes are maternally inherited in most of the cultivated crops, minimizing the chances of transgene escape through pollen.

Engineering of the chloroplast genome is accomplished by the integration of foreign genes into the intergenic spacer regions without disrupting its native chloroplast genes. Two chloroplast genes are used as flanking sequences for facilitating the integration of the transgene into the chloroplast. Transgene cassette includes a selection marker gene and gene of interest, both are regulated by chloroplast gene promoters and its UTR regions. Chloroplast genome sequences are essential in building a transgene because they provide both flanking and regulatory sequences.

Transgene cassettes are cloned into the bacterial plasmids called chloroplast vectors and they are bombarded into the plant cells or callus using gold particles and a projectile gene gun. Even though the target gene cassette is integrated into the nuclear genome of the plant cell, flanking chloroplast genes will limit the expression due to non-availability of their regulatory regions. Even if such integration occurs, the transgenes can be easily identified by the evaluation of their integration sites and eliminated.

4.5.5.1 Applications of Chloroplast Genome Engineering

One of the main focus of chloroplast genetic engineering is the overexpression of target genes for enhancing the abiotic and biotic stress tolerance, which is vital for the protection of plants against environmental cues. So far, over 50 transgenes oriented with the insect resistance in the edible crops such as cabbage, soybean, and eggplant have been successfully integrated into the chloroplast genome. In addition to improved stress resistance, chloroplast genome has also been engineered to produce useful enzymes, biomaterials, biofuels. Chloroplast genetic engineering also plays a key role in the manufacture of protein drugs, which overcomes the steps such as expensive fermentation systems, purification systems, storage, transport, and short shelf life. Protein drugs expressed in lettuce leaf chloroplasts are stored indefinitely at ambient temperatures without losing their efficacy. Oral delivery of several human therapeutic proteins, expressed in chloroplasts is one of the efficient ways of diagnosing several human diseases including diabetes, cardiovascular disorders, pulmonary hypertension and Alzheimer's. Therapeutic proteins were initially expressed in tobacco chloroplasts and were subsequently expressed in lettuce chloroplasts at the clinical level. Oral delivery of exendin-4, modulating the secretion of insulin in a glucose-dependent manner, decreased the glucose levels in diabetic animals by stimulating the production of insulin in a manner similar to that of injectable drug.

Chloroplast genome engineering is also ideal for producing low-cost booster vaccines against several infectious diseases. No vaccine is currently available against malaria and cholera. In a recent study, Davoodi-Semiromi (2010) has fused the cholera toxin-B subunit (CTB) of *Vibrio cholerae* with malarial vaccine antigen, apical membrane antigen-1 (AMA1) and merozoite surface protein-1 (MSP1), expressed in lettuce chloroplasts. With no suitable models exists to test human malaria, cholera toxin using mice immunized with chloroplast- expressed CTB was highly effective, providing the longest duration of protection in the published literature.

4.5.6 Chloroplast Genome of *Euglena gracilis*

Euglenaceae family comprises the genera of photosynthetic euglenids such as *Euglena, Cryptoglena, Monomorphina, Trachelomonas, Strombomonas, Colacium, Euglenaria*, and *Euglenaformis* exhibiting significant diversity in their chloroplast shape, number, position and presence of pyrenoids, etc. Hallick et al. (1993) have sequenced the chloroplast genome of *Euglena gracilis*, which was the first euglenophyceaean chloroplast genome to be sequenced and was used for long time as a representative chloroplast genome for all Euglenophyceae, till sequencing of *Euglena longa* chloroplast genome by Gockel GHW (2000), a taxon that has lost the ability of photosynthesis. In 2012, a series of euglenophyceaean chloroplast genomes were sequenced and published, belonging to *Eutreptia viridis*, *Eutreptiella gymnastica*, *Euglena viridis*, *Monomorphina aenigmatica*, *Colacium vesiculosum*, and *Strombomonas acuminata* and *Euglenaformis proxima*.

Euglena gracilis is a unicellular facultative photosynthetic organism, closely related to flagellate protists. Chloroplasts of *Euglena gracilis* share many common structural and functional features with that of chlorophytes and land plants. Chloroplast DNA of *Euglena gracilis* was the fourth chloroplast sequenced after tobacco, liverwort, and rice. This is also one of the well-characterized organellar genomes because of its low GC content (buoyant density), generating clear discrimination between nuclear and plastid DNA. Chloroplast DNA of this organism is one of the very first known examples of a circular chloroplast genome, which is quite different from the genomes of algae and higher plants with respect to its organization. Unique features of this genome include the presence of a variable number of tandem repeats serving as an origin of DNA replication, presence of group II introns and twintrons in the genes involved in PSII synthesis.

4.5.6.1 Genome Sequencing and Annotation

DNA sequences of a number of *Euglena gracilis* chloroplast genes had been reported previously, which are available in the databases. For completing the full-length genome sequence, all known regions of the genome were re-compiled and annotated, by making several corrections in the earlier annotated data, facilitating the identification of all unknown regions, by cloning with appropriate overlaps and sequencing of both the strands. The last 54 bp of the sequence represented a VNTR sequence. The size of Euglena cpDNAs will be more than 143,170 bp, depending on the number of 54 bp repeated segments. The sequence data were compiled and evaluated using the software from Genetics Computer Group, 575 Science Drive, Madison, Wisconsin, USA 53711. Gene identification was based on screening of the GenBank Release 75.0, EMBL Release 30.0, PIR-Protein Release 33.0, PIR-Nucleic Release 36.0, and SwissProt Release 22.0 databases with the FASTA and BLITZ algorithms from EMBL, Heidelberg, and the BLAST algorithm available through the BLAST network service at the NCBI, USA.

4.5.6.2 Chloroplast Genome Organization

The 143,170 bp physical map of *Euglena gracilis* chloroplast genome is numbered from the first nucleotide after the 54 bases VNTR in a clockwise manner to the last nucleotide before the VNTR region 143,116 base (Fig. 4.2). Annotation of the genome has been provided in the EMBL database with accession number X70810. The *Ori* locus is in close proximity to the VNTR region. The base composition of the genome is 26.1% G + C and 73.9% A + T. The genome comprises of up to 19.6 kb repeated rDNA sequence, accounting for 13.7% of the genome. This region is

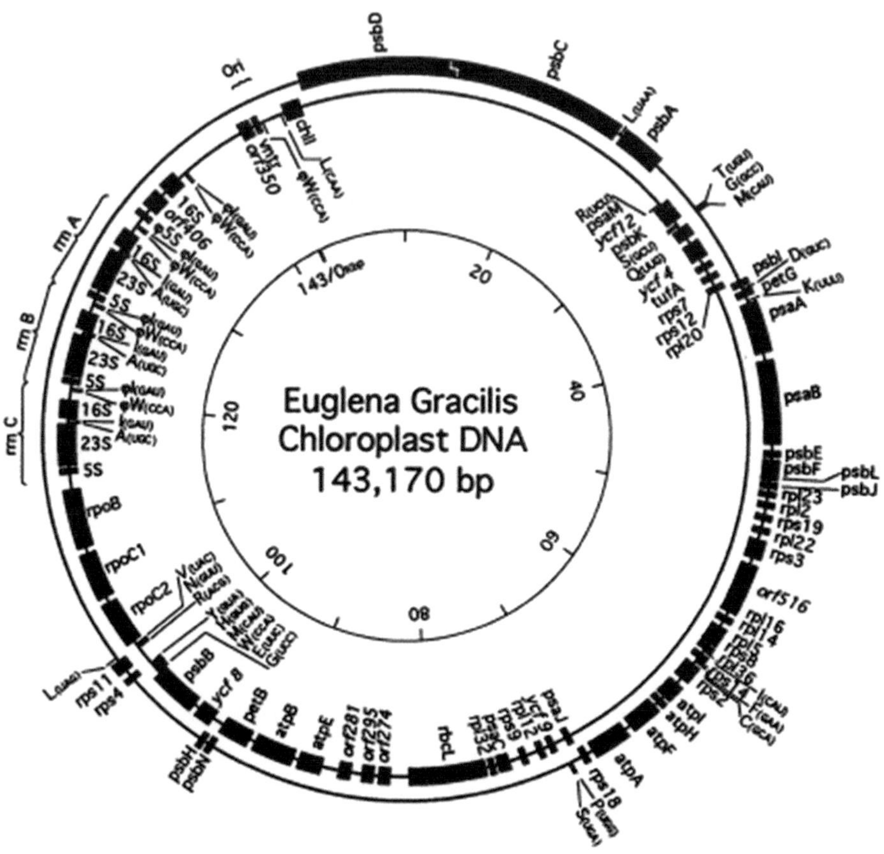

Fig. 4.2 Circular map of *Euglena gracilis* chloroplast genome. Genes are represented by filled boxes which are proportional to gene length, including exons and introns. Genes on the outer circle are transcribed clockwise. Genes on the inner circle are transcribed counterclockwise. Transfer RNA genes are identified by the single-letter code for the cognate amino acid, with the anticodon in parentheses (Courtesy, Richard B. Hallick et al. 1993)

particularly rich in G + C content (41.0%) compared to the entire genome. There are 3 copies of tandemly repeated 5918 nt ribosomal RNA operons and a fourth partial operon encoding a complete 16S rRNA gene positioning from 135,492 to 137,229. The DNA duplications are from 115,663 to 132,813 (2.9 repeats). Regions of small insertions and deletions in the genome are from 115,606 to 133,549. The remainder of the genome is a single copy sequence, densely packed with the genes encoding for polypeptides and tRNAs. None of the tRNA genes contain introns. Except for 8 of 21 ribosomal protein genes and 6 of 27 photosynthesis-related genes, all genes are interrupted by more than one intervening sequences.

One of the significant features of the *Euglena gracilis* genome organization is the arrangement of coding and non-coding DNA strands with respect to the *Ori*. Replication of the DNA is bi-directional beginning from *Ori* to the terminator on the opposite side of the circular DNA. Most of the gene clusters are transcribed away from the *Ori* bi-directionally towards the presumptive terminator. The strong bias of gene polarity away from the origin of replication could be an indication that replication and transcription are closely linked in the chloroplasts of *Euglena gracilis*.

4.5.6.3 Genes Encoding for Chloroplast Transcription and Translation Apparatus

Fifty five genes of the chloroplast 70S ribosomes, tRNAs, and translation factors including 16S, 23S, and 5S rRNAs, 27 different tRNA species, 11 ribosomal proteins of the 30S subunit, 10 ribosomal proteins of the 50S subunit and the gene for elongation factor EF-Tu are known to be constitutively expressed. Three genes encode for the subunits of chloroplast DNA-dependent RNA polymerase. The high density of introns in *Euglena gracilis* chloroplast DNA mask the location of protein-coding regions, making cDNA sequence analysis necessary for the identification of chloroplast genes. All exons of RNA polymerase subunit genes and ribosomal proteins are been confirmed by cDNA analysis. Most of these exons are very small in size. Among 168 exons of intron-containing genes, 54 encode for the exons of the length less than 20 amino acids.

4.5.6.4 Genes for tRNAs and Pseudo-tRNAs

All the 61 codons of the universal genetic code are found in the protein synthesis of *Euglena gracilis* chloroplast genome. The *trnl-trnA* genes are co-transcribed with the rRNA operons and are the only tRNA genes present in multiple gene copies. Usage of codons in the protein synthesis of chloroplasts and their corresponding tRNA anticodons for the translation of each codon have been identified in *Euglena gracilis* chloroplast (Table 4.1). The codon usage frequency indicates a very high A + U content in this chloroplast genome.

tRNAs with potential U:N pairing are tmnA-UGC, trnL-UAG, tmP-UGG, tmR-ACG, trnS-UGA, trnT-UGC, and trnV-UAC. Isoaccepting tRNAs are present only for leu, ile, arg, ser and gly codons. 4.8:1 ratio of codons is ending in either A or U compared with G or C. In codons ending with purines, there is a 3.6:1 bias of A over G. In codons ending in pyrimidines, there is a 7.4:1 bias of U over C. Few of the codons such as Leu-CUC, Leu-CUG, and Arg-CGG are used very rarely. Leu-CUC is used three, Leu-CUG twelve, and Arg-CGG nine times respectively. 9 pseudo-tRNA genes have been identified in this genome. 5 copies of pseudo-*trnW-CCA* genes precede the transcription imitation site of all four 16S rRNA genes. This pseudogene is also located near the *Ori*, adjacent to the VNTR sequences. 4 copies of a pseudo-*tmI-GAU* gene were also identified, each preceding a 16S rRNA gene.

4.5.6.5 Genes Encoding Chloroplast Ribosomal Proteins

DNA of *Euglena gracilis* chloroplast encodes for at least 21 chloroplast specific ribosomal protein genes, including 11 for the 30S small subunit and 10 for the 50S large subunit. 10 of these genes are present in a single ribosomal protein operon. Ribosomal protein-coding capacity is similar to that of land plant chloroplast

Table 4.1 Codon usage frequency in the protein encoding genes of *Euglena gracilis* and their corresponding tRNA anticodons

S. No.	Amino acid	Codon	Frequency	tRNA anticodon
1	Phe	UUU	627	
2	Phe	UUC	92	*tmF-GAA*
3	Leu	UUA	677	*trnL-UAA*
4	Leu	UUG	214	*trnL-CAA*
5	Leu	CUU	231	
6	Leu	CUC	3	*trnL-UAG*
7	Leu	CUA	75	
8	Leu	CUG	12	
9	Ile	AUU	620	
10	Ile	AUC	58	*tml-GAU*
11	Ine	AUA	372	*tntl-CAU*
12	Met	AUG	236	*trnM-CAU*
13	f-Met	AUG	48	*trnM-CAU*
14	Val	GUU	475	
15	Val	GUC	31	*tmV-UAC*
16	Val	GUA	233	
17	Val	GUG	43	
18	SER	UCU	320	
19	Ser	UCC	43	*trnS-UGA*
20	Ser	UCA	189	
21	Ser	UCG	52	
22	Pro	CCU	295	
23	Pro	CCC	33	trnP-UGG
24	Pro	CCA	142	
25	Pro	CCG	23	
26	Thr	ACU	294	
27	Thr	ACC	28	trnT-UGU
28	Thr	ACA	277	
29	Thr	ACG	65	
30	Ala	GCU	383	
31	Ala	GCC	39	trm4-UGC
32	Ala	GCA	233	
33	Ala	GCG	61	
34	Tyr	UAU	335	
35	Tyr	UAC	56	trnY-GUA
36	End	UAA	40	
37	End	UAG	6	
38	His	CAU	234	
39	His	CAC	28	tmH-GUG
40	Gln	CAA	311	trnQ-UUG
41	Gln	CAG	47	
42	Asn	AAU	497	

(continued)

Table 4.1 (continued)

S. No.	Amino acid	Codon	Frequency	tRNA anticodon
43	Asn	AAC	105	trnN-GUU
44	Lys	AAA	771	trnK-UUU
45	Lys	AAG	139	
46	Asp	GAU	379	
47	Asp	GAC	70	trnD-GUC
48	Glu	GAA	470	trnE-UUC
49	Glu	GAG	107	
50	Cys	UGU	98	
51	Cys	UGC	35	trnC-GCA
52	End	UGA	2	
53	Trp	UGG	189	trnW-CCA
54	Arg	CGU	206	
55	Arg	CGC	47	trnR-ACG
56	Arg	CGA	89	
57	Arg	CGG	9	
58	Asn	AAU	497	
59	Asn	AAC	105	trnN-GUU
60	Lys	AAA	771	trnK-UUU
61	Lys	AAG	139	
62	Ser	AGU	173	
63	Ser	AGC	28	trnS-GCU
64	Arg	AGA	233	trnR-UCU
65	Arg	AGG	58	
66	Gly	GGU	480	
67	Gly	GGC	65	tmG-GCC
68	Gly	GGA	319	trnG-UCC
69	Gly	GGG	60	

genomes (18 of 21 genes). This genome has small subunit genes *rps2, rps3, rps4, rps7, rps8, rpsll, rpsl2, rpsl4, rpsl8*, and *rpsl9*, common in all known chloroplast genomes of land plants and large subunit genes *rpl2, rplJ4, rplJ6, rpl20, rp122, rpl23, rpl32* and *rpl36*. Gene *rpl5* is specific to chloroplast DNA of *Euglena gracilis*, also found in *Astasia longa*, red alga *Porphyra purpurea* and *Cyanophora paradoxa*, but not found in the chloroplasts of land plants. Gene *rps9* is also specific to the chloroplasts of *Cryptomonas, Cyanophora paradoxa* and *Cyanelle*.

4.5.6.6 Genes Involved in Photosynthesis

The chloroplast genome of *Euglena gracilis* encodes 27 genes for components of the thylakoid membranes, the chloroplast ATP synthase complex and RUBISCO. There are 5 genes encoded for photosystem I polypeptides, 10 for photosystem II and 2 for the cytochrome b6/f complex. Six genes encoding the *Euglena* ATP synthase subunit are organized in two operons similar to land plants. Genes for the subunits of NADH dehydrogenase complex and cytochrome-f are not found in the *Euglena gracilis* chloroplast genome. Euglena chloroplast DNA along with red alga *P. purpurea* contains *chlI* gene in their genome, might be involved in the chlorophyll biosynthesis. Till date, *chlI* gene is not reported in the chloroplast genomes of land plants.

4.6 Mitochondrial Genomes

Mitochondria are the energy-transducing organelle of the cells, which fuels the conversion of cellular metabolisms into ATP through oxidative phosphorylation. Structurally, mitochondria have two membranes, outer and inner. An outer membrane separates the mitochondrion from the cytosol and the inner membrane is invaginated to form the cristae, which protrude to form the matrix. Five enzyme complexes involved in the oxidative phosphorylation system are embedded in the mitochondrial inner membrane. Mitochondria contain its own genome called mitochondrial DNA (mtDNA), located in its matrix.

Mitochondrion might have originated from an incorporated proteobacterial endosymbiont K-purple bacteria. During its evolution, it has either transferred many of its essential genes to the nucleus or has lost them completely. Few eukaryotes have secondarily adapted to anaerobic lifestyle with consequent modification of their mitochondria to form hydrogenosomes (membrane-bound organelle, which releases molecular hydrogen (H_2) as a by-product of energy generation under anaerobic conditions) and mitosomes (highly reduced cryptic mitochondria found in the anaerobic eukaryotic parasites). These modifications are due to the partial or complete loss of the mitochondrial genome. Still, mitochondria carry the hallmarks of their bacterial ancestor such as utilizing N-formylmethionyl-tRNA (fMet-tRNA) as an initiator of protein synthesis. Mitochondrial genetics differ considerably from Mendelian genetics. Uniparental inheritance, cellular polyploidy, and deviation from the standard genetic code are few peculiarities in the mitochondrial genetics, which strongly dictate the functional consequences of the mtDNA.

4.6.1 Sequencing Technologies for Mitochondrial Genomes

Mitochondrial DNA has major features such as maternal inheritance, replicative segregation, threshold expression, high mutation rate, and heteroplasmy. New mtDNA mutations might also arise in cells, co-existing with wild-type mtDNAs that segregate randomly during cell division. These features make mtDNA a valuable genetic biomarker and a prognostic/diagnostic indicator for a number of diseases as well as in forensic sciences for human identification, where the biologic evidence contains too little or no nuclear DNA. Sanger's dideoxy has been the most standard method for mtDNA sequencing. But this approach has certain limitations with throughput, scalability, speed, and resolution, necessitating the throughput sequencing approach. Massively Parallel Sequencing Technology (MPST) provides platforms for more comprehensive coverage of the genome per sample than currently is possible with Sanger's sequencing. The number of samples can be distinguished by barcoding and sequencing simultaneously. Genome sequencers available for this purpose are the Ion Torrent Personal Genome Machine (PGM™) (LifeTechnologies, San Francisco, CA) and the MiSeq™ (Illumina, Inc., San Diego, CA). The PGM exploits non-optical sequencing on CMOS integrated circuits by detecting small changes in the pH, due to the release of H^+ during the addition of a nucleotide to the growing strand within 2 h of running time. The MiSeq uses fluorescently tagged terminator chemistry that requires 39 h for paired-end sequencing but has a higher throughput and an associated simpler, less labor intensive library preparation methodology than the PGM. Using MPS, it is possible to sequence 96 samples of mitochondrial genomes at a time. Sequencing of the entire mitochondrial genome provides a higher resolution and discrimination power than is currently possible with the sequencing of only a few portions of the non-coding region of the mitochondrial genome (for high information content) or by targeted analyses (for a few SNP or deletions noted in the coding region).

Each cell consists of hundreds to thousands of mtDNAs with different proportions of mutant and wild-type. Heteroplasmy is a condition in which the wild type mtDNA and mutant mtDNA co-exist and the degree of heteroplasmy is used to describe the ratio of the mutant mtDNA compared with the wild type. Mutations in the heteroplasmic mtDNA are responsible for many diseases and the degree of mutations may vary among different tissues, even in the same organism. The existing mitochondrial genome sequencing methods are neither sensitive nor specific enough for detecting the mtDNA heteroplasmy. Screening for the detection of mitochondrial genome-wide heteroplasmy includes denaturing HPLC, surveyor nuclease digestion and high-resolution melt (HRM) profiling. Quantification of mtDNA heteroplasmy levels can be achieved by polymerase chain reaction-restriction fragment length polymorphism (PCR-RFLP) analysis, allele-specific oligonucleotide dot-blot analysis, real-time amplification refractory mutation system quantitative PCR pyrosequencing and parallel sequencing. Through parallel sequencing, mitochondrial genome-wide heteroplasmy can be identified and quantified with extraordinary sensitivity.

4.6.2 Structure of Mitochondrial Genome

Mitochondrial DNA is a small, highly abundant DNA molecule, which was the target of many early genome sequencing projects. Mitochondrial genome sequencing of a large number of organisms has been determined. Human mtDNA was the first documented complete sequence of a mitochondrial genome, elucidated in 1981 and was further revised in 1999. It is a closed circular, double-stranded DNA molecule of 16,569 bp, distinguished into cytosine-rich light (L) and guanine-rich heavy (H) strands on the basis of G + T composition (Fig. 4.3).

H strand encodes the genes for two rRNAs, 14 tRNAs, and 12 polypeptides, whereas L strand codes for eight tRNAs and a single polypeptide. In total, the human

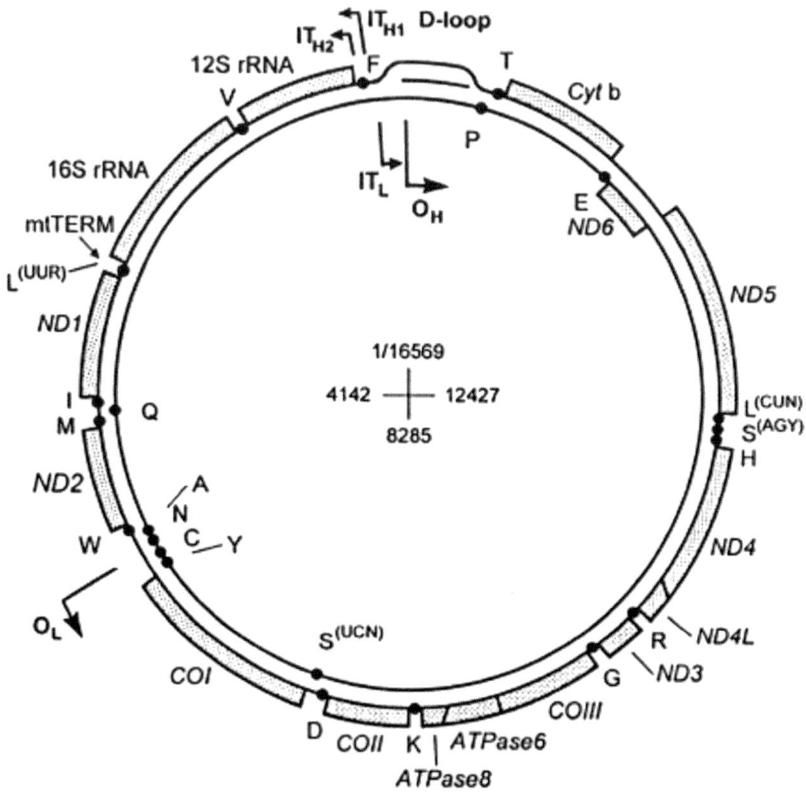

Fig. 4.3 Human mitochondria genome map. The outer circle represents the H-strand and inner circle represents L-strand. The D-loop is shown as a three-stranded structure. The origins of H-strand (OH) and L-strand (OL) replication and the direction of DNA synthesis are indicated by long bent arrows; the initiation of transcription sites (ITL, ITH1, ITH2) and the direction of RNA synthesis are denoted by short bent arrows. tRNA genes are depicted by dots and the genes coding for the two rRNA species (12S and 16S) and the 13 protein-coding genes are depicted by shaded boxes (*Courtesy Jan-Willem Taanman, Royal Free Hospital School of Medicine, University of London*)

mitochondrial genome contains 37 genes encoding 13 proteins, which are the main constituents of the enzyme complexes belonging to the oxidative phosphorylation system. Both rRNA and tRNA molecules are unusually small. Few of the protein-encoding genes are overlapping and part of the termination codons are not encoded but are generated post-transcriptionally by polyadenylation of the mRNAs. All core subunits of RC complexes I, III, IV and V and the RNA necessary for mtDNA translation are 2 rRNAs (12S and 16S) and 22 tRNAs. Nuclear genes encode the remaining ~70 OXPHOS components and all other proteins required for mitochondrial metabolism and maintenance, are imported *via* specialized systems. No introns and no intergenic non-coding nucleotides were identified in the genes of mitochondrial genome, with an exception of 1.1 kb displacement loop (D-loop), containing the transcriptional promoters and at least one of the proposed replication origins. 6–10 copies of mtDNA are organized in a stable protein–DNA macro-complexes termed as nucleoids, which are exchanged between mitochondria. MtDNA replication, maintenance, repair, and recombination are the main functions of the major constituent proteins associated with these structures. Human mtDNA is strictly inherited through the maternal lineage and structure and gene organization of mtDNA is highly conserved among mammals.

Comparison of the mitochondrial protein sequences revealed several deviations from the standard genetic code and even in the codon usage. In mitochondrial genomes, TGA is used as tryptophan codon rather than as a termination codon and AGA/AGG codon specifies a stop codon among vertebrates, serine in Echinodermata and arginine in yeast as the standard genetic code. Another unique and surprising feature of the mitochondrial genome is the usage of simplified decoding mechanism, allowing the translation of all codons with less than 32 tRNA species required as per Crick's wobble hypothesis. This reduction in the number of tRNAs in the mitochondrial genome is achieved by the usage of a single tRNA with U in the first anticodon position to recognize all codons. Mitochondria genomes of fungi use a modified U in the wobble position to read two codon families with a purine in the third position of the codon, preventing the misreading of two codon families with a pyrimidine in the third position, which is conserved among the mitochondria of vertebrates. Mitochondrial DNA of metabolically active vertebrate cells contains a short three stranded structure called displacement loop or D-loop, comprising of genes for tRNAPhe and tRNAPro, the origin of replication and the major promoters for transcription.

4.6.3 Mutations in the Human Mitochondrial DNA and Diseases

Human mitochondrial DNA has been a source of a very high rate of mutations, which are 10–17 folds higher than that of nuclear DNA. Repairing systems that are existing in the mitochondria are not sufficient to counteract the damage sustained by the mitochondrial DNA due to the presence of respiratory complex in its proximity and the amount of ROS generated are sufficient to alter the DNA present in the inner mitochondrial membrane. These alterations play a crucial role in human health. First mutations in the mt DNA were identified in 1988. Since then, over 250 mutations responsible for a wide variety of diseases have been identified and characterized by mitochondrial DNA (Fig. 4.4). But, the exact prevalence of mt DNA is difficult to ascertain due to the heterogeneity of the diseases and a plethora of known causative mutations. It is estimated that 1 in 10,000 humans have clinically manifesting DNA disease and 1 in 6000 are at risk. Recent studies have reported that the frequency of ANG mutation is 0.14% and that of MT-RNR1 mutation oriented with the aminoglycoside-induced sensorineural hearing loss is 0.2%, suggesting the seriousness of health impairers caused by the mutations in the mtDNA.

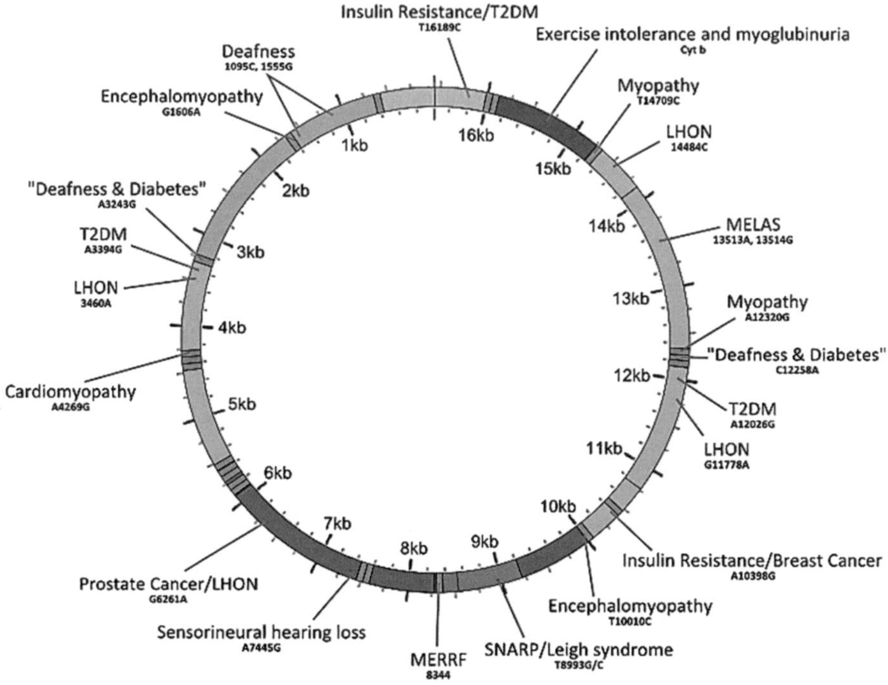

Fig. 4.4 List of disorders caused due to the mutations in the mtDNA. LHON: leber's hereditary optic neuropathy, MELAS: mitochondrial encephalopathy lactic acidosis with stroke-like episodes, MERRF: myoclonic epilepsy and ragged-red fiber, T2DM: type II diabetes mellitus (Courtesy Wallace D.C, University of California, Irvine)

4.6.3.1 Point Mutations and Rearrangements in the Mitochondrial Genome

Point mutations in the mitochondrial genomes are reported to occur in the tRNA, rRNA genes and even in their proteins. These mutations are maternally inherited. 50% of the point mutations that leads to diseases are reported to be occurring in the tRNA genes. Point mutations in the protein-coding genes of mtDNA will affect the respiratory complex of the corresponding protein and mutations occurring in the mitochondrial tRNA genes further impairs the overall protein function. Most of the point mutations occurring in the mitochondria are considered to be heteroplasmic, displaying clinical heterogeneity and are highly recessive. Majority of the re-arrangements occurring in mtDNA are deletions, whose size varies from 1.3 kb to 8 kb. Deletions in the mitochondrial DNA might be due to the inheritance of mutations in the nuclear DNA, whose products are involved in the maintenance and replication of mitochondrial nucleotides. A single deletion in the mtDNA occurs early in the development and multiple deletions occur in aged postmitotic tissues and also in the individuals with neurodegenerative disorders. These deletions occur between OH and OL regions, flanked by short direct repeats. The mechanism by which these deletions are formed in the mitochondrial DNA is not clear and is currently hypothesized to occur during the replication of mitochondrial DNA.

4.6.3.2 Functional Characterization of Mutations

Pathogenic mutations in mtDNA invariably affect the respiration, leading to a reduced ability in the production of cellular ATP. However, the clinical phenotypes of mitochondrial disorders are extremely heterogeneous, involving some molecular mechanisms along with reduced respiration rates, responsible for the variability. For developing the effective therapies to mitochondrial disorders, it is important to identify the biochemical and molecular mechanisms connecting the phenotype to the genotype.

4.6.3.3 Clinical Syndromes

The clinical syndromes associated with mtDNA mutations are highly variable and can be reflected at any stage of human life, depending on the level of mutation and the severity of the deficit. Genetic (nuclear) and environmental factors also affect the level and intensity of the disease to some extent along with the accumulated mtDNA mutations during aging, neurodegeneration, and tumorigenesis. Leigh syndrome is a progressive neurodegenerative condition reflected during the onset of infancy, particularly deteriorating the brainstem, diencephalon, and basal ganglia in a step-wise manner. Leigh syndrome occurs due to severe failure of oxidative metabolism in combination with a variety of different genetic defects such as m.8993 T N C/G, m.10158 T N C, m.10191 T N C affecting the mtDNA and *SURF1* gene in the nuclear genome. Depletion syndromes are severe disorders that occur in the organs with depleting mt DNA. These disorders occur in the childhood resulting in severe muscle weakness, progressive encephalopathy or liver failure, necessitating organ transplantation. Kearns-Sayre syndrome (KSS) is associated with the development of retinitis pigmentosa and progressive external ophthalmoplegia before the age of 20. It is a multisystem disorder caused by single, large-scale deletion resulting in the development of neurological disorders such as cerebellar ataxia, cognitive impairment, and deafness, as well as non-neurological features of cardiomyopathy, complete heart block, short stature, endocrinopathies, and dysphagia. Pearson syndrome is a rare disorder in infants occurs due to the deletion of large-scale single mtDNA, characterized by sideroblastic anemia and exocrine pancreatic failure, leading to an early death. Chronic progressive external ophthalmoplegia (CPEO) is one of the most common presentations of mtDNA disease in adults, characterized by the progressive paralysis of the eye muscles leading to impaired eye movement and ptosis. CPEO is caused by a single deletion. Neuropathy, ataxia and retinitis pigmentosa occurs mainly due to the MT-ATP6 m.8993 T N G mutation in the mtDNA.

4.6.3.4 Diagnosis of mtDNA Diseases

Lack of proper genotype-phenotype correlations and complex interactions between nuclear and mitochondrial genomes in many patients are the factors limiting the laboratory diagnosis of diseases generated due to the mutations in the mtDNA. Family pedigree history and histological changes in the affected tissues are the salient features that provide the basis for the diagnosis of mitochondrial disease, supported by the molecular defects such as single, large-scale rearrangements or point mutations in mtDNA or in nuclear genome, leading to the abnormalities such as loss of mtDNA copy number or multiple mtDNA deletions.

Most of the patients with mtDNA diseases show histological and histochemical changes in affected tissue such as ragged red fibers indicating the dysfunction of the respiratory complex. These fibers appear by staining with Gomori trichrome stain, showing abnormal mitochondria. Biochemical testing of clinically affected tissue also involves the determination of respiratory complex activities in mitochondria from the homogenates made with fresh or frozen tissue samples including measurements of mitochondrial substrate oxidation, oxygen consumption, and ATP synthesis.

Identification of isolated or multiple abnormalities further narrow down the options for identifying mtDNA re-arrangements and certain point mutations that are identified to be present only in clinically affected tissues. MtDNA rearrangements such as deletions and duplications have been traditionally assessed by Southern blotting hybridization, replaced by the long-range PCR. Deletions in mtDNA indicate the screening of appropriate nuclear mtDNA maintenance genes including POLG, POLG2, SLC25A4, PEO1, and OPA1. Quantitative loss of mtDNA determination by the real-time PCR with reference to a nuclear housekeeping gene is indicative of a recessive mtDNA depletion disorder leading to myopathic or hepatocerebral phenotype. Common point mutations in mtDNA such as m.3243A N G, m.8344A N G, m.8993 T N G/C, can be readily screened by the restriction fragment length polymorphism (RFLP) analysis by the addition of a radioactive or fluorescent label in the final PCR cycle to permit the accurate quantification of mtDNA heteroplasmy.

Due to its small size, screening of the entire mitochondrial genome can be undertaken easily for identification of rare or novel mutations in patients with suspected mtDNA disease. Denaturing gradient gel electrophoresis (DGGE) and denaturing high-performance liquid chromatography (dHPLC) were the main techniques for the identification of novel and rare mutations. With the emergence of re-sequencing microarrays such as the Affymetrix MitoChip 2.0, many laboratories are continuing to sequence the whole mitochondrial genome of affected tissues, using overlapping primers. Given the highly polymorphic nature of the mitochondrial genome, careful assessment of newly identified mtDNA sequence variants must be undertaken to establish a clear link with the human disease for the gene therapy or organ transplantation.

4.6.3.5 Gene Therapy for the Prevention of Diseases Due to mt DNA Mutations

Despite the considerable advance in the understanding of mitochondrial genetics, function, and diseases concerned with the pathogenesis of mtDNA, no effective treatment options are currently available for patients with mitochondrial dysfunction, except in rare cases where surgery or transplant may be indicated due to the following reasons.

(i) Identification of the diverse functional effects of mtDNA mutations has proved challenging.

(ii) Engineering mtDNA is extremely problematic due to its high polyploidy, inaccessibility being a double-membrane organelle, apparent lack of minimal recombinations. These features have prevented the successful transfection of autonomously replicating foreign DNA plasmids or modified mtDNA into mitochondria.
(iii) Development of suitable models that can accurately portray the pathogenesis of mitochondrial diseases is quite slow.

Manipulation of heteroplasmy levels for shifting the balance of mutant into wild-type genomes has been attempted by a number of approaches. Antigenomic therapy, involving the usage of sequence-specific nucleic acid derivatives that selectively target and inhibit the replication of mutated mtDNA, thus allowing the wild-type genome to propagate has initially shown promising results in vitro with peptide nucleic acids. However, peptide nucleic acids were unable to cross the inner mitochondrial membrane, forcing the design of molecules with a greater polarity such as cell membrane crossing oligomers (CMCO) and CMCO: PNA hybrid molecules for their entry into the into mitochondria in whole cells. Alternative strategies for manipulating mtDNA heteroplasmy levels include mitochondria-targeted restriction endonucleases, capable of distinguishing between mutated and wild-type genomes. Selective targeting of mutated mtDNA has also been successfully attempted with a targeted chimeric zinc finger DNA methylase, capable of binding and modifying in a sequence-specific manner. Even though the initial results look promising, the therapeutic use of this strategy is limited by the need for transfection and its successful expression in defective tissues.

4.6.4 Mitochondrial Genome of Neandertal Fossil

Even though the exact relationship of Neandertals with modern day human beings is a subject of debate, they are the hominid forms, which are considered to be most closely related with the human beings that are existing in the present day. Information regarding their relationship was made available in 1997 by sequencing 379 bp hypervariable I region (HVRI) of mtDNA obtained from Neandertal specimen fossil found in 1856 in Neander Valley, near Düsseldorf, Germany. Since then, 15 complete or partial HVRI sequences, as well as two HVRII sequences from the mitochondria of Neandertal have been described and their phylogenetic analyses suggest that the mtDNAs of Neandertal man fall outside the variation of modern human mtDNAs. As the mitochondrial genome is inherited maternally without undergoing any recombinations, these results indicated that Neandertals did not make any strong contribution to the modern human mitochondria gene pool.

With the availability of High-throughput 454 sequencing technology that can be applied to fossil DNA, retrieval of ancient DNA was made possible. The main application of the 454 sequencing technique is the sheer volumes of sequence data can be obtained from the ancient DNA sequencing projects, particularly from mitochondrial genomes due to their small size and abundance in cells. Data obtained from the 454 sequencers have also provided increased understanding of DNA modification during burial deposition. Degradation, cytosine deamination and chemical modification of DNA are known to occur in the fossil or ancient DNA, particularly in their ends as these are often single-stranded. Deamination of cytosine has resulted in the formation of uracil that is read as thymine by DNA polymerases, leading to a high rate of C to T transitions. A very high rate of G to A transitions have also been observed near the 3′ ends, which might be attributed to the deaminated cytosine residues on the complementary strands that were used as templates during the fill-in reaction for creating the blunt ends during library construction for sequencing. Through 454 sequencing, a 34.9-fold coverage of the Neandertal mtDNA genome was generated from a Neandertal bone excavated from Vindija Cave, Croatia in 1980. Hence, this fossil bone is called Vindija bone 33.16. This fossil has been dated to 38,310+/−2130 years before present. In the previous studies carried out on this fossil, the sequence of mtDNA HVRI has been determined and a 2414 bp of mtDNA sequence was also performed by 454. The complete sequencing of Neandertal mitochondrial DNA is mentioned, which provides an insight regarding the evolution of modern human being.

4.6.4.1 Extraction of Fossil Mitochondrial DNA, Sequencing

Mitochondrial DNA was extracted from 100–200 mg of Vindija bone 33.16, obtained from the Neandertal fossil. The extract was analyzed for the modern contamination with modern human mtDNA by PCR using primers in the HVRI that distinguishes the existing humans from Neandertal and amplifies both mtDNA types with similar efficiency. Following amplification, the PCR product was cloned and a total of 103–112 clones were made from the extract, sequenced to determine the contamination rate, which was ranging from 0% to 0.9%. Totally, 9 libraries were generated from the extract using 454 adapters with a unique Neandertal specific sequence key for unequivocally identifying each sequence. From these 9 libraries, 39 million sequence reads were generated by 147 runs on a GS FLX sequencing platform. Neandertal sequences were identified in each read by the virtue of its similarity with primate genome and mtDNA sequences were identified by the similarity to the human mtDNA. In 2100 mtDNA molecules per cell of Neandertal bone, 8341 mtDNA sequences were identified with an average length of 69.4 bp. Length of the longest mtDNA fragment was 278 bp and the shortest fragment was 30 bp (limited by the flow cycles performed on the GS FLX instrument). The complete genome sequence of human Neandertal mtDNA is available in the NCBI database under accession number AM948965.

4.6.4.2 Genome Assembly

Assembling the ancient DNA sequences which are archetypally short, and exhibiting a very high rate of nucleotide misincorporation poses a tough challenge. To solve these issues, a novel assembling procedure was identified. According to this procedure, each mtDNA sequence has been aligned to its full length with the human mtDNA sequence (UCSC build hg18) as a reference and all these alignments were completely merged. During this process, four regions with a total length of 20 bp without any sequence coverage was found to be existing. Eight regions on the Neandertal mtDNA, accounting to a total of 117 bp has been covered only by single reads. Nine positions on the genome, without any majority base, was also existed because of the low coverage. 31 homopolymers with insufficient data for determining their length remained in the assembly. These specific regions were PCR amplified from a Neandertal bone extract, cloned and sequenced by Sanger technology in order to complete the assembly process. The entire sequence was re-applied to the Neandertal mitochondrial detection pipeline, using the previously assembled Neandertal mtDNA as a reference, resulting in the detection of additional 721 mtDNA sequences missing previously. 8341 mtDNA fragments consisting of 578,733 nucleotides have yielded a final assembly of 16,565 nucleotides, with no position has less than a 9-fold sequence coverage.

4.6.4.3 Sequence Analysis

Alignment of the 16,565 bp Neandertal mtDNA with the 16,568 bp human reference mtDNA sequence have revealed 206 differences, among which 195 are transitions and 11 transversions. In the non-coding region of the Neandertal mitochondrial DNA, a sequence of four base pairs (CACA) was deleted at rCRS position 514 and one base pair was inserted at position 16,263. No major differences were identified in the 13 protein-coding genes, 22 tRNA genes and 2 rRNA genes of Neandertal mtDNA, human and chimpanzee mtDNA. Differences between the pairwise sequences 53 humans all over the world, Neandertal and chimpanzee mtDNA were around 99. These pieces of evidence keep the Neandertal mtDNA outside the variation of existing humans.

4.6.5 Structure and Organization of Plant Mitochondrial Genome

Due to the technical difficulties in sequencing the large and complex DNA molecules, the complete sequence of plant mitochondria could not be done till Brennicke's group reported the 366,924 bp of *Arabidopsis thaliana* complete mitochondrial genome in 1997. After that, over 269 plant mitochondrial genomes have been sequenced and the data indicates that flowering plants are containing the largest mitochondrial genomes reported so far with circular and linear shapes (maize cytoplasmic male sterility lines-S). Plant mitochondrial genomes exist in large circular forms denoted as master chromosome containing several large direct (>500 bp) and few inverted repeats. The genome replicates in a recombination-dependent manner, with intra and intermolecular recombinations between the repeats are responsible for different isomeric forms and sub-genomic versions of the master chromosome. Variations in the size of a master chromosome might be due to the presence of repeated sequences and large duplications (up to 1000 bp), representing up to 35% of the total genome size. Size of mitochondrial genomes in maize lines varies from 535,825 bp to 739,046 bp with genome complexities ranging from 506,760 bp to 537,180 bp. Unique sequences in the mitochondrial genomes differ intra-specifically by upto7%. Sequence loss from mitochondrial genomes can be compensated by transfer to the nuclear genome. Plant mitochondrial heteroplasmy covers a significant part of the mitochondrial genome. >20% of the plant mitochondrial genome is represented by known proteins and the genes that are encoding for rRNA and tRNA. The coding region of the plant mitochondria is highly conserved with a possibility of either mitochondrial or chloroplast origin. Additionally, every plant mitochondrial genome has 10% of putative conserved open reading frames (ORF). Recombinations between the short repeats might alter a few known coding sequences and ORFs, leading to cytoplasmic male sterility. The number of genes in the plant mitochondrial genome is 50–60. The differential number of genes is due to the differential gene content in the subunits of Complex II, especially the ribosomal proteins and tRNAs. Genes are translated according to the Universal Genetic Code. The composition of tRNA genes in plant mitochondria is quite unique. Among 15–21 tRNA genes encoded, 10–12 are orthologous to moss mitochondria. This class of tRNA genes is called as native. The remaining set of tRNA genes either show high homology to cp DNA or their sequences do not match with the tRNAs of any known source. Differential allocation of tRNA genes to these classes is also frequently found indicating that the recruitment of tRNA genes has occurred independently in plants. Few mitochondrial genes of angiosperms are interrupted by introns. The average number of introns per mitochondrial genome is 20–24, constituting up to 13% of the genome. All the introns in the mitochondrial genomes were identified as group II type except horizontally transferred group I introns, which are rarely documented in the *cox1* gene of Peperomia. One of the main striking features of angiosperm mitochondrial introns is the presence of trans-splicing, particularly in the genes encoding for Complex I subunits. A transition from cis to a trans-re-arrangement of introns during their evolution has been noticed, supported by the finding that all introns in non-angiosperms are with a *cis* arrangement and are trans-spliced in angiosperms.

The multipartite structure of plant mitochondrial genomes can be used to explain their heterogeneity. When plant mitochondrial (mt) DNA is examined by electron microscopy and gel electrophoresis, the morphology of the genome was not found to be consistent with the circular model and the observed molecules have shown variation with respect to their shapes and sizes. The shape varied from linear to circular and is of various sizes. Additionally, few genomes have shown Y, H and theta (θ) shaped branched forms presumably representing recombination intermediates, indicating the entity of plant mitochondrial genome as a mixture of various DNA molecules. It still remains unclear how the multipartite and branched mitochondrial DNA is replicating and is transmitting to the next generation.

4.7 Conclusions

Increased number of chloroplast and mitochondrial genomes has provided us with a better understanding of their structure, content, and organization. Chloroplast genomes are more conserved than the mitochondrial genomes and high variations in their size have been observed in the mitochondrial genomes, which makes them more structurally varied and capricious. The mechanisms of DNA maintenance and repair are quite common between these two genomes, accounting for convergent genome evolution of organelle. However, the organellar genome sequences of most of the model organisms are available in databases and increased sequence information and technology will further enhance the understanding of their genome diversity, mutations, and architecture.

Eukaryotic Genome Organization, Regulation, Evolution and Control

5

Abstract

This chapter aims to show the vital components of eukaryotic genomes such as chromosome structure, genome duplications, transposons, satellite DNA. The genomes of important and highly used model organisms yeast, *Caenorhabditis elegans*, Drosophila, Arabidopsis, rice, soybean, and human beings have been discussed in detail. Importance of comparative genomes, complexity, and evolution of eukaryotic genomes with special emphasis on the human genome was also added.

5.1 Introduction

Prokaryotic genomes are simpler with one or two circular or linear chromosomes, consisting of mostly coding DNA, whereas Eukaryotic genomes are more complex with linear, multiple chromosomes containing a large sea of non-coding DNA with small islands of coding regions. Eukaryotic genomes vary in the types of sequences, amount of DNA and the number of chromosomes. Different cells in the same organism can vary in the number of chromosomes (n, number). Somatic cells are diploid (2n) and the germ cells are haploid (1n). Size of the genome is always related to the haploid known as the C-Value, which was 3.5×10^3 bp in small viruses such as coliphage, MS2 and 2×10^5 bp in large viruses, 2×10^7 bp in bacteria and unicellular eukaryotes such as yeast, 4 times larger in primitive multicellular organisms, which increases up to four to tenfolds in higher organisms. Wide range of organisms has shown that the c-value of a particular phylum or family is related to their structural and organizational complexity and organisms which are evolutionarily more complex have great minimum genome size. Further, the complexity of the eukaryotic genomes is mostly due to the presence of repetitive elements, pseudogenes, jumping genes, etc. This chapter will discuss the details of the eukaryotic genomes with the genomes of model organisms including human genome as appropriate examples.

K. V. Chaitanya, *Genome and Genomics*,
https://doi.org/10.1007/978-981-15-0702-1_5

5.2 Organisation of Eukaryotic Genomes

Most of the completed and ongoing genome projects have provided a great deal of information about the eukaryotic genome organization, which refers about the linear arrangement of thousands of single or multiple copy genes, the spatial organization of euchromatin and heterochromatin regions and the position of chromosomes in the nucleus.

5.2.1 Organization of Chromosomes

Chromosomes are the first functional units that are subdividing the cell nucleus. Fluorescent In Situ Hybridization (FISH) of fixed eukaryotic cells reveals the occupation of chromosomes at distinct volumes termed as chromosome territories, defined as the volume of chromosomes that are not exclusively but preferentially localized in a nucleus. These chromosome territories are also non-randomly distributed within the nuclear space and gene-rich chromosomes, localized internally than that of gene-poor chromosomes, which are more peripheral. More intense studies on the chromosomes have identified certain variations in their size and centromere position. A variety of staining techniques are available for understanding the variations in chromosomes. For G banding, the chromosomes are initially digested with protease trypsin and then stained with Giemsa, which divides the chromosomes into positive dark G bands consisting of A-T rich regions and negative pale G-C rich R bands. At an average, each chromosome can have up to 2000 light and dark bands (Fig. 5.1). Alternatively, chromosomes can be stained with fluorescent quinacrine for R banding, which intercalates between the nucleotides.

Heterochromatin region of a chromosome can be identified by the treatment of the chromosome with an acid followed by an alkali before staining with Giemsa. The unstained region of the chromosome is referred to as euchromatin.

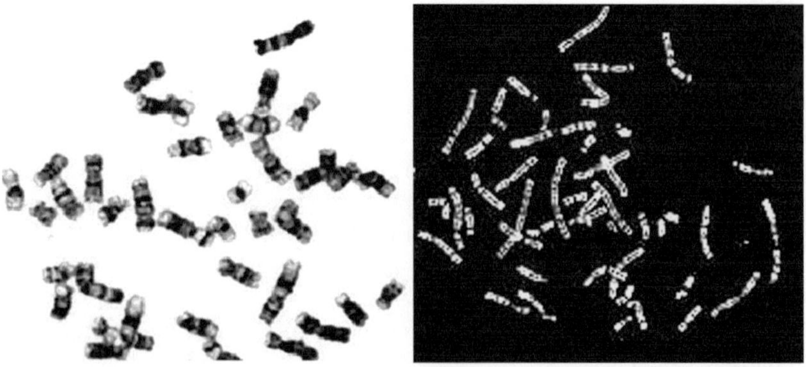

Fig. 5.1 G-banding and R banding of chromosomes

CpG islands are also identified in the human genome by using Giemsa stain. CpG islands (also called as HpaII Tiny Fragment (HTF) islands) are short regions of 1–2 Kb in the mammalian DNA, containing many sites for the restriction endonuclease HPaII. CpG islands account for ~2% of the mammalian genomes. These are the short interspersed sequences that deviate significantly from the rest of the genome by their un-methylated and GC-rich characteristic features. CpG Islands are generally found at the 5′ ends of all housekeeping genes and a proportion of genes with a tissue-restricted pattern of expression and are considered as the sites of transcription initiation, including the sites that are remote from the currently annotated promoters. 70% of the promoters from the vertebrate genomes are associated with the CpG islands.

5.2.2 Centromeres of Eukaryotic Chromosomes

The centromere is a distinctive region, which divides the chromosome into the long arm and short arm. This region is also required for the attachment of chromosome to the mitotic spindle during chromosome segregation. In unicellular eukaryotes like *Saccharomyces cerevisiae*, centromeres are defined by a c DNA sequence that is specifically found on all chromosomes, called point centromeres. In multicellular eukaryotes, centromeres are largely filled with arrays of satellite repeats. Many similarities have been identified between the centromeres of different species that highlight its unique nature and essential functions.

The centromere of budding yeast was the first eukaryotic centromere to be isolated and characterized. This centromere consists of three conserved DNA sequences that are common to all chromosomes, termed as Centromere DNA element I (CDE I), Centromere DNA element II (CDEII) and Centromere DNA element III (CDE III). CDEI is generally 8 bp long, essential for the high-fidelity chromosomal segregation. CDEII has a well conserved 78–86 bp AT-rich region, also participates in the chromosomal segregation and CDEIII is a 26 bp long sequence that contains seven invariant nucleotides, conserved among all chromosomes. These three elements together form 116 to 120 bp sequence for conferring mitotic stability and any mutation in these three elements will abolish centromere function. The centromeres of *Schizosaccharomyces pombe* are long with an approximate size of 40–100 kb, comprising of multiple microtubule attachment sites per each centromere. These attachments sites are composed of three central core regions cnt1, cnt2 and cnt3, flanked on both sides by inverted repeat sequences. Centromeres in *Drosophila melanogaster* consists of satellite DNA embedded with transposable elements. Mammalian centromeres are built with alphiod DNA, consisting of a 171 bp motif repeated in a tandem head to tail fashion. The centromere of *Arabidopsis thaliana* has a similar motif with 178 bp repeat. Centromeres of cereal plants contain 200 copies of a gypsy-like retrotransposon called cereba.

5.2.3 Telomeres

Telomeres are the ends of the chromosomes, discovered by Herman Muller in Drosophila and Barabara McClintock in corn during the 1930s much before the DNA was known to be the genetic material. The essential role of telomeres is to protect and maintain the chromosome ends, which are also the ends of linear duplex DNA. When the synthesis of DNA is initiated in 5′–3′ direction by the extension of an RNA primer, it cannot be preceded after the removal of the primer, resulting in shorter chromosomes after each cell division. Telomeres are the sequences with a short repeat of TTAGGG. Length of these repeats varies in the telomeres of higher organisms. In human telomeres, the repeats are 50 Kb and in mouse, telomeres are up to 100 Kb respectively. Telomeres in Drosophila is exceptional without any short conservative repeats of the telomeres, found in most of the organisms with at least one or sometimes more Long Interspersed Nuclear Elements (LINEs). The replacement of the telomeres occurs by the transposition of the telomere sequence to the chromosome ends. The composition of telomere repeated sequences varies considerably among organisms and a specific DNA sequence is not required for the telomere function. DNA strand, running from 5′ to 3′ towards the end, has more G residues arranged in clusters, than the other strand. This G strand can form non-Watson-Crick base-pairing structures, such as four-stranded helices and multiple G-G base pairs. The 3′ end of telomeres has a single-stranded overhang of 12–15 bases.

Telomeres of eukaryotic chromosomes use telomerase for their replication. Telomerase is technically called telomere terminal transferase is a ribonucleoprotein enzyme, whose RNA component binds to the single-stranded end of the telomere and an associated reverse transcriptase activity of the enzyme maintains the length and structure of the telomere. Telomerase activity solves the first end replication problem as it can elongate the 3′ single-strand G-tails in the absence of a DNA template and RNA primed DNA replication can further fill the strand. The activity of telomerase is highly progressive in *Tetrahymena,* allowing the addition of multiple telomeric repeats in a single round of replication, thus regulating its activity without being added at the end of every cell cycle. Further, telomerase specifically lengthens the short telomeres in the cell.

5.2.4 Spatial Organisation of Eukaryotic Genomes

Spatial organization is an important feature, which defines the life of an organism. In eukaryotes, cells are organized into tissues, which are further segregated from each other both spatially and functionally, performing biological activities related to growth, metabolism and cell division. Condensation and alignment of chromosomes during mitosis and their de-condensation during interphase or post mitosis are critical features providing the completed segregation as well as their spatial organization. DNA thread in the nucleus is folded hierarchically into several layers of higher-order structures that eventually form a chromosome. Chromosomes fold and occupy characteristic regions of the interphase nucleus. In unicellular eukaryotes such as yeast, as well as in metazoan cells, chromosomes are associated with the nuclear envelope and their centromeres, telomeres cluster together at opposite sides of the nucleus. In multicellular eukaryotes, chromosomes fold and tend to occupy distinct, non-overlapping territories. The position of chromosome within the nucleus will also determine its organization, with gene-rich chromosomes that are more frequently and highly transcribed are positioned towards the center of the nucleus and gene-poor chromosomes are positioned towards the periphery. DNA is compacted and can be accommodated in the limiting space of the cell nucleus called chromatin. Compaction begins with the wrapping of a 147 bp DNA fragment around a histone octamer to form the nucleosome. The histone octamer is composed of eight subunits that assemble as one histone H3–H4 tetramer and two histone H2A–H2B dimers. This nucleoprotein complex serves as a repeating unit of chromatin. Further, the nucleosomes are organized into a 10 nm beads-on-string chromatin fiber and their further condensation is yet to be completely established and ultimately the chromatin must be folded into highly-condensed interphase chromosomes, a process that still remains to be poorly understood.

5.2.5 Whole Genome Duplications and Segmental Duplications

Genomes are formed by a series of processes involving substitutions, insertions, deletions, transpositions, shuffling of exons or chromosomes, lateral gene transfer, gene fusion or fission, de novo origination, gene and genome duplications. The fate of the modified genes directly depends on the type of process. Duplication is the primary driving force involved in the adaptive evolution of genomes and may even be considered as a master mutation, which promotes the accumulation of subsequent mutations on duplicates. Duplication can be segmental, varying from a few nucleotides to several thousand bases or even whole genome (polyploidization). Both these duplications involve different evolutionary processes with widely different impacts.

Segmental duplication occurs more frequently in eukaryote lineages as a part of the continuous process. Whole genome duplication is a highly infrequent and spectacular mutational event that results either in the extinction or re-diploidization (a process involving non-homologous recombinations along with deletions and pseudogenization of genes, generating duplicated chromosomes differing in large segments). Most of the segmental duplications generate two gene copies and one of them loses its function by pseudogenization. Majority of the duplications in vertebrates are pseudogenized. In a few cases, both the duplicated genes will be fixed. Among them, one duplicated gene will shift towards a new function called as neofunctionalization. Rarely, two copies of the duplicated genes will have differences in their expression called subfunctionalization.

5.2.5.1 Mechanisms of Duplication

The possible mechanisms responsible for the segmental duplications are unequal crossing-over, primate major histocompatibility complex (MHC), transposition and retrotransposition events, that are associated with the lateral gene transfer. Whole genome duplication (polyploidization) in eukaryotes comprises of both autopolyploidy (polyploidization within a species) and allopolyploidy (hybridization between closely-related species). Genome duplication can also occur through other mechanisms such as genomic doubling, gametic non-reduction, and polyspermy. Genome doubling and gametic non-reduction involve the failure of cell division during mitosis and meiosis, respectively. Unreduced eggs seem common in both animals and plants, whereas unreduced spermatozoa seem to be common only in plants.

Baker's yeast (*Saccharomyces cerevisiae)* was the first eukaryotic organism to have its nuclear genome fully sequenced and is also the first eukaryotic genome to be completely investigated for genome redundancy and gene duplications. Wolfe and Shields (1997) proposed that the present day yeast genome is the derivative of an ancestral tetraploid genome, which underwent massive gene loss and translocations. Based on the orientation of the 55 duplicated regions present in the centromeres and the absence of any triplicated region (duplication of a region which was duplicated), it was hypothesized that the whole-genome duplication with a differential gene loss in the genome of *S. cerevisiae* was responsible for the present day yeast genome. The possibility of ancestral whole genome duplication was also reported in the genome of *Candida glabrata*. Following this whole-genome duplication, genes have been lost differentially between the duplicated species. Segmental duplications have also been experimentally reproduced in *S. cerevisiae*. Koszul et al. (2004) showed that large inter and intrachromosomal duplications, covering from 41 to 655 Kb and encompassing up to several hundreds of genes, occurred in the yeast genome with a frequency of close to 10^7 per cell/generation. Junctions of these segmental duplications frequently contain either microsatellites or transposable elements. The stability of these segmental duplications during meiosis and mitosis was shown to rely both on the size of the duplication and on its structure.

It has been hypothesized that the present day vertebrate genomes were the resulted from two rounds of whole genome duplications, that might have occurred early during their evolution. Comparative genomics of *Homo sapiens* with *Drosophila melanogaster* and *Caenorhabditis elegans have* revealed a number of significantly large paralogous regions in the human genome. Molecular clock analysis has revealed that a burst of gene duplications might have occurred in the early vertebrate ancestor. Comparison of human and chimpanzee (*Pan troglodytes*) genomes have shown that a large fragment of 32 Mb is duplicated in humans, but not in the chimpanzee. Further, 515 duplications were identified in the human genome exclusively on chromosomes 5 and 15. *Core duplicons* comprising of ancestral blocks with high gene density were also identified in the segmental duplications of humans, chimpanzees, and macaques. No detectable polyploidization event has occurred in the Drosophila lineage for the last 600 million years.

One more round of whole-genome duplication in diploids leads to polyploidy, which is one of the well-studied phenomena among flowering plants. In monocots,

Zea mays is an allotetraploid, generated by the fusion of two diverged ancestors ~11.4 million years ago. *Triticum aestivum* is an allohexaploid, containing three sets of homologous chromosomes. Among dicots, 11 plants such as tomato, potato, soybean, barrel medic, cotton (both *Gossypium arboreum* and *Gossypium herbaceum*), and *Arabidopsis thaliana* genomes have exhibited large scale gene duplications, indicated by their ploidy levels. An abundant amount of triplicate and quadruplicate sequences were found in the genome of soybean, which might be due to the occurrence of more than one round of whole genome duplication. *Arabidopsis thaliana* genome consists of 24 large duplicated regions of size 100 Kb, covering up to 58% of the genome. Three whole genome duplication events might have occurred in *Arabidopsis thaliana* during the last 150 million years.

5.2.6 Transposable Elements

Transposable elements are the DNA fragments, capable of inserting into new chromosomal locations and duplicate themselves during this process, responsible for the rapid increase in the genome size. These elements are the single largest component of genetic material in most of the eukaryotes accounting for 45% of the human genome and 50–90% of some plant genomes. Transposons were elegantly discovered as genetic agents responsible for the pigmentation by Barbara McClintock in maize kernels in 1950, several years before the structure of DNA has been elucidated.

5.2.6.1 Classification

Eukaryotic transposable elements are divided into two classes based on their transposition intermediate as Class 1 or retrotransposons with mRNA and Class 2 with DNA as intermediates. Retrotransposons utilize reverse transcriptase for their transposition through an RNA intermediate. They are divided into two classes. (1) Viral retrotransposons are the transposable elements whose properties are similar to retroviruses. Examples of Viral retrotransposons are *Ty* elements of yeast and the *Copia* elements of *Drosophila*. (2) Non-viral retrotransposons, which are further sub-divided into Long Interspersed Nuclear Elements (LINEs), Short Interspersed Nuclear Elements (SINEs), *Dictyostelium* Intermediate Repeat Sequence Elements (DIRS) and Penelope like elements (PLE). The abundance of LINES and SINES in the eukaryote genomes is high when compared with DIRS and PLE elements. Each of these groups contains autonomous and non-autonomous elements. Autonomous elements have open reading frames, encoding the products required for transposition. Non-autonomous elements do not encode transposition proteins but are able to transpose because they retain the *cis*-sequences necessary for transposition.

LINES and SINES comprise a major part of the animal genome. These elements are defined by the existence of a large number of relatively short sequences that are related to one another. LINEs and SINES occupy up to 50% of the higher animal genomes. Properties of LINES and SINEs are provided in Table 5.1. A common group of LINEs in the mammalian genomes are L1, with a size of ~6500 bp long and

Table 5.1 Properties of LINEs and SINEs

Properties	LINES	SINES
Common types	L1 (human)	Alu (human)
Length	6.5 Kb	< 0.3 Kb
Termini	Poly A repeats	Poly A repeat
Target repeats	7–21 bp	7–21 bp
Enzymatic activity	Reverse transcriptase/none	Endonuclease
Organisation	1 or 2 uninterrupted ORFs	None

terminates in an A rich track. This element consists of two open reading frames, ORF1 and ORF2, transcribed as a bicistronic mRNA. ORF1 encodes a trimeric nucleic acid chaperone protein that binds with L1mRNA to form ribonucleoprotein complexes, considered to be the intermediates for transposition. ORF2 encodes for an endonuclease and a reverse transcriptase, required for the retrotransposition in wild-type cell lines. LINE transcripts are initiated at a Pol II promoter within the 5′ end of the element and terminate at the downstream of the simple repeat sequence. The number of full-length elements is usually small and most of the copies are truncated. As implied by its presence of repetitive DNA, LINES family shows sequence variation among individual members. LINEs amplify by an interesting mechanism called Target Primed Reverse Transcription, TPRT) that appears to have played a critical role in the evolution of primates.

L1 elements transpose in possible three steps:

(i) Formation of a nick in the double-stranded DNA at AT-rich sequences, mediated by the ORF2endonuclease activity.
(ii) Annealing of the L1 mRNA poly(A) tail to the 5 poly(T) tail at the nick and cDNA synthesis by the reverse transcriptase activity of ORF2.
(iii) Degradation of the L1 mRNA and second-strand synthesis followed by ligation.

Human and mouse genomes consist of ~850,000 and ~660,000 LINEs representing up to 20% of their respective genomes. The genome of *Tetraodon nigroviridis* contains 700 times fewer transposable elements with half of them are LINEs. LINEs are involved in large chromosomal rearrangements such as segmental duplications. They also play an active role in evolution by regulating global genome transcription.

SINEs are a heterogeneous group of elements that vary in their lengths ranging from 90 to 300 bp. Their short length and a high degree of repetition makes them comparable with satellite DNA, with the only exception of their dispersal around the genome, instead of confined to tandem clusters. Both the human and the mouse genomes contain approximately 1,500,000 SINEs, which make them the most abundant repeated elements in these two genomes. In the human genome, a large part of the moderately repetitive DNA exists as sequences ~350 bp, interspersed with non-repetitive DNA. A large portion of this repetitive DNA was cleaved by the restriction enzyme Alu I at a single site, located at 170 bp along the sequence. The cleaved sequences are members of a single family known as Alu family and there are ~300,000 members of Alu family in the haploid human genome at the rate of at least 1 member per 6 Kb of the genome. Alu elements are also found in the genomes of primates along with the MIR elements. SINES also exist as B1, B2, ID elements in rodents. Alu elements are composed of two 130-bp monomers separated by a short A-rich region, which links these two monomers. Each monomer was ancestrally derived from the 7SL RNA, following duplication. They transpose through a mechanism similar to that of L1 LINES. They are classified into several families and subfamilies, based on sequence conservation. S and J families are the oldest ones, which might have originated 35–55 million years ago, followed by the Y family ~25 million years ago. Alu elements are more active in humans than in chimpanzees.

Class 2 DNA transposons have a simple structure with a short terminal inverted repeat of 10–40 bp and a single gene encoding for transposase. Transposase binds in a sequence-specific manner to the ends of its autonomous and non-autonomous family members. After binding, transposase initiates a cut-and-paste reaction, where the element is excised from its donor site by generating an empty site and gets inserted into a new site of the genome. Repair of the double-strand break at an empty site can be precise with leaving no trace of the element or imprecise by leaving a transposon footprint of a few to several base pairs or even deleting the adjacent host DNA. The human genome contains about 300,000 copies of DNA transposons, which was 100 times higher than that of the *C. elegans* genome and 700 times higher than the *Drosophila melanogaster* genome.

DNA transposons are classified into three major subclasses

(i) Elements that excise as double-stranded DNA and transpose by a classical "cut-and-paste" mechanism. Example; Drosophila P elements;
(ii) Elements that utilize a rolling-circle mechanism, such as Helitrons
(iii) Elements that utilize a self-encoded DNA polymerase. The mechanism of transposition is not well understood.

Some eukaryotic genomes such as *S. cerevisiae* and *S. pombe* do not contain any DNA transposons, although they do contain retroelements. However, the genomes of other ascomycetes such as *Y. lipolytica* and *C. albicans* contain several DNA transposons. Rice genome contains a 100–200 copies of a Miniature Inverted-repeat Transposable Element (MITE) called Micron.

5.2.7 Satellite DNA

Satellite DNA was first identified as a fraction of DNA that was sedimented as a strong and localized band above and below the main DNA band during the density gradient centrifugation using cesium chloride. Further molecular analysis of satellite DNA by restriction endonuclease digestion gave one or few distinct low molecular weight bands following electrophoresis. These bands were tandemly repeated sequences. The DNA bands were eluted from the gel for sequencing, which gave two types of sequence repeats named as mini-satellite and microsatellite DNA repeats. Satellite DNA is found in the heterochromatin and subtelomeric regions of eukaryote genomes such as mammals, plants, Drosophila, but is not found in hemiascomycetes and *Saccharomyces pombe*. Satellite DNA is characterized by the large tandem repeats of million nucleotides length and the repeat units show great variation in their size ranging from 2 to several hundred nucleotides. The human genome consists of several satellite repeats in their centromeric regions, with the length ranging from 5 nucleotide GGAAT repeat to 171 nucleotide α-satellite repeats. Plant genomes contain satellite repeats in centromeric and telomeric regions of the chromosomes and harbor repeat unit lengths ranging from 118 to 755 bp. Satellite DNA of *Drosophila melanogaster* has been extensively studied and its location was completely mapped on each chromosome. Size of the satellite DNA repeats in Drosophila ranges from 5 to 359 bp, with the larger units being found essentially within heterochromatin covering about half of the X chromosome. Shorter satellite repeats such as (AA GAGAG)n and (AATAT)n are localized on all chromosomes. The heterochromatic Y chromosome carries nine satellite repeats ranging from 5 to 7 nucleotides, three of them mapping only to the Y chromosome.

5.2.7.1 Minisatellites

Minisatellites are the non-coding, tandemly repeated regions of the genome with the size ranging from 6 bp to 100 bp. It was estimated that a haploid human genome consists of ~1500 minisatellite loci with many of them exhibiting extreme polymorphisms due to variation in the variable number of tandem repeats (VNTRs) and a mutation rate up to 10% per gamete. Alec Jeffreys et al. (1985) has detected and developed DNA probes, which can detect hypervariable minisatellite loci in larger numbers. Hybridization of digested and electrophoresed DNA with these core sequences at low stringency has detected a pattern of fragments, which was unique for unrelated individuals, providing the background for the fingerprint analyses. DNA fingerprinting have several important applications in forensic medicine, as markers for linkage studies in genetic analyses, and as a means for establishing kinship between individuals, including paternity determination. Through the development of minisatellite variant repeat polymerase chain reaction (MVR-PCR), it is possible to analyze single pairs of minisatellite alleles, which has enabled measurement of changes in the number of repeats and also their occurrence, frequency, and location of point mutations along the sequence of repeats in a single allele. Minisatellites in the human genome are unevenly distributed, and are mainly localized in the sub-telomeric regions of the chromosomes, implying their limitation in the linkage analyses. Sub-telomeric localization of the human minisatellite repeats can be correlated with a high density of chiasmata during meiosis, indicating a possible association with the meiotic crossing over.

5.2.7.1.1 Mini Satellite Fingerprinting and its Applications

Extreme individual variations in the mini satellites have provided an exceedingly efficient tool for the recognition of individuals by their electrophoretic pattern. The possibility of amplifying DNA by PCR has further made it possible to use an extremely small amount of DNA isolated from single hairs or tiny bloodstains for minisatellite typing. Fingerprinting of satellite DNA-therefore has lent itself to analyses in forensic medicine and also in historical and archeological samples. Although the reliability of the fingerprinting method has been questioned in some conspicuous legal cases, the usage of this tool nevertheless has become more and more a routine procedure in forensic sciences. The occurrence of minisatellites and other repetitive DNA sequences is wide throughout the organism world. Use of satellite typing has become an important and highly valuable tool in the population ecology. Investigations by Burke and Bruford (1987) have shown a pronounced variation in the fingerprints among and between the bird species. Fingerprinting analysis of house sparrows through their blood samples demonstrated mapping of population structure, mating selection, and various polygamous pairing strategies. Repetitive DNA sequences insufficiently stable minisatellites are also useful for the phylogenetic investigations of evolutionary processes by comparisons of species, subspecies, and populations.

5.2.7.1.2 Minisatellites Association with Human Diseases

Along with regular minisatellites, some cases of pathogenic minisatellites have also identified. One of the best studied pathogenic minisatellites is the *Ha-ras* protooncogene locus *HRASI* VNTR, located 1000 bp downstream of the polyadenylation signal, consisting of repetitive units up to 28 bp, forming ~30 alleles. Four of these alleles comprise up to 94%, believed to have given rise to the other alleles. Rare alleles of pathogenic minisatellites are found to be three times more common in cancer patients than in controls and these alleles are associated with the 11 forms of cancer. Mutations in the minisatellites might affect the insulin gene (*INS*). Minisatellite *INS* VNTR, located 600 bp upstream of the transcription initiation site is composed of 14 bp repeat units arranged in three allelic classes with modal lengths of 600 bp for Class I, 1200 bp for Class II and 2200 bp for Class III respectively. The presence of Class I minisatellite is also associated with a doubling of the relative risk for type I diabetes mellitus (IDDM). At least six genes named as IDDM 1–6 are main contributors to the risk for diabetes and IDDM2 has been mapped to the INS VNTR minisatellite.

5.2.7.2 Microsatellites

Microsatellites, also known as simple sequence repeats (SSRs) or simple sequence length polymorphisms (SSLPs), are the repetitive sequences of 2–4 nucleotides length, which are widely interspersed in the genomes of multicellular organisms. Microsatellite repeats are inherited through the Mendelian process of inheritance, hence are used for identifying the genetic relationships. They are one of the main mechanisms by which polymorphisms are generated, hence can be used as markers for forensic analysis and food authenticity. They comprise 3% of the human genome with an average of one repeat per every 2 Kb of the genome. Among the microsatellites, dinucleotide repeats are the most common form, comprising of 0.5% of the human genome with dAC, dTG, dAT and dTA repeats dominating the human genome than dGC and dCG repeats. They occur on average of every 30,000 bp, constituting a significant part of human DNA. Trinucleotide repeats are a little rare.

The significance of these repeats in the human genome is oriented with several disorders. Microsatellites in recent years have attracted a lot of attention because of the establishment of a direct connection between expanded arrays of CG rich trinucleotides and several human neurological diseases. Tandem duplications of the short sequences that build microsatellites can occur during DNA replication, leading to a number of disorders as identified in Fragile X syndrome patients. They exhibit hundreds or even thousands of the trinucleotide repeats CGG, whereas healthy individuals have about 44 repeats. Tandem repeats can also occur in coding regions as identified in the Drosophila *glue* protein gene, which contains 19 direct tandem repeats of a 21 bp long sequence encoding 7 amino acids. Another example is the gene for α2(1) collagen found in chicken, mouse, and humans, comprising of 52 exons with introns varying in length from 80 to 2000 bp. All exons are the multiples of 9 bp with either 54 or 108 bp length. Huntington's disease is an autosomal neurodegenerative disorder whose symptoms depends on the expansion of CAG tandem repeats, translated to polyglutamine. This trinucleotide repeat is located in the coding region of the Huntington's disease gene IT15, codes for huntingtin protein. The instability CAG repeats are also influenced by another trinucleotide repeat CCG downstream to the CAG repeat, coding for proline. Huntington's disease usually occurs in the late stages of human life, but with the increased number of CAG repeats through male gametes, the symptoms become more severe and onset earlier in the subsequent generations.

5.3 Complexity of the Eukaryotic Genomes

Till date, the genome sizes of more than 10,000 organisms from plants, animals, etc. have been estimated, helping us in understanding the genetic complexity of an organism, which is completely contrasting with its organism complexity. For example, the marbled lungfish, *Protopterus aethiopicus*, contains more than 40 times the amount of DNA per cell than human beings. In fact, it has the largest recorded genome of any eukaryote comprising of 132.8 billion base pairs haploid genome, whereas haploid copy of the human genome per cell is 3.5 billion bp (Size of a genome is measured in pictograms, which gets converted into approximate nucleotide number. On average, one pictogram of genomic DNA is approximately equivalent to 1 billion bp). The similar contrasting relationship has been observed in the protein-encoding genes and the organism complexity, with a human genome consisting of 30,000 protein-coding genes, when compared with that of *Trichomonas vaginalis*, a unicellular parasitic organism comprising of 60,000 and *Drosophila melanogaster* consisting of 13,000 protein-coding genes.

Size of the genome and the protein-coding genes not only transform eukaryotic organisms into complex ones as they have evolved other mechanisms such as genome behavior and expression of the genes for generating biological complexity. Alternative splicing, RNA editing, trans-splicing, and tandem chimerism are important phenomena, which makes the genome more complex. During alternative splicing, the introns are removed and the exons are spliced together to make a continuous mRNA. However, all introns are necessarily spliced out. Depending on how the introns are spliced and the exons are joined back, a single gene, which is also a transcription unit can code for the multiple proteins. It has been estimated that 30,000 genes of the human genome can code for more than 100,000 proteins or even more proteins. RNA editing is the alteration of an mRNA molecule after its transcription, such as modification of its cytosine to uracil before it undergoes translation. The impact of RNA editing varies between the organisms and their genes, with some of them are detrimental (associated with a disease). RNA editing that leads to changes in its protein structure could be advantageous to that organism. Trans-splicing is a mechanism by which separate transcripts are spliced together to form an mRNA molecule. Tandem chimerism occurs when adjacent transcription units are transcribed together to form a single chimeric mRNA molecule.

The complexity of an organism will also depend on the number of genes expressed at the same level. If all genes of *T. vaginalis* are operated at the same level, its genome would have become more complex than Homo sapiens. It was suspected that the increased number of genes in *T. vaginalis* might be due to their duplication and more than half of the genes could be pseudogenes and de-functional genes. Many genomes are also rich in non-coding RNA sequences, that coordinate the gene expression with regulatory elements such as promoters.

5.4 Yeast Genome

Saccharomyces cerevisiae, commonly known as brewer's yeast, baker's yeast is the principal form of the yeast that is most frequently used in food fermentations and industrial processes. It is perhaps, one of the most useful organisms owing to its usage since ancient times in bakery and brewery. Yeast is believed to be originally isolated from the grape skins as a thin white film on the grape skin and the wax of the cuticle. Since prehistoric times, human beings have exploited the capability of *Saccharomyces cerevisiae* for the conversion sugars into alcohol and other desirable flavored compounds, which can enhance the shelf life of foods and beverages and improve the digestibility. Ancient brewers, winemakers, and bakers have learned that inoculation of unfermented foods with a small portion of a fermented product resulted in fast and rapid fermentation, termed as back slopping. This could be due to the continuous growth of yeast lineages in these man-made environments and has lost contact with their natural niches, thus providing a perfect definition for domestication. However, still, it is not clear whether the yeast diversity is due to selection and domestication or because of the neutral divergence caused by geographic isolation and limited dispersal.

The position of *S. cerevisiae* as a eukaryotic model system provides many intrinsic advantages. *S. cerevisiae* cells are round to ovoid, 5–10 μm in diameter. It generally reproduces by a vegetative process called budding. *S. cerevisiae* has a short generation time (doubling time of 1.5–2 h at 30 °C) and can be easily cultured in a basic medium. *S. cerevisiae* can also be easily transformed permitting either addition or deletion of new genes by homologous recombination. Furthermore, the ability to grow *S. cerevisiae* as a haploid organism simplifies the creation of gene knockout strains. As a simple eukaryote, *S. cerevisiae* shares its complex internal cell structure with plants and animals without the high percentage of non-coding DNA. Many human protein homologs related to cell cycle, signaling, protein processing were first discovered in yeast. Mutations leading to respiratory chain deficit (p mutation) were first studied in yeast. Another yeast species close to *S. cerevisiae* is *S. pombe,* which was predicted to have got diverged approximately before 300 to 600 million years has also been well studied. The life cycle of *S. cerevisiae* is most ideally suitable to the classical genetic analysis, which enabled the construction of a detailed genetic map, defining the full set of haploid chromosomes. Using the present technology, any of its nuclear genes can be completely deleted or even can be replaced with a mutant allele with absolute accuracy. The yeast genome is the combination of a large number of chromosomes with small genome size.

5.4.1 Genome Sequencing

Sequencing of the *Saccharomyces cerevisiae* genome was possible by the combined effort of more than 600 scientists belonging to Europe, North America, and Japan. It is the first complete genome sequence of a eukaryote and is one of the largest genomes to be completely sequenced. Sequencing of this genome was done by dividing the responsibilities among the international groups. *S. cerevisiae* contains a haploid set of 16 well-characterized chromosomes, ranging in size from 200 to 2200 kb. In May 1992, the complete nucleotide sequence (315 kb) of chromosome III was published by 35 European laboratories, which formed as a European Union for yeast genome sequencing project, revealed many interesting structural features. It also revealed many new genes representing unknown functions and one-third of these genes have no known homologs. In 1994, sequencing of chromosome II (820 kb) and chromosome XI (666 kb) was published. By the end of 1995, more than 80% of the yeast genome has been sequenced under the European Union project and the entire sequence of the yeast genome was known by 1996. Using specific computer programs, potentially transcribed and protein-coding regions were identified. Few selected open reading frames (ORFs) were further analyzed by genetic and biochemical methods. The total sequence of *S. cerevisiae* chromosomal DNA, constituting of 12,052 kb, was released in the public domain on 24th April 1996. Since then, regular updates have been maintained at the *Saccharomyces* Genome Database (SGD). This is a highly annotated and cross-referenced database for the yeast researchers. Another important *S. cerevisiae* database is maintained by the Munich Information Center for Protein Sequences (MIPS).

5.4.2 Genome Overview

Saccharomyces cerevisiae genome is composed of 12,156,677 bp and 6275 genes, compactly organized on 16 chromosomes. Among them, 5885 genes are believed to be truly functional with specified protein products. It has been estimated that every 2 Kb of the yeast genome encodes a protein synthesizing gene and 70% of the total genome consists of coding regions. It is estimated that yeast shares about 23% of its genome similarity with humans. A larger number of ORFs were predicted by considering shorter proteins. In contrast to the genomes of multicellular organisms, the yeast genome is highly compact, with genes representing 72% of the total sequence. The average size of yeast genes is 1.45 kb, or 483 codons, with a range from 40 to 4910 codons. A total of 3.8% of the ORF contains introns. All of the genes were characterized experimentally and most of their functions have been predicted. Genes with unknown function either contain a motif of a characterized class of proteins or corresponds to the genes encoding proteins that are structurally related to functionally characterized gene products from yeast or from other organisms. Ribosomal RNA is coded by approximately 140 copies of a single tandem array on chromosome XII and 40 genes encoding snRNAs scattered throughout the genome. Yeast genome consists of 275 tRNA genes belonging to 43 families, of which 80 have introns. In addition, chromosomes contain movable DNA elements, retrotransposons that vary in number and position in different strains of *S. cerevisiae*, with most of the laboratory strains having approximately 30.

5.4.3 Chromosomal Organization

Most of the chromosomes in yeast consist of GC-rich and GC-poor DNA regions, correlating with their variations in gene density. In chromosome III, periodicity of the base composition is paralleled by variations in recombination frequency along the chromosome arms. GC-rich regions on the chromosome arms are coinciding with regions of high recombination and AT-rich troughs coinciding with the recombination-poor centromeric and telomeric regions. Four smallest chromosomes of yeast nucleus I, III, VI, and IX exhibit average recombination frequencies of 1.3 to 1.8 times higher than the average of the whole genome. It is well known that Yeast Artificial Chromosomes (YAC) of 150 kb in size are mitotically unstable, raising the questions related to the minimal size of yeast chromosomes and process by which these small chromosomes have achieved their current size. It has been suggested that the 31 kb gene-poor region at each of the chromosome ends may act as "fillers" for increasing their size and also their stability. Genetic redundancy is most frequently noticed at the ends of yeast chromosomes. Two terminal domains of chromosome III shows nucleotide sequence homology with each other and also with the terminal domains of chromosomes V and XI. The duplicate of chromosome I right terminal region is found at the left terminal region of chromosome I and at the right terminal end of chromosome VIII.

The left telomere of chromosome III contains a repeated sequence element X and a pseudo-X element at an internal site about 4 Kb from the true X. These two elements represent long terminal repeats (LTRs) of a new class of yeast transposon called Ty5. Yeast genome also contains 19 telomere-associated highly conserved repeats called Y′ elements with an ORF, producing a protein product, reminiscent of the RNA helicases. No Y′ helicase ORFs are found on chromosomes I, III and XI. Synthesis of new telomeric repeats by telomerase is essentially a reverse transcription process. Current mechanisms of telomere biogenesis in many eukaryotes had their origins in the activities of retrotransposons or retroviruses. Many of the polymorphisms observed between the homologous chromosomes of different *S. cerevisiae* strains are either due to transposition or recombination between transposons or their LTRs. Yeast genome contains 52 complete Ty elements, 264 LTRs and other remnants that are the footprints of previous transposition events. 11 out of 13 Ty2 elements are found in the sites of previous transposition activity (old sites) and 16 out of 33 of the Ty1 elements are found in "new" sites. Thus, yeast transposons appear to get inserted preferentially into specific chromosomal regions that may be termed as transposition hot spots.

5.4.4 Organellar, Plasmid, and Viruses in Yeast Genome

Yeast mitochondrial DNA encodes the components of mitochondrial translational machinery and ~15% of the mitochondrial proteins. Yeast r^o mutants completely lack mitochondrial DNA and are deficient in the respiratory polypeptides synthesized on mitochondrial ribosomes such as cytochrome *b* and subunits of cytochrome oxidase and ATPase complexes. The 2-mm circle plasmids, present in most of the *S. cerevisiae* strains, apparently function solely for their own replication. *cir*^{o} strains, which lack 2-mm DNA, have no observable phenotype. However, a certain chromosomal mutation, *nib1*, causes a reduction in the growth of *cir*$^+$ strains, due to an abnormally high copy number 2-mm DNA.

All *S. cerevisiae* strains contain dsRNA viruses, constituting up to 0.1% of total nucleic acid. Five families of dsRNA viruses, L-A, L-BC, M, T, and W replicate in yeast but so far have not been shown to be viral. M dsRNA encodes for a toxin, and L-A encodes for a major coat protein and components required for the viral replication and maintenance of M. The two dsRNA, M and L-A, are packaged separately with the common capsid protein encoded by L-A, resulting in virus-like particles that are transmitted cytoplasmically during vegetative growth and conjugation. L-B and L-C have an RNA-dependent RNA polymerase similar to L-A and are present in the intracellular particles. *KIL*-o mutants, lacking M and consequently, the killer toxin, are readily induced by growth at elevated temperatures, chemical, and physical agents. Yeast also contains a 20S circular single-stranded RNA that appears to encode an RNA-dependent RNA polymerase, which acts as an independent replicon and is inherited as a non-Mendelian genetic element.

5.4.5 Genome Evolution

The existence of two or more sets of genes with identical sequences, encoding proteins provides the basic raw material for the evolution of novel functions. Complete genome sequence analysis of *S. cerevisiae* suggests that it has undergone duplication at some point during its evolution as evidenced in the redundancy of pericentric regions and arms of chromosomes. Although redundancy does occur close to the ends of chromosomes, exchanges between such regions are probably too frequent and too recent to help us discern the overall history of the yeast genome. In *S. cerevisiae* genome, simple direct repeats were seen in several forms with the most frequent one is 23 PAU genes, which specifies seripauperines (a set of identical serine-poor proteins of unknown function), whose ORFs show a very high codon bias and an NH2-terminal signal sequence. These PAU genes, reside in the subtelomeric regions of yeast chromosomes. Clustered gene families are less common in the yeast genome, with an exception of YAR family occurring on the chromosome I. YAR family consists of six related but non-identical ORFs, YAR023-YAR033, specifying the membrane proteins of unknown function. 16 members of this family are dispersed on six chromosomes II, III, V, VIII, XI, and X respectively.

Analysis of numerous Cluster Homology Regions (CHRs) in the yeast genome provides a better understanding of its genome evolution. CHRs are large regions containing homologous genes arranged in the same order, with the same relative transcriptional orientations, on two or more chromosomes. A 7.5 kb CHR region has been identified on chromosomes V, X and a 15 kb region identified on chromosomes III and XIV, consisting of four ORFs with similarly ordered homologs in the centromeric regions of both chromosomes. Chromosomes IV and II shared the longest CHR region, comprising of two pericentric regions with varying lengths 120 kb and 170 kb respectively, sharing 18 pairs of homologous genes (13 ORFs and five tRNA genes). A homologous set of genes encodes for citrate synthase were identified in the yeast genome, with CIT2 on chromosome III is the peroxisomal enzyme and CITJ on chromosome XIV is mitochondrial. This is one of the very good examples of evolution through gene duplication. A recently identified third citrate synthase gene (CIT3) has been discovered on chromosome XVI, making the evolution of yeast genome more interesting.

5.4.6 Comparative Genomics of Yeast

Ever since the yeast genome has been sequenced, researchers believed the best way to identify the true genes in the genome is by determining the DNA sequences that code for proteins. With the availability of more than three yeast genomes, the current list of genes needs to be well revised. By comparing the new genome sequences with the original, 50 new genes and 70 stretches of DNA that regulates the yeast genes were uncovered. They also proposed that around 500 DNA sequences, which were previously thought to be genes should be crossed off from the list and the final number of genes should be 5538.

It was also proposed to check if there was any evolutionary pressure preserving these 500 sequences, which were not considered as genes. Using comparative genomics studies, more than 70 new sequences were identified, which are involved in the regulation of gene activity. These regulatory sequences are of two types. (1) Sequences act like tiny traffic lights, indicating the gene when to express and when not to express. (2) Sequences that are involved in the signaling, as a response to a stimulus. Most of the variation in yeast genes occurs in the telomeres as they get exchanged rapidly. It was hypothesized that similar telomere exchange is happening in the human genome, where most of the gene rearrangements are likely to occur.

5.4.7 Yeast Proteome

Availability of yeast complete genome sequence denotes the accessibility of the complete proteome of a eukaryotic cell for the first time. During this process, 50% of its proteins have been identified and classified on the basis of their amino acid sequence homology and other proteins of known function. However, homology-based searches often provide only a general description of their biochemical functions with no indication of their specific biological function. An attempt to classify yeast proteins according to their function conservatively, predicted by such computer analyses have been carried out by Munich Information Center for Protein Sequences (MIPS). It has been identified that yeast cell devotes 11% of its proteome to metabolism, 3% to energy production and storage, 3% to DNA replication, repair, and recombination, 7% to transcription and 6% to translation. Totally 430 proteins are involved in intracellular trafficking or protein targeting and 250 proteins play structural roles. More than 200 transcription factors, 250 transporters (both primary and secondary) have been identified. However, these statistics refer only to the yeast proteins for which significant homologs were found. Another approach that is expected to be greatly facilitated by the availability of all yeast protein sequences is 2-D gel electrophoretic analysis, which permits the resolution of more than 2000 soluble protein species. With this technique, many of these membrane proteins are not resolved and the reproducibility of electrophoretograms from one laboratory to another is still poor.

Using yeast proteome, many proteins of this organism were uncovered, whose existence was doubtful. Earlier, it was believed that yeast does not have an H1 histone protein. In reality, yeast contains H1 histone protein, whose gene was found on chromosome XVI. Another example is the discovery of a yeast gamma-tubulin gene on chromosome XII, which had previously eluded yeast geneticists, despite their intensive efforts. It has been hypothesized previously that complete availability of the yeast proteome and its understanding is a prerequisite for the complex human proteome as half of the proteins known to be defective in human heritable diseases show some amino acid sequence similarity to yeast proteins and a majority of the yeast proteins have human homologs. These human proteins are in the process of being classified on the basis of their structural or functional similarity to that of the yeast proteome.

5.4.8 Future Research

The availability of the *S. cerevisiae* genome along with its complete set of deletion mutants has enhanced *S. cerevisiae* as a powerful model system for understanding the regulation of eukaryotic cells. Further, analysis of its genetic interactions through the synthetic genetic array of double mutants will take this one more step further, whose approaches can be applied in many divergent fields of biological and medicinal science.

5.5 *Caenorhabditis elegans* Genome

5.5.1 Importance of *C. elegans* as a Model Organism

Caenorhabditis elegans is a free-living, transparent nematode that grows up to 1 mm length, lives in temperate soil environments. Research on the molecular and developmental biology of *C. elegans* has begun in 1974 by Sydney Brenner. Since then, it has been extensively used as a model organism. *C. elegans* is symmetrical roundworm, which feeds on bacteria that develop on decaying vegetable matter. This organism has two sexes, hermaphrodites, and males. Most of the individuals in the population are hermaphrodite, with males comprising just 0.05% of its total population on an average.

C. elegans is studied as a model organism for a variety of reasons.

1. It is a simple multicellular eukaryotic organism that can be studied in great detail. Strains are cheap to breed and can be frozen. When subsequently thawed they remain viable, allowing long-term storage
2. *C. elegans* is a transparent organism, facilitating the study of cellular differentiation and other developmental processes in the intact organism.
3. The fate of development of every single somatic cell (959 in the adult hermaphrodite; 1031 in the adult male) has been mapped out. In both the sexes, a large number of additional cells (131 in the hermaphrodite, most of them would otherwise become neurons), are eliminated by programmed cell death. This concept has been thoroughly studied in this organism, specifically because of this apoptotic predictability, which has contributed to the elucidation of some apoptotic genes, mainly through observation of abnormal, apoptosis-surviving nematodes.
4. *C. elegans* is one of the simplest organisms with a nervous system, comprising of 302 neurons, whose pattern of connectivity has been completely mapped out and shown to be a small-world network. Research has explored the neural

mechanisms responsible for several of the more interesting behaviors shown by the *C. elegans* including chemotaxis, thermotaxis, mechanotransduction, and male mating behavior.

5. *C. elegans* is relatively straightforward to disrupt the function of specific genes by RNA interference (RNAi). Silencing the function of a gene in this way can sometimes allow a researcher to infer what the function of that gene may be. The nematode can either be soaked in or injected with a solution of double-stranded RNA, the sequence of which is complementary to the sequence of the gene that the researcher wishes to disable. Alternatively, worms can be fed on genetically transformed bacteria which express the ds RNA of interest.
6. *C. elegans* has also been useful in the study of meiosis. As sperm and egg nuclei move down the length of the gonad, they undergo a temporal progression through meiotic events. This progression means that every nucleus at a given position in the gonad will be at roughly the same step in meiosis, eliminating the difficulties of heterogeneous populations of cells.
7. The organism has also been identified as a model for nicotine dependence as it has been found to exhibit differential behavioral responses to nicotine such as acute response, tolerance, withdrawal, and sensitization, similar to those observed in mammals, especially human beings.

5.5.2 Genome Sequencing

The genome sequence of *C. elegans* was based on the pre-existing physical map, initiated by the development of YACs and cosmid libraries, covering most of the genome and rapidly closed the gaps. By the end of 1989, 20% of the genome was represented only in YACs, generating the possibility of completing the sequence. In 1990, Joint funding from the National Institutes of Health and the UK Medical Research Council (MRC)] for a pilot study was arranged for sequencing the *C. elegans* genome, facilitating the sequence of the first 3 Mb. Success in this venture resulted in full funding and the expansion of the two groups of the consortium in 1993.

Sequencing of the *C. elegans* genome was initiated in the centromeres of chromosomes, where cosmid coverage and the density of genetic markers are high. Sequencing process was divided into two major parts:

1. Shotgun phase, which acquires the sequence from random sub-clones
2. Finishing phase, a directed sequence acquisition for closing the gaps and to resolve ambiguities in the low-quality areas.

Numerous improvements during the shotgun phase have increased the sequencing efficiency, improved data quality and decreased the expenditure. Restriction digests with several enzymes were performed on most cosmids, which provided many valuable checks regarding the sequence assembly, ambiguous because of the repeats. When available cosmids were exhausted, fosmids were screened by maintaining a single copy per cell and one-third of the gaps were bridged in the central regions of the chromosomes along with a very few gaps bridged in the outer regions. Most of the central gaps were recovered by using long-range PCR and the remaining gaps in the outer regions were recovered by sequencing the YACs. Comparison of the assembled YAC sequences with the overlapping cosmids and fosmids showed a few discrepancies due to a rearrangement in the cosmid. In the cases where a shotgun phase failed to yield a spanning subclone, plasmid clones that were obtained from cosmids were used to bridge the gaps. *C. elegans* was the first multicellular organism to have its genome completely sequenced. The sequence was published in 1998, although a number of small gaps were present; the last gap was finished by October 2002.

5.5.3 Overview of the Genome

The prediction of coding regions in multicellular genomes is challenged by the interruption of introns, generation of alternatively spliced forms and the relatively low gene density. This problem is more complex in *C. elegans* genome whose genes are trans-spliced and 25% of the genes are organized into operons. GENEFINDER was used to identify the putative coding regions and to have a basic overview of gene structure. Using GENEFINDER, it was observed that 92% of the predicted intron regions had an exact match with the experimentally confirmed introns and 97% of the introns were overlapped.

The *C. elegans* genome sequence is approximately 97 Mb consisting of 1,90,099 protein-coding genes with a gene density of 1 per every 5 kb. At an average, each gene contains 5 introns and 27% of the genome resides in exons. The number of predicted genes is much higher than the number of essential genes estimated from the classical studies. ~42% of the predicted proteins have distant matches outside Nematoda, with most of them contains functional information. 34% of predicted proteins matched only with other nematode proteins and a few of them have been functionally characterized. The fraction of genes with informative similarities is far lower than the 70% seen for microbial genomes, reflecting that a small proportion of nematode genes are devoted to core cellular functions.

Protein set of *C. elegans* was compared with that of *E. coli*, *S. cerevisiae,* and *Homo sapiens* to highlight the qualitative differences in the predicted protein sets. It was found that smaller genomes had matches to a larger fraction of their protein sets and larger genomes had a higher number of matching proteins. High protein similarities were found between *C. elegans* and *H. sapiens* and a substantial number of proteins common to *C. elegans* and *E. coli* which were not found in the yeast. *C. elegans* lacked the proteins that were found in yeast and *E. coli*. Genes encoding proteins with distant matches outside Nematoda were more likely to have a matching EST (60%) than those without such matches (20%), suggesting that conserved genes including housekeeping ones are likely to be highly expressed. In addition to the protein-coding genes, the genome contains at least several hundred non-coding RNA genes. There are 659 widely dispersed tRNA genes and at least 29 tRNA-derived pseudogenes. 44% of the tRNA genes are found on the X chromosome. Several other noncoding RNA genes occur in dispersed multigene families. U1, U2, U4, U5, and U6 spliceosomal RNA genes occur in 14, 21, 5, 12, and 20 dispersed copies, respectively. 5 dispersed copies of signal recognition particle RNA genes have been identified with at least four dispersed copies of splice leader 2 (SL2) RNA genes. The ribosomal RNA genes occur solely in an array at the end of chromosome I. 5S RNA genes occur in a tandem array on chromosome V, with array members separated by SL1 splice leader RNA genes. The number of known RNA genes in the genome has increased greatly due to the discovery of a new class of 21 U-RNA gene, and the genome is now believed to contain more than 16,000 RNA genes, up from as little as 1300 in 2005. Scientific curators continue to appraise the set of known genes, such that new gene predictions continue to be added and incorrect ones modified or removed.

5.5.3.1 Repetitive Sequences

A significant fraction of the repetitive sequences are found in the *C. elegans* genome, which was classified into local comprising of tandem, inverted, simple sequence repeats and dispersed. Tandem repeats account for 2.7% of the genome and are found at least once in every 3.6 kb. Inverted repeats account for 3.6% of the genome, found once per 4.9 kb. Many repeat families are distributed non-uniformly with respect to the coding regions and are found within the introns than between genes. Intronic region of the *C. elegans* genome contains 51% of the tandem repeats and 45% of the inverted repeats. The intergenic region accounts for 47% of the genome consists of 45% tandem repeats and 55% inverted repeats. Tandem repeats overlap with only 27% of the genome encoding proteins. Few of the simple repeat sequences such as CeRep11, CeRep26, and CeRep27 occur in families. CeRep26 is a tandemly occurring hexamer repeat TTAGGC, seen at the multiple sites internal to the chromosomes in addition to the telomeres. 711 copies of the CeRep11 were distributed over the autosomes and only one copy has been located on the X chromosome. 38 dispersed repeat families were identified in the *C. elegans* genome, with most of them are associated with transposition. However, these repetitive elements may not explicitly encode an active transposon.

5.5.3.2 Chromosome Organization

GC content of the genome is 36% and is unchanged across all the chromosomes. No centromeres were found in the chromosomes as in most of the metazoans. The gene density is fairly constant across all the chromosomes. Both inverted and tandem repetitive sequences are more frequent on the autosome arms than in the central regions of the chromosomes or even on the X chromosome. A fraction of the genes with high similarities to the organisms other than nematodes tends to be lower on the arms, as does the fraction of genes with EST matches. The difference between autosome arms and central regions indicates that there is a high rate of meiotic recombinations on the arms facilitating a rapid evolution of DNA in the arms region compared with that of the central region.

5.5.3.3 Comparative Genomics *C. elegans* with *C. briggsae*

In 2003, the genome sequence of the related nematode *C. briggsae* was also determined, allowing researchers to study the comparative genomics of these two organisms. The official version of the *C. elegans* genome sequence continues to change as and new evidence reveals errors in the original sequencing. Most of the changes are minor, adding or removing only a few base pairs (bp) of DNA. For example, the WS169 release of Worm Base (December 2006) lists a net gain of six base pairs to the genome sequence. Occasionally more extensive changes are made, as in the WS159 release of May 2006, which added over 300 bp to the sequence. The comparative analysis of the related nematodes *Caenorhabditis elegans* and *C. briggsae* offers a powerful approach toward understanding the genetic basis for the form and function of these simple animals. Studies to date have already yielded valuable insights into the evolution and role of particular sequences, genes, and pathways. Morphologically, these two species are almost indistinguishable, despite the fact that their most recent common ancestor (MRCA) existed about 100 million years ago. Both are soil-dwelling, self-fertilizing hermaphrodites, with facultative males. Both have a 100-megabase (Mb) genome apportioned into six chromosomes. Genes isolated in one species will frequently rescue mutants in the other. Despite these similarities, nucleotide alignments of the complete genome sequence of *C. elegans* with the draft sequence of *C. briggsae* strain AF16 reveal that 52.3% of the *C. elegans* genome and 50.1% of the *C. briggsae* genome aligns between the two species with the bulk of this in the coding sequence. The substantial body of knowledge procured on *C. elegans* over the past few decades will help in interpreting the sequence similarities and differences. Much less is known about *C. briggsae*.

It all be gained with the construction of a genome map by first generating a genetic map using molecular-based single nucleotide polymorphism (SNP) markers. This more detailed genetic map based on SNPs would be of use in its own right, for example, simplifying positional cloning of genetically defined genes. But, it would also provide long-range continuity, which would, in turn, allow the placement of much of the assembled sequence along the chromosomes. This long-range map of the genome allows a direct comparison of chromosomal organization in *C. briggsae* to the distinctive features of *C. elegans* organization. Using other wild isolates of *C. briggsae*, thousands of SNPs were discovered. By genotyping selected SNPs across recombinant inbred (RI) lines between the sequenced strain (AF16) and the SNP source strains, a genetic map was generated. The resultant genetic map was then combined and the sequence assembly information to place 91.2 Mb of sequence onto the six linkage groups, with another 9.9 Mb was tentatively associated with chromosomes.

The integrated map allowed to correct several misassemblies in the initial *C. briggsae* sequence. Of broader interest, chromosomal scale phenomena were also explored. Like in *C. elegans*, rates of recombination appear much higher on arms than in central regions for the autosomes. Autosome arms and centers also differ in their repeat content, coding density, and a fraction of highly conserved genes, as seen in *C. elegans*. Unexpectedly, the comparison has also revealed extensive conservation of synteny between these two organisms, with the vast majority of genes with 1:1 ortholog that resides on one chromosome in one species lying on a single chromosome in the other. Investigation of the entire sequence for clusters of discordant orthologs suggests five regions of more than 50 kilobases (kb) that are likely candidates for misassembly. In addition to the marked variation in recombination rates along the autosomes in *C. elegans*, repeat density and gene density were also found to vary by regions. Similar variation in the density of these features in the *C. briggsae* autosomes was observed, with the repeat density higher and intron length greater on the arms and exon density greater in the centers. As seen in *C. elegans*, telomere-related repeats (TTAGGC) show a particularly marked difference in their distribution. Strikingly, 1:1 orthologs, even after accounting for the greater exon density, are more common in the centers.

5.6 The Genome of *Drosophila melanogaster*

Genus Drosophila also called fruit flies, pomace flies, vinegar flies, wine flies belongs to the family Drosophilidae. Entire Drosophila genus consists of more than 1500 species and are highly diverse in appearance, behavior, and breeding habitat. Drosophila species are found all around the world, with most of them in the tropical regions. Drosophila species such as *Drosophila melanogaster*, *Drosophila immigrans,* and *Drosophila simulans* are closely associated with humans and are considered as domestic species, whereas *Drosophila subobscura* have been accidentally introduced by human activities such as fruit transports. Many species of this family linger around the baits of fermented bananas or mushrooms. Among them, *Drosophila melanogaster* has been highly used in the genetics and developmental biology research.

Drosophila was chosen as a test system to explore the applicability of whole-genome shotgun (WGS) sequencing for large and complex eukaryotic genomes.

5.6.1 *Drosophila melanogaster* as a Model Organism

Drosophila melanogaster is one of the most studied model organisms in biological research, particularly in genetics and developmental biology.

1. Fruit flies are small and hence, easy to grow in the laboratory. Their morphology can be easily studied after getting anesthetized either with CO_2 gas or Ether by rapidly cooling them through a fly nap.
2. Drosophila generation time is ~10 days at room temperature, making the possibility of studying several generations within a few weeks of time.
3. This organism is having a high rate of fecundity with females lays up to 100 eggs per day, and more than 2000 eggs in a lifetime.
4. Male flies can be readily distinguished from the females and virgin females are easily isolated, which facilitates genetic crossing.

5. The mature larvae of Drosophila contain giant polytene chromosomes in their salivary glands.
6. Males do not show any meiotic recombinations, which facilitates advanced genetic studies.
7. Recessive lethal balancer chromosomes comprising of visible genetic markers are used for identifying the lethal alleles in a heterozygous state without recombination due to multiple inversions in the balancer.

5.6.2 Genetic Markers

Drosophila genes have been traditionally named after their phenotypes resulting due to mutation. For example, the absence of a particular gene in Drosophila will result in a mutant embryo that does not develop a heart. Scientists have named this gene and the mutant as *tinman*. This system of nomenclature resulted in a wider range of gene names than in other organisms. Genetic markers are commonly used in Drosophila research, as most of the phenotypes are easily identifiable either with the naked eye or under a microscope. A few of the common genetic markers are listed below.

The allele symbol is followed by the name of the gene affected and a description of its phenotype. Recessive alleles are in lower case, while dominant alleles are capitals

- Cy^1: curly; The wings curve away from the body, the flight may be somewhat impaired.
- e^1: ebony; Black body and wings (heterozygotes are also visibly darker than wild type).
- Sb^1: stubble; Hairs are shorter and thicker than wild type.
- w^1: white; Eyes lack pigmentation and appear white, vision may be somewhat impaired.
- y^1: yellow; Body pigmentation and wings appear yellow.

5.6.3 Genome Sequencing

Because of its importance as a model organism in genetics studies used by the large research community and its modest genome size, Drosophila was the first eukaryotic organism to be considered for its applicability of whole-genome shotgun (WGS) sequencing for large and complex eukaryotic genomes. The basic groundwork of this project was laid by the fly research community, which in turn has been supported and funded by the federally funded Human Genome Project, genome projects in the United States, Europe, and Canada. The Berkeley Drosophila Genome Project (BDGP) and the European Drosophila Genome Project (EDGP) initiated the genomic sequencing and had completed 29 Mb of the genome sequence. The BAC map and other genomic resources of Drosophila served both as an independent confirmation of the assembly of data from the shotgun strategy and as a set of resources for further biological analysis of the genome. The Drosophila genome is ~180 Mb in size, distributed in 4 chromosomes, an X/Y pair and three autosomes labeled as 2, 3 and 4. 120 Mb of the genome is presented as euchromatin region, distributed on two large autosomes, X chromosome, and up to 1 Mb on the small fourth chromosome. The heterochromatin region consists of short, simple sequence elements repeated for many megabases, occasionally interrupted by the transposable elements and tandem arrays of rRNA genes. Heterochromatin region is also embedded with small islands of unique sequences, such as MAPKinase gene rolled on chromosome 2, flanked on both sides by at least 3 Mb of heterochromatin.

Whole genome shotgun sequencing of the Drosophila genome was initiated with two major goals:

(i) To test the WGS strategy on a large and complex eukaryotic genome as a prelude to sequencing the human genome.
(ii) To provide a complete, high-quality genomic sequence to the Drosophila research community, so as to advance research on this important model organism.

WGS sequencing is an effective and efficient way of sequencing the prokaryotic genomes, whose size ranges from 0.5 to 6 Mb. Through WGS strategy, whole genomic DNA of an organism is sheared into segments of a few thousand bp in length and cloned directly into a plasmid vector suitable for DNA sequencing. DNA is sequenced sufficient times so that each base pair is covered numerous times, in a fragment size of 500 bp. After sequencing, the fragments are assembled in overlapping segments to reconstruct the complete genome sequence. Along with large size, most of the eukaryotic genomes often contain substantial amounts of repetitive sequences with the potentiality to interfere with correct sequence assembly. Weber and Myers in 1997 presented a theoretical analysis of the whole genome shotgun sequencing in which they examined the impact of repetitive sequences, discussed experimental strategies to mitigate their effect on sequence assembly and suggested that the WGS method could be applied effectively to large eukaryotic genomes. A vital component of the whole genome shotgun strategy is obtaining the sequence data from each end of the cloned DNA inserts and the juxtaposition of these end-sequences (also called mate pairs), critical for producing a correct assembly.

Libraries for whole genome shotgun were prepared with 2 kb, 10 kb and 130 kb insert sizes of cloned DNA. End-end sequences of the cloned BACs provided long-range linking information, which was used to confirm the overall structure of the assembly. Totally, three million sequence reads were obtained from the whole-genome libraries. A BAC-based physical map comprising 95% of the euchromatic region was constructed by screening the BAC libraries with sequence-tagged site (STS) markers. The clone-based draft sequence was used for two purposes: (1) For improving the likelihood of accurate assembly and (2) For the identification of templates and primers, which can be used for filling the gaps that remain after the genome assembly. Assembly of the sequenced BAC clones comprising the euchromatin region was performed using the data obtained by WGS. Only 2% of the sequenced region contained heterochromatin in the form of sequence repeats, indicating that the heterochromatin DNA is not stably cloned in the small-insert vectors used for the WGS libraries.

Due to the unclonable repetitive DNA surrounding the centromeres, it is very much unlikely that the genome sequence of entire eukaryotic chromosomes such as Drosophila will ever be completed. It is, therefore, necessary to provide an assessment of the continuity and accuracy of the sequence. Sequencing of Drosophila was completed in the March 2000, named as Release 1, curated at the database called Flybase. The size of the genome was 165 million base pairs (Mbp) and more than 60% of the genome appears to be functional. Sequencing the Drosophila genome represents the first demonstration of a multicellular organism through the WGS approach. Completion of this project has made a significant milestone for several reasons

1. It is the culmination of a century of genetic and molecular studies on this model organism.
2. This genome project has permitted the applicability of WGS to complex multicellular eukaryotic genomes
3. Sequencing of the Drosophila genome allowed many meaningful comparisons with the already sequenced genomes of *C. elegans* and *S. cerevisiae*.
4. The gene sequences identified and annotated through the genome sequence will provide the basis for many future studies.

5.6.4 Gene Prediction

For the prediction of the genes in this genome, different gene-finding approaches were made including the gene-finding programs such as Genscan and EST prediction using Genie. Genscan predicted 17,464 genes and Genie predicted 13,189 genes, which were closer to the number of experimentally determined ones. Results of the computational analyses were presented to the annotators by means of a custom visualization tool, that allowed annotators to define transcripts on the basis of EST and protein sequence similarity information, Genie and Genscan predictions. The present annotation of the Drosophila genome predicts 13,601 genes, encoding 14,113 transcripts through alternative splicing in some genes. More than 10,000 genes with database matches were reviewed manually and the remaining 3000 predicted genes have no database matches that can be used to refine intron-exon boundaries. Genes predicted by Genscan that did not overlap with the predictions made by the Genie.

Identified genes in the Drosophila genome were classified according to a functional classification scheme called Gene Ontology, built with the collaboration among FlyBase, the *Saccharomyces* Genome Database and Mouse Genome Informatics. Gene ontology consists of a set of parameters that provides a consistent description of the genes in terms of their molecular function, biological role, and cellular location. Proteins encoded by 1539 Drosophila genes have been annotated by FlyBase using 1200 Gene Ontology classifications. Additionally, 718 proteins from *S. cerevisiae* and 1724 proteins from mouse had also been annotated and were placed into Gene Ontology categories. These predicted Drosophila genes were used as queries by BLASTX or BLASTP and grouped on the basis of the GO classification. 7400 transcripts were assigned to 39 major functional categories and 4500 were assigned to 47 major process categories. The largest predicted protein is Kakapo, a cytoskeletal linker protein required for adhesion between and within the cell layers, with 5201 amino acids and the smallest is the 21 amino acid ribosomal protein L38. There are 56,673 predicted exons with an average of four per gene, occupying 24.1 Mb of the 120 Mb euchromatin. The size of the average predicted transcript is 3058 bp. There was a systematic underprediction of 59 and 39 untranslated sequences mainly due to less than complete EST coverage and the inability of gene-prediction programs to predict the non-coding regions of transcripts. There are 41,000 introns whose size is ranging from 40 bp to more than 70 kb, occupying 20 Mb of sequence. 292 tRNA genes and 26 genes for spliceosomal small nuclear RNAs (snRNAs) were identified. Average gene density in the Drosophila genome is one per every 9 kb. Determination of sex in *Drosophila* occurs by the ratio of X chromosomes to autosomes, not because of the presence of a Y chromosome as in the case of human beings. Although the Y chromosome is entirely heterochromatin region, it contains at least 16 genes, many of which are thought to have male-related functions.

5.6.5 Gene Similarity with Humans

Approximately, 75% of the genes, oriented with the known human diseases have a considerable match with the genetic code of Drosophila and 50% of this fly protein sequences have mammalian homologs. An online database named Homophila has been generated and is available to search for human disease gene homologs in Drosophila and vice versa. *Drosophila* is being used as a genetic model for several human diseases including the neurodegenerative disorders such as Parkinson's, Huntington's, Alzheimer's and spinocerebellar ataxia. This organism is also being used as a model for studying the mechanisms underlying aging, oxidative stress and drug abuse.

5.7 Genome of *Arabidopsis thaliana*

Arabidopsis thaliana is a small flowering plant belongs to the family Brassicaceae, which is a native of Europe, Asia, and North-Western Africa. This plant is discovered by Johannes Thal in the sixteenth century. Arabidopsis is a spring Annual with a short life cycle. Around 750 natural accessions of *Arabidopsis thaliana* have been collected so far, which are available from the two major stock centers ABRC and NASC. *Arabidopsis thaliana* had begun to be used in the developmental studies since the beginning of 1900 and its mutants were developed in 1945. Since then, it is being used as a model system for studying genetics, plant development and evolution. Even though *Arabidopsis thaliana* has very less importance in agriculture, it has several traits that make it a useful model for understanding the physiology, cellular and molecular biology of flowering plants.

5.7.1 *Arabidopsis thaliana* as a Model Organism

The small size of the plant and its rapid life cycle are highly advantageous for research. *Arabidopsis thaliana* is also called spring ephemeral, which takes 6 weeks to complete its life cycle. Its small size is convenient for the cultivation in a small space. The selfing nature of this plant assists genetic experiments. An individual plant produces several thousand seeds. Gene transformation in Arabidopsis, through *Agrobacterium tumefaciens,* can be done by "floral-dip", (dipping a flower into a solution containing Agrobacterium + the gene of interest and a detergent) avoiding the need for tissue culture. The site of T-DNA insertions has been determined in more than 300,000 independent *Arabidopsis thaliana* transgenic lines. Through the collection of t-DNA insertion lines, insertional mutants are available for almost all the functional genes in Arabidopsis. Young seedlings and their root system are translucent, facilitating the live cell imaging in this plant highly suitable for the microscopy analysis using fluorescent stains. The genome of this plant is one of the smallest (1.25 Mb) among plants, distributed in five chromosomes. It was the first plant genome to be sequenced, useful for the genetic mapping, molecular analysis, and identification of superior traits for plant breeding programs. Sequencing of this genome was taken up by Arabidopsis Genome Initiative (AGI) comprising of the research groups from Europe, Japan, and the United States in 1996 and was completed in the year 2000. The updation and maintenance of this genome are being done by The Arabidopsis Information Resource (TAIR).

5.7.2 Genome Sequencing

Large insert bacterial artificial chromosomes (BAC), P1 phage and Transformation Competent Artificial Chromosome (TAC) libraries were used as the primary substrates for the sequencing. Physical maps of the Arabidopsis genome, obtained by the sequencing of 79 cosmid clones were assembled by using the Restriction Fragment Finger Print analysis of the BAC clones by hybridization or through PCR of sequence-tagged sites (STS). These physical maps were integrated for obtaining the genetic map, which provided a foundation for the assembly of contigs into sequence-ready tiling paths. End sequence comprising of 47,788 BAC clones was used from the BACS to extend as well as for the integration of contigs. 10 contigs representing the chromosome arms and heterochromatin located in the centromere were assembled from 1569, BACs, TACs and cosmid libraries with an average insert size of 100 Kb. 22 genomic DNA products were PCR amplified, cloned and sequenced to link the regions that were not covered by the cloned DNA region. Telomere sequence was obtained from Yeast Artificial Chromosome (YAC), phage clones and products derived from the inverse polymerase chain reaction (IPCR) of genomic DNA. Selected clones were sequenced on both the strands and their overlapping sequences were compared and assembled, with an accuracy of 99.999%.

Arabidopsis genes were predicted by using a combination of algorithms, optimized with the parameters based on the known Arabidopsis genes and their similarities with known proteins and ESTs. 80% of the predicted genes were completely consistent with the algorithms, 93% of the ESTs matching with the predicted genes and 1% of the ESTs matched with the non-coding regions, indicating the identification of almost all potential genes of this genome. A total of 25,498 genes are predicted in this genome was the largest gene set predicted and published till date. The gene density of the Arabidopsis is similar to that of *C. elegans*. Arabidopsis genome also has a great number of tandem gene duplications and segmental duplications, accounting for its large gene set. The rDNA repeat regions on chromosomes 2 and 4 could not be sequenced because of their repetitive structure and content, which is already known. The centromeric regions are not fully sequenced because of the large blocks of monotonic repeats such as 5S rDNA and 180-bp repeats.

5.7.3 Gene Prediction

Among the identified 25,498 genes of the Arabidopsis genome, functions of 17,593 (69%) genes were predicted and classified based on their sequence similarity with the proteins of known function and only 9% of them have been experimentally characterized. Similar proportions of the Arabidopsis mitochondrial gene products were targeted to the secretory pathway and 14% of the gene products are targeted to the chloroplast. Comparing the genes of unicellular and multicellular eukaryotes with that of Arabidopsis indicates that the genes involved in cellular communication and signal transduction in Arabidopsis have more counterparts in multicellular eukaryotes than prokaryotes. Significant redundancy is also evident in the Arabidopsis genome in the form of segmental duplications, tandem arrays, and many highly conserved genes are also found to be scattered over the genome. Segmental duplication is responsible for 6303 gene duplications.

A total of 11,601 protein types were identified in the Arabidopsis genome, among which 35% are unique and the proportion of proteins belonging to families with more than 5 members is 37.4%. The absolute number of Arabidopsis gene families and their types is similar to other multicellular eukaryotes, indicating that a proteome of 11,000–15,000 types is sufficient for a wide diversity. Most of the domain functions identified in Arabidopsis are also conserved in, *S. cerevisiae*, Drosophila and *C. elegans* genomes, indicating ubiquitous pathways in eukaryotes. Arabidopsis

Table 5.2 Arabidopsis genes with high similarity to human genes

S. No.	Arabidopsis genes	Human genes
1	Putative calcium ATPase	Xeroderma Darier±White
2	Putative DNA repair protein	Xeroderma Pigmentosum
3	DNA excision repair cross-complementing protein	Xeroderma pigment
4	Multidrug resistance protein	Hyperinsulinism
5	Probable H + -transporting ATPase	Renal tubul. acidosis
6	Putative ABC transporter	HDL deficiency 1
7	ATP-dependent copper transporter	Wilson, ATP7B
8	DNA ligase	Immunodeficiency, DNA ligase
9	Putative ABC transporter	Stargardt's, ABC
10	Ataxia telangiectasia mutated protein	Ataxia telangiectasia
11	Niemann-Pick C disease protein-like protein	Niemann-Pick
12	ATP-dependent copper transporter, putative	Menkes
13	Deafness, hereditary	Putative unconventional myosin
14	Putative myosin heavy chain	Fam, cardiac myopathy
15	Repair endonuclease	Xeroderma Pigmentosum
16	Glucose-6-phosphate dehydrogenase	Cystic fibrosis
17	Putative glycerol kinase HNPCC	Glycerol kinase defic
18	Putative DNA mismatch repair protein	ABC transporter-like protein

gene set has also shown some significant similarity with that of humans. Using BLASTP, it was identified that out of 289 human disease genes, 139 (48%) had similarity with Arabidopsis. Among them, 17 human disease genes have more similarities with Arabidopsis genes than yeast, Drosophila or *C. elegans* genes (Table 5.2). Even though numerous protein families with similar functions are shared between eukaryotes, plants contain only 150 unique protein families, including transcription factors, structural proteins, enzymes and proteins of unknown function. Members of the gene families common to all eukaryotes have undergone substantial increase or decrease in their size in the Arabidopsis genome. Finally, the transfer of a relatively small number of cyanobacteria-related genes from a putative endosymbiotic ancestor of the plastid has added to the diversity of protein structures that are found exclusively in plants.

5.7.4 Genome Organization and Duplication

Sequencing of the Arabidopsis genome has provided a long insight of its evolutionary history and complete view of its chromosomal organization. Arabidopsis genome is organized into 1528 tandem arrays, consisting of 4140 (17%) genes. These arrays are raging up to 23 adjacent members. Large segmental duplications were identified either by the direct alignment of the chromosomal sequences or by aligning the proteins for the conserved gene order. All the five chromosomes were aligned in both orientations using MUMer 30, which has revealed 24 large segmental duplications of the size above 100 Kb comprising up to 65.6 Mb (58%) of the genome. Much of the duplications appeared to have undergone shuffling after the duplication. Altogether, the amount of Arabidopsis genome that was duplicated encompasses to 67.9 Mb (60%). The sequence conservation of the duplicated genes varies greatly, with 6303 (37%) of the 17,193 genes in the segments are highly conserved. The proportion of the highly conserved genes in each duplicated segment varies between 20 and 47%. The segmental duplication of the Arabidopsis genome reveals that polyploidy is a key factor involved in the evolution of Arabidopsis from a tetraploid ancestor like maize and the duplication might have occurred approximately 112 million years ago. The conservation of duplicated segments might be due to their divergence from an autotetraploid form or might even reflect the differences present in an allotetraploid ancestor. There is also a possibility of several independent duplications events that might have taken place instead of a tetraploid formation. However, the diploid genetics of Arabidopsis and the genetic divergence of the duplicated segments have masked its complete evolutionary history.

5.7.5 Telomeres and Centromeres

Telomeres of Arabidopsis chromosomes are composed of CCCTAAA repeats with an average of 1 repeat per 2–3 kb of the telomere. All telomeres are separated from coding sequences by repetitive subtelomeric regions, measuring less than 4 kb, except telomere 4 North (TEL4N), whose consensus repeats are adjacent to the Nucleolus Organizer Regions (NOR). Imperfect telomere-like arrays make up to 24 kb are found near centromeres of the genome, might have arisen due to the ancient rearrangements such as inversions. Centromere DNA mediates chromosomal attachment to the meiotic and mitotic spindles and often forms dense heterochromatin. Arabidopsis centromeres, like those of many the higher eukaryotes, contain numerous repetitive elements such as retroelements, transposons, microsatellites, and middle repetitive DNA. These repeats are rare in the euchromatic arms, but are highly abundant in the pericentromeric DNA and have a high affinity for the DNA binding dyes and dense methylation patterns. Inhibition of homologous recombination in centromeric regions indicates that they are highly heterochromatic and are considered as very poor regions for the gene expression. At least 47 expressed genes were encoded in the genetically defined centromeres of Arabidopsis, residing on the unique sequence islands flanked by the repetitive arrays such as 180-bp or 5S rDNA repeats. These genes are identified as the members of 11 functional categories that comprise the Arabidopsis proteome. The centromeres are not subjected to recombinations. Consequently, genes residing in these regions might exhibit unique patterns of molecular evolution. The function of these centromeres may be specified by the proteins that bind to centromeric DNA, by their epigenetic modifications or by their secondary or higher order structural analysis. Pairwise comparison of these non-repetitive portions of all five Arabidopsis centromeres showed that they share very limited sequence similarity of 1–7%. Most of the centromeric DNA is not shared between chromosomes, complicating efforts to derive their evolutionary relationships. Apart from this, 41 families of small conserved centromere sequences (AtCCS) are enriched in the centromeric and pericentromeric regions that differ from the sequences that are found in the centromeres of other eukaryotes. Molecular and genetic assays needed to be performed for confirming the role of these motifs in performing the centromere activity.

5.7.6 Transposable Elements

Transposons have been found in all eukaryotic and prokaryotic genomes. In the Arabidopsis genome, transposons account up to 10%, with a wealth of 2109 class I and 2203 class II elements, including several new groups (1209 elements). Mobile histories for many of these elements were obtained by identifying regions of the genome with significant similarity to empty' target sites (RESites), thus providing high-resolution information concerning the termini and target site duplications. These regions were readily detected because of the propensity of transposons to integrate into repeats and duplications in the sequence. Transposable elements such as Copia, CACTA-like elements, gypsy-like long terminal repeat (LTR) retrotransposons, long interspersal nuclear elements (LINEs), short interspersed nuclear elements (SINEs), hobo/Activator/Tam3 (hAT)-like elements, and MITEs are found in the Arabidopsis genome. Some MITEs contain ORFs with similarity to the transposases of bacterial insertion sequences. *Basho* and Mutator Like Elements (MULEs) were first discovered in the Arabidopsis, representing a structurally unique transposon. Basho elements have a target site preference for mononucleotide 'A' and wide distribution among plants. Members of the MULEs lack long terminal inverted repeats (TIRs).

In the Arabidopsis genome, Class I transposons are less abundant and has primarily occupied the centromere. In contrast, Basho elements and class II transposons such as MITEs and MULEs are predominately present in the periphery of pericentromeric regions. MULEs and CACTA elements are clustered near centromeres and heterochromatic knobs. The distribution pattern of Arabidopsis transposable elements reflects different types of pericentromeric heterochromatin regions similar to those found in animals. Usually, transposon-rich regions are gene-poor and have lower rates of recombination, indicating a correlation between low gene expression, high transposon density and low recombination After elucidation of the Arabidopsis genome sequence, the role of transposons in genome organization and chromosome structure can now be completely addressed.

5.7.7 Gene Regulation

Arabidopsis genome is having 3000 proteins involved in the gene regulation along with the additional DNA methylation potentially mediating gene silencing and parental imprinting. Arabidopsis possesses SNF2-type chromatin re-modeling ATPases, which regulates the expression of almost all genes. DDM1, a member of the SNF2 superfamily and MOM1, a gene with similarity to the SNF2 family are involved in the transcriptional gene silencing. Consistent with its methylated DNA, Arabidopsis genome possesses eight DNA methyltransferases (DMTs), with two types, are orthologous to mammalian DMT whereas the third one, chromomethyltransferase, is unique to the plants. Arabidopsis also encodes eight proteins with methyl DNA-binding domains (MBDs). The Arabidopsis genome encodes typical eukaryotic transcription machinery comprising of three nuclear DNA-dependent RNA polymerase systems. Transcription by RNA polymerases II and III involves the similar machinery used by the other eukaryotes. However, the transcription factors for RNA polymerase I are yet to be identified. As of now, two polymerase I regulators, which are homologs of yeast RRN3 and mouse TTF-1 are said to be identified in Arabidopsis along with four genes encoding single subunit plastid or mitochondrial RNA polymerases. Based on the analysis, similarity searches, domain matches, a total of 1709 proteins were identified in Arabidopsis with significant similarity to known classes of plant transcription factors. Any of the members belonging to the transcription factors, such as REL, homology region proteins, nuclear steroid receptors, and fork head-winged helix and POU (Pit-1, Octand Unc-8b) domain families of developmental regulators were not identified in Arabidopsis genome. Among 29 classes of Arabidopsis transcription factors, 16 classes such as AP2/EREBP-RAV, NAC, and ARFAUX/IAA families, which contains unique DNA-binding domains are specific to plants. Other transcription factors contain plant-specific variants of more widespread domains such as the DOF and WRKY zinc-finger families and the two-repeat MYB family. The number of transcription factors identified in Arabidopsis is double to that of Drosophila or *C. elegans*. This increase might be attributed to the segmental, chromosomal and tandem duplications in the Arabidopsis genome generating large gene families and transcription factors, which were expanded with the addition of the genes involved in metabolism, defense, and environmental interaction.

5.7.8 Developmental Regulation

Developmental regulation in Arabidopsis involves cell to cell communication, governed by a set of transcription factors and the regulation of chromatin state. Arabidopsis genome analyses the evolution of many processes contributing to the development of both plants and animals, which have converged on similar processes of pattern formation by utilizing and expanding different transcription factor families as major regulators. These processes can be best studied in insects and mammals. Segmentation in insects and differentiation of anterior and posterior limb axes in mammals involve the activation of a series of genes belonging to the homeobox family and this activation is vital for the later differentiation of body and limb axis regions. In plants, the pattern of whorls (sepals, petals, stamens, carpels) is also achieved by the activation of transcription factors belonging to the MADS-box family. Homeobox genes were reported in plants and many animals, and also reported to have MADS-box genes, implying that each lineage has separately invented its own mechanism of spatial formation, converging on the actions and interactions with the transcription factors. Other examples which demonstrate the greater divergence of plant and animal development are AP2/EREBP and NAC families of transcription factors, which have important roles in flower and meristem development of plants. Similarly, receptor tyrosine kinases were not found in plants, but the Arabidopsis genome contains 340 genes for receptor Ser/Thr kinases, belonging to many different families, defined by their putative extracellular domains.

Several families of transcription factors are involved in cell signaling, such as CLV1 receptor, S-glycoprotein homologs involved in signaling from pollen to stigma and the BRI1 receptor necessary for the brassinosteroid signaling. The Leucine-Rich Repeat (LRR) family of Arabidopsis receptor kinases shares its extracellular domain with many animal and fungal proteins that do not have associated kinase domains. At least 122 genes were identified in the Arabidopsis genome, that codes for LRR proteins without a kinase domain. Other Arabidopsis receptor kinase families have extracellular domains that are not found in animals. Several genes belonging to the Arabidopsis genome with a developmental function such as ethylene receptors, phytochrome receptors, and light receptors appear to have been derived either from cyanobacteria or a chloroplast genome, without any close relationship to animal or fungal proteins.

5.7.9 Photomorphogenesis and Photosynthesis

Light is the primary driving source of plant life on this planet, which serves as an energy source as well as triggers and modulates complex developmental pathways. Light stimulates the chlorophyll production, leaf development, cotyledon expansion, chloroplast biogenesis and the coordinated induction of many nuclear and chloroplast encoded genes, inhibiting the stem growth, by a process called photomorphogenesis, defined as the establishment of a body plan that allows a plant to be an efficient photosynthetic machine under varying light conditions.

Even though many of the vital components involved in this process were previously defined, the sequencing of the Arabidopsis genome has provided an opportunity to comprehensively identify the genes involved in the photomorphogenesis and establishment of photoautotrophic growth. More than 100 genes were identified in the Arabidopsis genome, involved in the light perception, signaling and nuclear-encoded genes that potentially function in photosynthesis. Among them, the exact functions of 35 genes have been defined. All light receptors of Arabidopsis have been identified along with 11 proteins of photosystem I, 8 proteins of photosystem II, 26 chlorophyll a/b binding proteins and many new proteins involved in the regulation of photomorphogenesis. 10 genes related to the electron transfer chain including two plastocyanins, five ferredoxins, and three ferredoxin/NADP oxidoreductases were identified. 40 genes with a possible role in CO_2 fixation, all enzymes involved in the Calvin-Benson cycle, 16 genes involved in chlorophyll biosynthesis and

31 genes in carotenoid biosynthesis were also identified. Arabidopsis genome sequencing has also revealed several components related to the complex distribution of the photosynthetic apparatus between nuclear and plastid genomes.

5.7.10 Metabolism

A large portion of the Arabidopsis genome encodes for the enzymes supporting the metabolic processes such as photosynthesis, respiration, intermediary metabolism, mineral acquisition, synthesis of lipids, fatty acids, amino acids, nucleotides, and cofactors. With respect to these processes, the Arabidopsis genome contains a complement of genes similar to a photoautotroph cyanobacterium *Synechocystis*. Arabidopsis has seven genes for the pyruvate kinase and additionally five more for pyruvate kinase-like proteins, whereas *Synechocystis* consists of only one gene for pyruvate kinase. 11 enzymes involved in the Arabidopsis glycolysis pathway are encoded by 51 genes present in eight copies each. Similarly, among the 59 genes involved in the glycerol lipid metabolism, 39 are represented by more than one gene. Genome duplications and expansion of gene families by tandem duplication might have contributed to this diversity. But, this high degree of structural redundancy does not necessarily have any implications on functional redundancy, as studied in mitochondrial serine hydroxymethyltransferase consisting of seven genes. Still, a mutation in this gene completely blocks the whole photorespiratory pathway.

The metabolome of Arabidopsis differs from any organism sequenced to date by the presence of many genes encoding enzymes for pathways unique to the vascular plants. For example, 420 genes have been assigned probable roles in the pathways responsible for the synthesis and modification of cell-wall polymers. 12 genes encode for cellulose synthase, 29 genes encode for 6 families of structurally related enzymes, synthesizing major polysaccharides. 52 genes encoding polygalacturonases, 20 encoding pectate lyases and 79 encode pectin esterases, indicates a massive gene investment for the pectin. Similarly, the presence of 39 β-1,3-glucanases, 20 endoxyloglucan transglycosylases, 50 cellulases, and other hydrolases, 23 expansins reflects the importance of wall re-modeling during the growth of plant cells. 69 genes with significant similarity to known peroxidases and 15 laccases in Arabidopsis genome indicates the importance of oxidative processes in the synthesis of lignin, suberin, and other cell-wall polymers. The high degree of redundancy in the genes involved in cell wall metabolism reflects the differences in substrate specificity by some of the enzymes. It is well known that different cell types in plants have different cell wall compositions and each cell type requires relevant enzymes for the synthesis of the biomolecules required for their cell wall construction. Even though 40 different cell types that plants make has been identified, a large number of genes involved in their wall metabolisms are yet to be defined.

Higher plants synthesize more than 100,000 secondary metabolites. An important factor in the rapid evolution of metabolic complexity is cytochrome P450, a superfamily of heme-containing proteins, which catalyzes NADPH and O_2 dependent hydroxylation reactions. They participate in vital biochemical pathways devoted to the synthesis of plant secondary metabolites, such as phenylpropanoids, alkaloids, terpenoids, lipids, cyanogenic glycosides and Glucosinolates. They are also involved in the synthesis of plant hormones such as gibberellins, jasmonic acid, and brassinosteroids. Arabidopsis genome consists of 286 P450 genes, whereas Drosophila, *C. elegans,* and yeast are having 94, 73 and only 3 P450 genes respectively. Such a low number in the yeast genome indicates a few reactions of basic metabolism, catalyzed by P450s. Even though many genes related to the plant P450s have been identified, only a few dozen P450 enzymes from plants have been characterized till date. The discrepancy might be due to the production of a large number of metabolites which are yet to be identified. In addition to the high P450 gene number, Arabidopsis has many other genes whose pathways and their function is currently unknown. Discovery of these genes and many other intriguing genes in the Arabidopsis genome has created a wealth of new opportunities for understanding the metabolic diversity of higher plants.

5.7.11 Comparative Genomics with *B. oleracea*

Comparing the genome sequences of closely related species is one of the highly useful approaches for the identification of important regions in a genome as demonstrated in *B. oleracea*. By using a whole-genome shotgun approach, over 400,000 sequences were generated from *B. oleracea* accounting for 274 Mb of its genome. These sequences were aligned by comparing with the Arabidopsis genome using BLAST Z and conserved segments from *B. oleracea* were identified by filtering the overlapping alignments.

For identification of the novel genes, conserved sequences of both genomes that are falling into the intergenic regions were chained together with an assumption that each chain represents a potential gene. Using the stringent filtering mechanisms, over 2000 potential genes were identified and manual curation of 500 among these resulted in the curation of new gene models and the extensions of existing genes, which were positively tested for their expression using PCR and cDNA libraries. Overall, comparative genomics of Arabidopsis and *B. oleracea* have led to the identification of a significant number of novel genes.

5.8 The Soybean Genome

5.8.1 Soybean and its Importance

Legumes play an important role in the world's agriculture by fixing atmospheric nitrogen through symbioses with microorganisms. Soybean (*Glycine max* L. Merr.) is an important legume with a high nutrient value of proteins, oils, carbohydrates, primary and secondary metabolites, vitamins, minerals, that is consumed worldwide. Based on the usage and consumption, soybean seeds are classified into two types, i.e., food beans and oil beans. In eastern countries, soybeans are normally harvested, dried and processed into various commercial products such as oils, lubricants, soaps, etc. whereas countries like China and Korea prefer soybean for consumption as whole bean vegetable as well as in the form of sprouts, soy milk, etc. The people of South-East Asian countries preferably use vegetable soybean in their diet and are commonly known as "mao dou" in Chinese; "poot kong" in Korean and "edamame" in Japanese languages. Apart from the high protein and oil contents, soybeans are supplemented with carbohydrates comprising of both soluble and non-soluble sugars. Soybean considered as the highest vegetable protein yielding legume, constituting all the essential amino acids that cannot be synthesized by the human body and need to be supplemented from a food source. The fiber content of the soybean helps in the lowering cholesterol, blood sugar and curtails the exposure of colon cancer. Soybean also forms an excellent source of nutraceutical constituents such as fiber, protease inhibitors, phytic acid and polyphenols including flavonoids, isoflavones, lignins, glycosides, and tannins, which are also possessing antioxidant, antimutagenic and anticarcinogenic properties. Soybean consumption reduces the formation and progression of cancer, cardiovascular diseases, Alzheimer's disease, and osteoporosis. The phytohormone genistein, an important isoflavone and is also a major constituent of soybean that plays a vital role in cancer inhibition and in the suppression of cancer cell growth. The soluble phenolic compounds are abundant in soybean. These nutritional constituents along with saponins, phytic acid, fiber, protease inhibitors, etc., are beneficial for human health and thus soybean is considered as a functional food. The phytoestrogen abundance in soybean extracts culminates its extensive use for the preparation of diet supplements that can be used effectively as a natural alternative to hormonal replacement therapy in the woman during menopause. The demand for the production of nutrient-rich soybean can be met only by the identification of superior cultivars, and sequencing of its genome will provide more information about its economically valuable traits.

5.8.2 Genome Sequencing and Assembly

Soybean (*Glycine max* var. Williams 82) genome is the largest whole-genome shotgun sequenced plant genome so far, comprising of 950 Mb assembled and anchored genome embedded in 20 nuclear chromosomes. Seeds of soybean cv. Williams 82 were cultivated in a growth chamber and were etiolated for 5 days before the harvest. DNA was extracted by standard phenol-chloroform method, treated with RNase and Proteinase K, precipitated with ethanol. Sequencing was performed by the whole genome shotgun method using ABI 3730XL capillary sequencing machines, at the Joint Genome Institute in Walnut Creek, California.

15,332,163 sequence reads were assembled using Arachne v.20071016 formed into 3363 scaffolds covering 969.6 Mb of the soybean genome. The resulting assembly was integrated with the genetic and physical maps previously built for soybean and a newly constructed genetic map to produce 20 chromosome-scale scaffolds covering 937.3 Mb. Additional 17.7 Mb sequence present in 1148 unanchored sequence scaffolds consists of highly repetitive sequences, with less than 450 predicted genes. Positions of scaffolds were determined with extensive genetic maps composed of 4991 single nucleotide polymorphisms and 874 simple sequence repeats. Among 397 scaffolds, 377 were unambiguously oriented on the chromosomes and 20 unoriented scaffolds are in repetitive regions, with a scarcity of recombinations and genetic markers. Telomeric repeats TTAGGG or CCCTAAA are present on both the arms of 8 chromosomes and on the single-arm of 11 chromosomes. 19 of the 20 chromosomes contain a large block of 91–92 bp centromeric repeats.

5.8.3 Gene Annotation

Gene annotation of the soybean genome was done using Fgenesh and Genome Scan based on the EST alignments and peptide matches of Arabidopsis, rice, and grapevine. Models were reconciled with EST alignments and were filtered for high confidence by eliminating the genes related to the transposable elements, with low sequence entropy, short introns, incomplete start or stop codons, low C-score. One of the striking features of the soybean genome is that 57% of the genome is repeat-rich, with low-recombination heterochromatic regions surrounding the centromeres. The average ratio of genetic-to-physical distance is 1 cM per 197 kb in euchromatic regions and 1 cM per 3.5 Mb in the heterochromatic regions. 93% of the recombinations are occurring in the repeat-poor, gene-rich euchromatic genomic region, accounting for 43% of the genome. 21.6% of the genes are found in the repeat-rich and transposon rich regions of the centromeres. A total of 46,430 protein-coding loci were identified in the soybean genome, using a combination of full-length cDNAs, ESTs, homology, and ab initio methods. ~20,000 more protein-coding loci were also identified, enriched with hypothetical, partial and/or transposon-related sequences, and possess shorter coding sequences and fewer introns. The exon-intron structure of genes is highly conserved in soybean, poplar and grapevine, consistent with a high degree of position and phase conservation found more broadly across angiosperms. Intron size is also highly conserved in soybean, indicating that a few insertions and deletions might have accumulated within or among introns over the past 13 million years.

Among the 46,430 protein-coding loci, 34,073 (73%) are found to be clearly orthologous with one or more sequences in other angiosperms, which can be assigned to 12,253 gene families. Soybean is particularly enriched in genes containing nucleotide-binding-site-APAF1-R-Ced) (NBARC) and leucine rich-repeat (LRR) domains, associated with the plant immune system. From the protein families in the sequenced angiosperms, 283 putative legume-specific gene families containing 448 high-confidence soybean genes were identified with AP2 domain, a protein kinase domain, cytochrome P450, and PPR repeat are the top most domains found in the soybean genome. An additional 741 putatively soybean-specific gene families each consisting of at least two or more soybean genes were also identified, with protein kinase and protein tyrosine kinase, AP2, LRR, MYB-like DNA binding domain, as well as GDSL-like lipase/acylhydrolase and stress-upregulated Nod19 as top genes in this set.

5.8.4 Repetitive Elements

A combination of structure-based analyses and homology-based studies have resulted in the identification of 38,581 repetitive elements, covering almost all types of plant transposable elements, which make up 59% of the soybean genome. Long terminal repeat (LTR) retrotransposons are the most abundant class of transposable elements in the soybean genome containing up to 42% of the repetitive elements. 510 families containing 14,106 intact elements were found with sizes ranging from 1 to 21 kb, with an average size of 8.7 kb. 69% of the intact families are Gypsy-like, and 17% of the elements are divided into *Tc1/Mariner, haT, Mutator, PIF/Harbinger, Pong, CACTA* superfamilies, Helitrons, and the remainder 14% of the elements are Copia-like. Most of these families are present in low copy numbers, with 78% of them are lesser than 10. 18,264 solo LTRs and 4552 nested insertion events were also identified in soybean (Table 5.3). *Tc1/Mariner* and *Pong*, comprising 0.1% of the genome seem to be inactive and relatively old. Other transposon families appear to have undergone amplification recently and may still be active as indicated by their high similarity with multiple elements (up to 98%). Table 5.3 displays the list of repetitive elements in soybean genome.

Table 5.3 Repetitive elements in the soybean genome

S. No.	Repetetive element	Copy number	% Of DNA
Retro transposon			
LTR retro transposon			
1	Ty1/copia	124,516	12.47
2	Ty3/gypsy	185,189	29.52
Non-LTR retro transposon			
	LINE	3420	0.25
DNA transposon, subclass I			
3	Tc1/Mariner	536	0.03
4	hAT	938	0.04
5	Mutator	100	4.53
6	PIF/Harbinger	10,207	0.29
7	Pong	1755	0.09
8	CACTA	127,467	10.16
9	MITE	46,335	0.83
10	Tourist	19,168	0.33
11	Stowaway	27,167	0.50
Subclass II			
12	Helitron	7128	0.53
13	Satellites repeats	11,004	1.18

5.8.5 Structural Organization of the Genome

Homologous blocks were identified in the soybean genome. Because of the multiple duplications, diploidizations and chromosomal rearrangements in the genome, multiplicons or blocks between chromosomes can occur, involving more than two chromosomes. 61.4% of the homologous genes are found in blocks involving only two chromosomes, 5.63% of the genes spanning three chromosomes and 21.53% traversing four chromosomes. Chromosome 14 is the highly fragmented one with a block matching with 14 other chromosomes, which is the highest number of all chromosomes. Chromosome 20 is highly homologous with the long arm of chromosome 10 and has a few more matches in other parts of the genome.

5.8.5.1 Root Nodulation Genes

One of the unique features of leguminaceae members is their ability to establish nitrogen-fixing symbioses with soil bacteria of the family Rhizobiaceae. Sequence comparisons with the previously reported nodulation genes have identified 28 nodulin genes and 24 key regulatory genes in the soybean genome, which might be the true orthologues of known nodulation genes in other legume plants. 25% of the nodulin genes produce transcript variants, among them nodulin gene 24 is solely responsible for the production of 10 transcript variants. However, none of the soybean regulatory nodulation genes are found to produce transcript variants.

5.8.5.2 Oil Biosynthesis Genes

Identifying the genes involved in triacylglycerol biosynthesis of soybean might benefit the soybean oil composition as well as an increase in its yield. A comparison of the soybean genome with 614 Arabidopsis genes that are involved in the acyl lipid biosynthesis has resulted in the identification of 1127 putative orthologous and paralogous genes in the soybean genome. This could probably a low estimate owing to the high stringency conditions used for gene mining. The number of genes involved in storage lipid synthesis, fatty acid elongation, and wax/cutin production was similar in both Arabidopsis and soybean. A number of genes involved in other subclasses such as in lipid signaling, degradation of storage lipids and membrane lipid synthesis, were two-to-three fold higher in soybean than Arabidopsis, indicating the complexity of acyl lipid synthesis in soybean. The number of genes involved in plastid de novo fatty acid synthesis was 63% higher in soybean than that of Arabidopsis.

5.8.5.3 Transcription Factors

5671 putative transcription factor genes distributed in 63 families were identified in soybean, which represents 12.2% of the 46,430 predicted protein-coding loci in the soybean genome. These transcription factor genes are homogeneously distributed across the chromosomes with an average abundance of 8–10% on each chromosome. Among the transcription factor genes identified, 538 are tandemly duplicated.

The overall distribution of transcription factor genes in soybean is very similar to that of Arabidopsis with a very few exceptions such as the members of ABI3/VP1 family, which are 2.2 times more abundant in Arabidopsis and members of the TCP family, 4.4-times more abundant in soybean. Additionally, FHA, HD-Zip (homeodomain/leucine zipper), PLATZ, SRS and TUB transcription factor genes are more abundant in soybean and HTH-ARAC (helix–turn–helix araC/ xylS-type) genes are exclusively identified in the soybean genome.

5.8.6 Importance of Soybean Genome

Hundreds of traits inherited by a single gene (also called as qualitative traits) have been characterized in soybean and many of them are genetically mapped. However, QTL mapping studies for the most important crop production traits and seed quality are in the process for more than 90 distinct traits of soybean such as plant developmental and reproductive characters, disease resistance, seed quality, and nutritional traits. In most of the cases, a functional gene or a transcription factor responsible for the QTL is not known. Integration of the whole genome sequence with the genetic marker map will permit us to study the association of mapped phenotypic effectors with the causal DNA sequence. Using the soybean genome map, mutation in the rsm1 (raffinose synthase) gene was identified, that can be used for the selection of low stachyose containing soybean lines for improving the ability of animals and humans to digest the soybeans. Through comparative genomics approach between soybean and maize, a single-base mutation responsible for the reduction of phytate production in soybean was identified. Phytate reduction could result in the reduction of a major environmental runoff contaminant from swine and poultry waste. Another remarkable application of the soybean genome is the cloning of the first resistance gene against the devastating Asian soybean rust (ASR) and its confirmation with viral-induced gene silencing. This will greatly benefit the worldwide soybean production.

Soybean is one of the most important global sources of protein and oil, has become the first legume species with a complete genome sequence. Sequencing of the soybean genome has not only become a key reference for more than 20,000 legume species but also for the remarkable evolutionary innovation of nitrogen-fixing symbiosis. This genome sequence is also an essential framework for vast new experimental information such as tissue-specific expression and whole-genome association data. With the knowledge of this genome sequence, an approach for understanding the plant's capacity to turn CO_2, water, sunlight along with elemental nitrogen and minerals into concentrated energy, protein and nutrients for human and animal usage has been made. Further, this genome sequence opens the door for crop improvements that are needed for sustainable food production, energy production, and environmental balance.

5.8.7 Database

The outcome of the soybean whole-genome shotgun project is available at DDBJ/EMBL/GenBank under the accession ACUP00000000, ACUP01000000 as version 1. Full annotation of this genome is also available at http://www.phytozome.net.

5.9 The Rice Genome

5.9.1 Importance of Rice Crop to the World

Rice (*Oryza sativa L.*) is one of the most important food crops in the world, which feeds more than half of the global population. Rice plays a central role in human nutrition and this is one of the oldest crops, which is under cultivation since 10,000 years. With an increase in the global population, the demand for increased rice production is also increasing and it was estimated that there should be an increase of 30% global rice production in the next 20 years, in order to meet the projected demands from the population. Rice is being cultivated in the most productive irrigated lands. With increasing levels of atmospheric pollution, unseasonal rains, a rapid increase in the night temperatures due to global warming, reductions in the suitable land and water are posing a threat for the rice cultivation with labor and energy-dependent fertilizers providing the additional constraints. Increasing the rice yield and productivity will be possible only with improved conventional breeding combined with applications of novel plant biotechnology tools, which needs a high-quality rice genome sequence. Possessing the smallest genome among the major cereals, dense genetic maps, ease of genetic transformation and extensive genome co-linearity with other Gramineae members makes rice a wonderful model system not only among the cereal plants but also among the crop plants.

With the genome size of 400–430 Mb, the rice genome is the smallest and a well mapped genome among the major cereals. The molecular map of rice plant genome with 6000 markers, has been useful in the alignment of physical chromosome maps. A library comprising of over 40,000 rice ESTs have been reported and many of them have been mapped. A rice YAC library has also been fingerprinted and ordered with mapped markers covering up to 60% of its genome. Several BAC libraries have also been described in early studies. After the introduction of novel methods for transformation, rice plant has become the easiest of all cereal plants for the genetic transformation, permitting the complement mutations or conferring the dominant phenotypes for verifying the gene function.

5.9.2 International Rice Genome Sequencing Project (IRGSP)

Following the progress made in understanding the rice molecular genetics, an effort to sequence its whole genome has been made in Japan during 1997, with 12 countries eventually decided to contribute in the International Rice Genome Sequencing

Table 5.4 Sharing the chromosomes of rice among different countries for the genome sequencing

Chromosome number	Country
1	Japan, Korea
2	United Kingdom (EU), Canada
3	USA
4	China (indica variety Guang Lu Ai 4)
5	Taiwan
6	Japan
7	Yet to be claimed
8	Yet to be claimed
9	Thailand
10	USA
11	India, USA
12	France

Project (IRGSP) (Table 5.4). After Wide-range of discussions, IRGSP has chosen a single Japanese rice variety (*Oryza sativa* L. ssp. japonica cv. Nipponbare) to be the source of DNA for sequencing, which prevents the allelic polymorphisms generated by the cultivated varieties and impedes the accurate compilation or integration of sequences. Members of the IRGSP have also fixed the standards for sequence quality similar to that of Human Genome Project, with less than one base-pair (bp) error in 10,000 bp. This standard is achievable through a combination of high-quality shotgun sequence reads, a seven-fold redundancy and the insistence that 97% of all bases are sequenced on both the strands. The level of accuracy can be gauged by the computational software such as phred/phrap/consed.

5.9.3 Physical Map and Sequencing of the Rice Genome

The physical map of rice has gained utmost importance as a basic tool for bridging the information in the nucleotide sequence to phenotypic traits.

1. The first step of mapping involves the understanding of rice DNA by making a linkage map based on polymorphisms within the DNA sequences, RFLPs, SSRs and CAPSs (Cleaved Amplified Polymorphic Sequences). Till date, around ten genetic maps have been made on rice by positioning the genetic markers indispensable for assembling the large DNA fragments for sequencing and for ascertaining the chromosomal locations of these fragments.
2. The next step is the construction of a rice genome-wide physical map. So far, only one physical map covering up to 60% of the rice genome assembled using YACs has been constructed. Due to the disadvantages of using YACs as templates for DNA sequencing, BAC/PAC (P_1 derived Artificial Chromosome) vectors were used for constructing the rice genomic libraries. Utilization of several restriction enzymes, such as Sau3AI for PACs, Hind III and EcoR I for BACs have addressed the uneven distribution of restriction sites along the rice genome, resulting in the genome-wide physical map of the rice.

3. Another strategy is to map EST markers on the YAC physical map to generate a dense EST map. Around 40,000 cDNA clones have been partially sequenced and were assembled based on their 3′-end sequences. Using this process, 5000 ESTs, each of them representing an independent group, had been mapped on the YAC physical map. These markers are thought to partly reflect the distribution of genes along each chromosome and should help in making sequence-ready contigs for gene-rich regions. Sequence-ready PAC contigs have been identified by EST selection and sequencing of several PACs identified by this strategy indicates a higher gene density of one gene per every 5000 bp. The PAC contigs constructed in this way covers up to 30% of the rice genome. The gaps where no EST markers or YACs can be filled either by flanking the gap to design PCR primers to 'walk' into the gaps or to use the end-sequence and fingerprint information from the BAC library to search for the clones that fill the gaps. The completion of a deep, sequence-ready physical map of rice is a basic requisite for the generation of a reliable sequence map composed of predicted genes, supplemented with the positional information from RFLP or EST markers and phenotypic traits. This will become the skeleton of a database used for establishing the rice genome.

5.9.4 Genome Annotation

The sequences generated by the Rice Genome Research Program are annotated by searching for a non-redundant protein database using BLASTX, searching the rice EST database using BLASTN, and scanning the sequences with GenScan for predicting the open reading frames and with Splice Predictor to identify the exon/intron splice sites. All these results were combined to make a final genome annotation, used not only for the gene prediction but also in characterization of repeat sequences (Inverted and Tandem, flanking and long terminal repeat) and transposable elements. For integrating the rice genome information, a new database named as Integrated Rice Genome Explorer (INE) has been developed. In this database, DNA markers on the genetic map play key roles in linking the YAC physical map, the EST map and the PAC/BAC physical map to define the chromosomal locations of clones on each map. Annotated genome sequences are available *via* PAC/BAC clones on a physical map. Similarities between the rice genome and other cereals will be made apparent by linking INE to databases of each of the other important cereal crop species. A preliminary comparison is currently being made between rice and maize databases.

5.9.5 Components of Rice Genome

The size of rice (*O. sativa* ssp. japonica cv. Nipponbare) haploid nuclear DNA is determined to be 394 Mb on the basis of flow cytometry and after addition of BAC contigs and gaps to the sum of the non-overlapping sequence, the total length of the rice nuclear genome was calculated to be 388.8 Mb. An independent measure for the pseudomolecules in the rice genome was obtained by searching for unique EST markers. Among 8440 ESTs searched for the pseudomolecules, 8391 (99.4%) were identified in the pseudomolecules. Hence, the pseudomolecules are expected to cover 95.3% of the entire genome with a specific estimate of 98.9% in the euchromatin region.

5.9.5.1 Centromere Location

Centromeres of rice chromosomes are typical eukaryotic type, consisting of repetitive sequences, such as satellite DNA at the center, retrotransposons and other transposons in the flanking regions. All centromeres contain a highly repetitive 155–165 bp CentO satellite DNA, along with centromere-specific retrotransposons. The CentO satellites with 59 kb and 69 kb CentO repeats are located within the functional domain of the rice centromeres, identified by the completely sequenced chromosomes 4 and 8. These CentO satellites repeats are flanked by numerous retrotransposons.

5.9.5.2 Gene content and their Expression

After masking the pseudomolecules for repetitive sequences, ab initio gene finder FGENESH was used to identify the genes, which are exclusively non-transposable elements related. A total of 37,544 non-transposable element protein-coding sequences were predicted in the rice genome, with a density of one gene per every 9.9 kb. Among the 37,544 gene models, 2927 genes were aligned well with ESTs of other cereals. 19,675 protein sequences had matches with entries of SwissProt database and 4500 protein sequences having no expression support at all. 63% of the predicted proteins are with one motif or domain and a total of 3328 different domains present in the predicted rice proteome. It was predicted that 51% of the predicted proteins might be associated with a biological process, 29.1% are involved in metabolism and other cellular physiological processes. 26,837 (71%) predicted rice proteins have homologues in the Arabidopsis proteome and 26,004 (88%) of the Arabidopsis proteome have homology with the rice proteome. Rice proteins also carry 38.1% homology with Drosophila, 40.8% with human, 36.5% with *C. elegans*, 30.2% with yeast, 17.6% with *Synechocystis* and 10.2% with *E. coli*. There are profound differences in plant architecture and biochemistry between monocots and dicots. 2859 rice genes are lacking homologues in the Arabidopsis genome. These genes could be monocot specific, such as seed storage proteins class belonging to prolamins that are not found in dicots (Table 5.5).

5.9.5.3 Tandem Gene Families

Analysis of the Arabidopsis genome has yielded 17% of genes that are arranged in the tandem repeats and a similar analysis on rice genome has produced 14% of the tandemly repeated gene sequences. However, manual curation of the rice chromosome 10 has shown one gene family encoding a glycine-rich protein with 27 copies and a TRAF/BTB domain protein with 48 copies. These tandemly repeated families are interrupted with other genes and are not included in strictly defined tandem repeats.

Table 5.5 Monocot specific proteins in rice genome

Protein name	Number
Abscisic stress ripening protein	4
Chitinase precursor	3
Citrate binding protein precursor	1
Endonuclease	1
Glucan 1,3-β-glucosidase precursor	3
Heterogeneous nuclear ribonucleoprotein	1
Jasmonate-induced protein	4
Mannosyl transferase	1
Pathogenesis related protein PR-10a	5
Phytosulfokines precursor	1
Prolamine	31
Proteinase inhibitor	10
Queuine tRNA ribosyl transferase	2
Ribosome inactivating protein	1
SAM-dependent methyltransferase	1
Seed allergen	5
Starch branching enzyme	1
Wound-induced protease inhibitor	1

5.9.5.4 Non-coding RNA Genes

763 tRNA genes and 14tRNA pseudogenes were detected in the rice genome with chromosome 4 having a single tRNA cluster and chromosome 10 has two large clusters derived from an inserted chloroplast DNA. MicroRNAs (miRNA) and small nucleolar RNAs (snoRNA) are a class of eukaryotic non-coding RNAs, believed to regulate the gene expression by interacting with the target mRNA. 158 miRNAs, 215 snoRNA, and 93 spliceosomal RNA genes were mapped in the rice genome (Table 5.6).

5.9.5.5 Organellar DNA Insertions into the Rice Nuclear Genome

A continuous transfer of chloroplast and mitochondria DNA into the nucleus of rice cells has resulted in the presence of chloroplast and mitochondrial DNA inserted into the nuclear chromosomes. Even though these two organelles contain the genome of several Mb when they were internalized, their genomes diminished and the present size of the mitochondrial genome is less than 600 kb and that of the chloroplast is only 150 kb. Homology searches have detected 421–453 chloroplast insertions and 909–1191 mitochondrial insertions, in the rice nuclear genome. Thus, chloroplast insertions contributing to 0.24% and mitochondrial insertions contributing to 0.19% of the rice nuclear genome. The distribution of chloroplast and mitochondrial insertions over the 12 rice nuclear chromosomes indicates that chromosome 12 contains ~1% of the mitochondrial DNA and 0.8% of the chloroplast DNA. It has been hypothesized that both of these transfers might have occurred independently.

Table 5.6 Distribution of non-coding RNA genes among the 12 chromosomes of the rice genome

Chromosome	miRNA	snoRNA	Spliceosomal RNA
1	18	13	6
2	23	17	25
3	12	64	13
4	19	12	9
5	12	19	1
6	15	16	11
7	10	22	10
8	19	14	6
9	7	5	0
10	6	14	4
11	7	7	5
12	10	12	3

5.9.5.6 Transposable Elements and Class 1 Simple Sequence Repeats

The rice genome is populated by the representatives from all known transposon superfamilies. Previous estimates of the 25% transposon content in the rice genome were confirmed by using the profile hidden Markov models, which allows the identification of more divergent elements. The present study had indicated that the transposon content *O. sativa* ssp. japonica cv. Nipponbare genome is at least 35%. Chromosomes 8 and 12 have the highest transposon content of 38% and 38.3%, respectively, and chromosomes 1 (31.0%), 2 (29.8%) and 3 (29.0%) have the lowest proportion of the transposons. Elements belonging to the IS5/Tourist and IS630/Tc1/mariner superfamilies are prevalent on the first three chromosomes. Class II transposable elements including the IS 5/Tourist and IS630/ Tc1/mariner superfamilies, outnumbered class I elements, which include long terminal-repeat (LTR) retrotransposons and non-LTR retrotransposons by more than twofold. However, the nucleotide contribution of class I transposable elements is greater than that of class II, mostly due to the large size of LTR retrotransposons and the small size of IS5/Tourist and IS630/Tc1/mariner elements. Most of the class I transposable elements are concentrated in the gene-poor, heterochromatin regions such as the centromeric and pericentromeric of the chromosomes. Members of some transposon superfamilies such as IS5/Tourist, IS630/Tc1/mariner, and LINEs, had a significant positive correlation with the recombination rate and gene density in rice.

Class 1 simple sequence repeats (SSRs) are the repeats of 20 nucleotides in length, that behave as hypervariable loci, providing a rich source of markers for their usage in genetics and plant breeding. 18,828 di, tri, and tetra-nucleotide SSRs of class 1, representing 47 distinctive motif families were identified and annotated on the rice genome, which makes an average of 51 hypervariable SSRs per every megabase. The highest density of these markers was found to be occurring on chromosome 3 and the lowest was found on chromosome 4. Several thousands of these SSRs markers have already been reported, known to amplify well and are highly polymorphic.

5.9.6 Outcomes of the Rice Genome Project

- 389 Mb of the rice genome sequence has been determined, which is ~260 Mb larger than fully sequenced dicot plant *Arabidopsis thaliana*. 370 Mb of the completed sequence, representing 95% coverage of the genome belonging to euchromatic regions has been generated.
- A total of 37,544 protein coding sequences, which are exclusively non-transposable element-related were detected, with a lower gene density of 1 per 9.9 kb. A total of 2859 genes were identified to be unique to rice and the other cereals.
- Gene knockouts are useful tools for determining gene function and relating genes to the phenotypes. 11,487 *Tos17*retrotransposon insertion sites were identified in the rice genome, among which 3243 are in the coding regions.
- 0.38–0.43% of the rice nuclear genome consists of chloroplast and mitochondrial DNA fragments, representing a repeated and ongoing transfer of organellar DNA to the nuclear genome.
- The transposon element content of rice is 35% and is represented by all known transposon superfamilies.
- 80,127 polymorphic sites that distinguish between two cultivated rice subspecies, *japonica* and *indica* were identified, resulting in the construction of a high-resolution genetic map for rice.
- Frequency of SNPs varies from 0.53–0.78%, which is 20 times more the frequency observed between the Columbia and Landsberg *erecta* ecotypes of *Arabidopsis*.

5.10 The Human Genome

Decoding of the DNA constituting human genome is one of the highly anticipated contributions that will help in understanding the evolution and diversity of human beings, the genetic basis for the causation of disease and studying the role of environment and heredity in defining their condition. A project has been proposed in 1985 for determining and decoding the complete nucleotide sequence of the human genome, which gave mixed reactions in the scientific community. After adjustments to the draft proposal, the Human Genome Project (HGP) was officially initiated in the United States during 1990, under the supervision of the National Institutes of Health and the U.S. Department of Energy. Proposed duration of the project was 15 years and the budget plan for completing the genome sequence was $3 billion. This project took 7–8 years to get into an active stage as in the 7 years of duration only 5% of the genome was sequenced.

In 1997, Weber and Myers proposed the whole-genome shotgun method for sequencing the human genome, which was not well received. In 1998 Craig Venter and his team of TIGR involved in the Human Genome Project has announced to build a unique genome facility for determining the sequence of complete euchromatin region in the human genome. In order to achieve this, In early 1998, Applied

Biosystems has developed an automated, high throughput capillary DNA sequencer, called ABI PRISM 3700 DNA Analyzer. Successful discussions between Applied Biosystems and TIGR scientists resulted in a plan to undertake the human genome sequencing through a whole-genome shotgun technique using 3700 DNA Analyzer. As a result, many of the operation facilities for genome sequencing were established in the TIGR. Some severely opposed the feasibility of the required 150-fold scale-up from the *H. influenzae* genome to the human genome with its complex repeat sequences. In order to achieve the required scale-up, Just sequenced genome of *Drosophila melanogaster* was chosen as a test case for whole-genome assembly on a large and complex eukaryotic genome. 120 Mb region of the Drosophila genome was determined for 1 year in collaboration with the Berkeley *Drosophila* Genome Project, resulted in two salient findings.

(i) Algorithms for chromosome assemblies have generated a highly accurate order of assembly with an orientation of less than tenfold coverage.
(ii) Performing multiple interim assemblies in place of one comprehensive and final assembly did not serve any purpose.

These findings, have led to a modified whole-genome shotgun sequencing approach to the human genome. Initially, it was proposed to perform tenfold sequence coverage of the genome for 3 years and then to in termly assemble the available sequence data quarterly. Sequencing of the human genome was started on 8th September 1999 and was completed on 17 June 2000. Completion of the first assembly was reported on 25 June 2000 and the final assembly of the 3 billion bp human genome was completed 1 October 2000.

The details of the Human Genome Project has been divided into seven sections.

1. Sources of DNA and Sequencing Methods
2. Genome Assembly Strategy and Characterization
3. Gene Prediction and Annotation
4. Genome Structure
5. Genome Evolution
6. A Genome-Wide Examination of Sequence Variations
7. An Overview of the Predicted Protein- Coding Genes in the Human Genome

5.10.1 Sources of DNA and Sequencing Methods

The initial version of the human genome was proposed to be a composite, derivative of multiple donors belonging to highly diversified ethnic backgrounds. 21 donors belonging to different ethanogeographic categories such as African-American, Chinese, Hispanic, Caucasian, etc. were volunteered for this project. Age, sex and a self-designated ethnogeographic group of each donor were recorded, linked by a confidential code to the donated sample 130 ml of whole heparinized blood was collected separately from males and females, along with the five specimens of semen from males, over a 6-week period. Permanent lymphoblastoid cell lines were created by Epstein-Barr virus immortalization. DNA from two males and three females comprising of one African-American, one Asian-Chinese, one Hispanic-Mexican, and two Caucasians were isolated. The decision of which ethnic group DNA to the sequence was taken based on a set of complex factors and technical issues such as quality of the DNA libraries and availability of the immortalized cell lines.

High-quality plasmid libraries with three insert sizes; 2 kbp, 10 kbp, and 50 kbp were obtained from each donor with equal representation from all parts of the genome and without any contamination from mitochondrial genome and *E. coli* genomic DNA. Process of whole-genome shotgun sequencing was fully automated and highly modular. Intermodule sample backlogs have allowed four principal modules such as (i) library transformation, plating and colony picking (ii) DNA template preparation (iii) dideoxy sequencing reaction set-up and purification (iv) sequence determination with the ABI PRISM 3700 DNA Analyzer, to be operated independently: As the inputs and outputs of each module have been carefully matched and sample backlogs are continuously managed, sequencing of the genome has proceeded without any interruption even for a single day, since the initiation of the *Drosophila* project in May 1999. The ABI 3700 is a fully automated capillary array sequencer, that can be operated with minimal hands-on time, ~15 min per day. The capillary system also facilitated a few correct associations of sequencing traces with samples through the elimination of manual sample loading and lane tracking errors associated with the slab gels. 65 members of the production staff were hired, trained and were rotated on a regular basis. A central laboratory information management system (LIMS) tracked all sample plates by unique bar code identifiers. The facility was supported by a quality control team that performed raw material and in-process testing and a quality assurance group with responsibilities including document control, validation, and auditing of the facility.

An automated trace-processing pipeline has been developed to process each sequence file. After quality and vector trimming, the average trimmed sequence length was 543 bp and the sequencing accuracy was exponentially distributed with a mean of 99.5% and with less than 1 in 1000 reads being less than 98% accurate. Each trimmed sequence was screened for matches to contaminants including sequences of vector alone, *E. coli* genomic DNA, and human mitochondrial DNA. The entire read for any sequence with a significant match to a contaminant was discarded. A total of 713 reads matched *E. coli* genomic DNA and 2114 reads matched the human mitochondrial genome. The importance of the base-pair level accuracy of the sequence data increases as the size and repetitive nature of the genome to be sequenced increases. Each sequence read must be placed uniquely in the genome, and even a modest error rate can reduce the effectiveness of the assembly.

5.10.2 Genome Assembly Strategy and Characterization

Two independent sets of data were used for the genome assembly. The first was a random shotgun data set of 27.27 million reads of 543 bp average length produced at

Celera, largely consisting of mate-pair reads obtained from 16 libraries constructed from DNA samples of five different donors. Libraries with insert sizes of 2, 10, and 50 kbp were used. The range of insert sizes in each library was characterized by the position of the mate pairs in that library, supported by a mean and standard deviation. The clone coverage is considered as the coverage of the genome in cloned DNA, considering the entire insert of each clone that has a sequence from both ends. It is a measure of the amount of l DNA coverage of the genome. Assuming the human genome size of 2.9 Gbp, trimmed sequences of Celera has given a 5.1× coverage of the genome in which the clone coverage was 3.42×, 16.40×, and 18.84× respectively for the 2, 10 and 50-kbp libraries for a total of 38.7× clone coverage. The second data set was from the publicly funded Human Genome Project (PFP), which was primarily derived from BAC clones comprising of 4443.3 Mbp sequence. The data for each BAC was deposited at one of four levels of completion. Phase 0 was the data of unassembled sequencing reads. Phase 1 was the data of unordered assemblies of contigs, called as BAC contigs or bactigs. Phase 2 was the data of ordered assemblies of bactigs. Phase 3 was the data of complete BAC sequences.

5.10.3 Whole-Genome Assembly

The genome was assembled using two different approaches. (1) A whole-genome assembly process using both Celera and the PFP data and (2) A compartmentalized assembly process that initially partitioned the Celera and PFP data into sets localized on large chromosomal segments and then ab initio shotgun assembly on each set was performed. The algorithms used for the whole-genome assembly (WGA) of the human genome were enhancements of the same used for the *Drosophila* genome, consisting of four principal stages: Screener, Overlapper, Unitigger, and Scaffolder.

1. **Screener:** finds, marks all microsatellite repeats of the size less than a 6 bp and screens out all known interspersed repeat elements, such as Alu 1, Lines and ribosomal DNA. These marked regions will be searched for overlaps, whereas screened regions do not get searched, but can be part of an overlap, involved in the matching of unscreened segments.
2. **Overlapper:** compares every read against every other read in search of complete end-to-end overlaps of at least 40 bp and with no more than 6% differences in the match. As all the data has been scrupulously vector trimmed, the Overlapper can find the genuine overlap matches. Overlapper is divided into two sections; true overlaps and repeat-induced overlaps.
3. **Unitigger:** Further processes the repeat-induced overlaps. These contigs are called unitigs, which are the uncontested interval subgraphs of the graph of all overlaps. Even though many of these assemblies are true overlaps, some of them are the collections of reads from the several copies of repetitive elements, collapsed into a single sub-assembly. These collapsed unitigs were identified by a discriminator, that gives the logarithm of the odds ratio, whether an unitig is composed of unique DNA or of a repeat consisting of two or more copies. The

result of running the Unitigger was a set of correctly assembled sub-contigs covering up to 73.6% of the human genome.

4. **Scaffolder:** The final step in assembling the genome was to order and orient the scaffolds on the chromosomes. These scaffolds were together on the basis of their order in the components from the compartmentalized shotgun assembly (CSA). These grouped scaffolds were re-ordered by examining the residual mate-pairing data. The scaffold groups were mapped onto the chromosome, using physical mapping data. This step is performed using reliable high-resolution map information such that each scaffold will overlap with multiple markers. Two genome-wide types of map information were available: high-density STS maps and fingerprint maps of BAC clones developed at Washington University.

The assembly of Celera's data, along with the bactig data has produced the scaffolds of 2.848 Gbp. The set of reads not incorporated in the assembly were 11.27 million (26%). More than 84% of the human genome was covered by scaffolds >100 kbp long. There were 93,857 gaps among the 1637 scaffolds >100 kbp. The average size of scaffold, contig, and gap were 1.5 Mbp, 24.06 kbp and 2.43 kbp respectively. More than 50% of all gaps were >500 bp, among them, 62% are <1 kbp long and no gap was >100 kbp. Similarly, more than 65% of the human genome sequence is in contigs of >30 kbp and the largest contig was 1.22 Mbp long. The final stage of resolving gaps was to fill them with assembled BAC clones data that covers the gap, called external gap walking.

In addition to the Whole Genome Assembly approach, a localized assembly approach known as Compartmentalized Shotgun Assembly (CSA) was also implemented for increasing the resolution of inter-chromosomal duplications. CSA is intended to subdivide the genome into large segments, each of which can be shotgun assembled individually, by clustering the Celera reads and bactigs into large, multiple megabase regions of the genome and then running the WGA assembler on the Celera data and shredded, faux reads obtained from the bactig data.

5.10.4 Gene Prediction and Annotation

The number of protein-coding genes in humans has been a subject of speculation. Initial estimation based on the re-association data was placed between 30,000–40,000, whereas later estimates exclusively from the brain were more than 100,000. Recent data obtained from both the corporate and public sectors estimated 142,634 genes based on the combination of EST data with CpG islands. Another prediction of 28,000–34,000 genes was derived from a methodology involving sequence conservation between humans and the pufferfish *Tetraodon nigroviridis*.

A rule-based expert system called Otto has been developed for the identification and characterization of genes in the human genome, which annotates a gene in one of the two ways. (1) By a high-quality match to the sequence of a known gene and (2) By evaluating a broad spectrum of evidence for the determination of a gene. A different gene-prediction strategy has also been used after along with Otto. Based on these two annotations, the total number of genes has been predicted to be 39,114 as an upper limit. If the requirements for other supporting pieces of evidence are made more stringent, this number drops rapidly to 26,383. Chromosome 19 is having the highest gene density.

5.10.4.1 Human Gene Transcripts

Based on the RefSeq transcriptional analysis, the average span of a gene in the human genome was estimated to be ~27,894 nucleotides. The transcripts predicted by Otto are longer, with 7.8 exons, whereas those promoted from gene-prediction programs were predicted to have 3.7 exons. The largest number of exons in a transcript is 234 in the titin gene mRNA.

5.10.5 Genome Structure

5.10.5.1 Cytogenetic Mapping

The basic element of the genome structure is the banding pattern produced by Giemsa stain. Human chromosomal banding studies have revealed that 17–20% of the human chromosome complement consists of C-bands, or constitutive heterochromatin, which is highly polymorphic, consisting of different alpha satellite DNA families with various higher order repeat structures. Many chromosomes have complex intra and interchromosomal duplications in their pericentromeric regions. ~5% of the human genome sequence was identified as alpha satellite sequences, and the remaining 80% of the genome belongs to the euchromatic component, divided into G, R and T bands. These cytogenetic bands have been presumed to differ in their nucleotide composition and gene density, whose boundaries were not predicted. T bands are the most G1C and gene rich, G bands are G1C poor. The density of genes was also greater in the regions of high G1C than in regions of low G1C content. Higher proportions of genes are located in the G1C poor regions. Chromosomes 17, 19, and 22, which also have a disproportionate number of H3 containing bands, had the highest gene density and chromosomes, X, 4, 18, 13, and Y also have the fewest H3 bands with low gene density. Chromosome 15 with few H3 bands has particularly high gene density and chromosome 8 with high H3 banding, still has a low gene density.

5.10.5.2 Linkage Mapping

Linkage mapping of the human genome had provided the genetic basis for studying the inheritance of traits and also in the positional cloning of the required genes. It has been observed that the rate of recombination in females is greater than that in males, which is the main reason for the non-uniform mapping across the genome. One of the outcomes of the complete human genome sequence of the generation of an ultimate physical map and to completely analyze it by corresponding with linkage and cytogenetic maps, which would narrow down the loop between the mapping and sequencing phases of the genome project. The location of the markers was mapped, constituting the Genethon linkage map to the genome and the rate of recombination

Table 5.7 Chromosome wide average rates of recombinations in the human males and females

Chromosome no.	Average recombinations in males	Average recombinations in females
1	1.12	1.76
2	0.78	1.40
3	0.86	1.30
4	0.67	1.40
5	0.67	1.43
6	0.71	1.67
7	1.16	1.21
8	0.73	1.36
9	0.99	1.66
10	1.03	1.51
11	0.72	1.32
12	0.76	1.55
13	0.55	1.19
14	0.98	1.63
15	0.94	1.56
16	1.00	2.32
17	0.87	1.83
18	1.37	2.24
19	0.97	1.75
20	0.89	2.15
21	1.26	1.90
22	1.10	2.08
X	NA	1.64
Y	NA	NA

was calculated and expressed as centimorgans (cM) per Mbp (Table 5.7). From this map, it was estimated that high rates of recombination occur in the telomeric regions of the chromosomes. A difference of 4.99 cM was identified between the lowest and the highest recombination rates with the largest difference in chromosome 16. These variations in the recombination rates across the human genome exceed the differences in recombination rates between males and females. The human genome is the hotspot of recombinations, with an average rate of fivefold or even more per every 1 kbp (Table 5.7).

5.10.5.3 CpG Islands

CpG islands are the stretches of unmethylated DNA with a higher frequency of CpG dinucleotides when compared with the entire genome. CpG islands are believed to preferentially occur at the transcriptional start of genes and it has been predicted that most of the housekeeping genes have CpG islands at their 5′ ends. Experimental evidence also indicates that CpG island methylation is correlated with gene inactivation and has been shown to be more important during gene imprinting and tissue-specific gene expression. Based on the experimental methods, it was estimated that

30,000–45,000 CpG islands are in the human genome and the number of CpG islands exclusively on chromosome 22 is 499.

5.10.5.4 Repetitive Elements

Around 35% of the human genome is full of repetitive sequences, with chromosome 19 is having the highest repeat sequence density of 57%.

5.10.6 Evolution of Human Genome

5.10.6.1 Retrotransposons

Retrotransposition of a processed mRNA in the human genome results in functional genes, called as intron less paralogs or inactivated genes or pseudogenes. A paralog occurs when a gene appears in more than one copy probably due to the duplication. Even though the existence of genes in both intron-containing and intron-less forms encoding a functionally similar proteins has been identified, cataloging those evolutionary events on the genome will provide a better understanding of the functional consequences leading to the gene-duplication events. Identification of conserved intronless paralogs in the mouse and other mammalian genomes provides the basis for obtaining the information regarding the evolution of these transpositions, leading to the gene loss and accretion in the mammalian radiation. Proteins corresponding to 901 single-exon genes in the genome were subjected to BLAST analysis and obtained 298 sequences of single to multi-exon correspondence. Among these 298 sequences, 97 might be the intronless paralogs of known genes.

Most of these sequences are flanked by the direct repeat sequences and have poly(A) tails, which are the characteristic features of retrotransposons found in the human genome. Retrotransposition from a single chromosome to multiple target chromosomes has also been identified in the human genome during the retrotransposition of ribosomal L21 gene from chromosome 13 onto chromosomes 1, 3, 4, 7, 10 and 14 respectively. Retrotransposition followed by the subsequent changes in the coding and noncoding regions will lead to a change in their function and these fragments, as evidenced in the 31-exon diacylglycerol kinase zeta gene present on chromosome 11 having an intronless paralog on chromosome 13.

5.10.6.2 Pseudogenes

A pseudogene is defined as a non-functional copy of a normal gene with a slight alteration leading to its non-expression. There are of two types of pseudogenes, (1) Processed pseudogenes that occur as a result of retrotransposition and (2) Duplicated or unprocessed pseudogenes arising from the segmental genome duplication. The general characters of the processed pseudogenes are (1) complete lack of intervening sequences, which are found in the counterparts of the functional genes, (2) presence of a poly(A) tract at the $3'$ end and (3) direct repeat sequences, which are flanking with the pseudogene sequence. The complete set of Otto predicted coding regions in the human genome were searched against the human genomic sequence by means of BLAST and has identified 2909 regions with an identity of 70% over the length of the transcripts, could be the processed pseudogenes. The exact number of pseudogenes in the human genome is still not known.

5.10.6.3 Gene Duplication and Large Scale Duplications

A graph-based algorithm called Lek was developed for grouping the predicted human protein set into protein families. The complete clusters that result from the Lek clustering provide one basis for comparing the role of whole-genome or chromosomal duplication in protein family expansion as opposed to other means, such as tandem duplication. Large scale duplications in the human genome were searched by using two independent methods; (1) Identification of highly conserved blocks of duplication and (2) Describing the most comprehensive method for identifying the interchromosomal block duplications. Application of the second method has identified a large number of duplicated chromosomal segments in all 24 chromosomes.

5.10.7 Sequence Variations in the Human Genome

Computational methods were used to identify single-nucleotide polymorphisms (SNPs) by comparison of the Celera sequence to other SNP resources. The SNP rate between two chromosomes was 1base per 1200–1500 bp. SNPs are distributed non-randomly throughout the human genome. Among them, only 1% will potentially hamper the protein function and coding regions. This results in an estimate that only thousands, not millions, of genetic variations, may contribute to the structural diversity of human proteins.

5.10.8 Analysis of Predicted Protein-Coding Genes

A preliminary analysis of the predicted human protein-coding genes was conducted by applying two separate methods for analyzing and classifying the molecular functions of 26,588 predicted proteins representing 26,383 gene predictions. (1) A method, based on the analysis of protein families from publicly available databases such as Pfam and Celera's Panther Classification (CPC). (2) Analysis of protein domains, with both these databases. With these two methods, functions of 12,809 proteins (~41%) could not be classified and termed as the proteins with unknown functions. 59% of the functions of the proteins were identified by classifying these proteins into broad classes based on their similarity with the known ones. In the analysis of 12,731 additional low confidence predicted genes, only 636 (5%) were assigned molecular functions by the automated methods. 212 these 636 predicted genes represented endogenous retroviral proteins and the majority of the 424 low confidence predicted genes with unknown function are not real genes. 12,095 genes appear to be unique among the genomes sequenced till date and many of them might represent false-positive gene predictions. The most common molecular functions

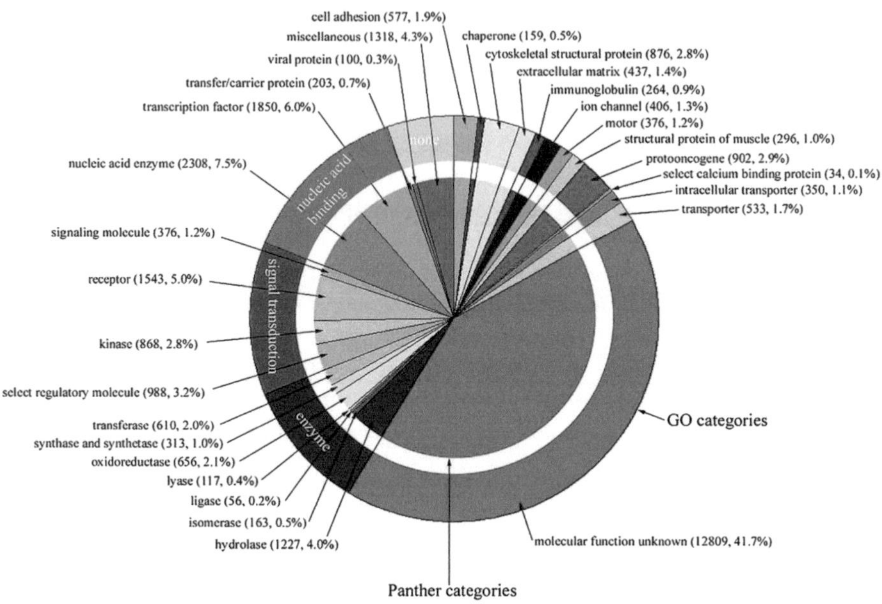

Fig. 5.2 Distribution of the molecular function 26,383 genes in the human genome

predicted include the transcription factors involved in nucleic acid metabolism, receptors, kinases, and hydrolases. Proteins that are members of proto-oncogene families, GTP-binding proteins and cell cycle regulators, proteins that modulate the activity of kinases, G proteins, and phosphatases are also identified (Fig. 5.2).

5.10.9 Evolutionary Studies and Comparative Genomics

With most of the genome-sequencing projects of model organisms have been completed, reasonable comparative information is available to begin the analysis of the evolution of the human genome. The genomes of *S. cerevisiae*, *C. elegans*, *D. melanogaster,* and *A. thaliana*, will provide a diverse background for the comparative genomics. Orthologs that are conserved between human and *D. melanogaster*, between human and *C. elegans* were identified for studying the core functions that appear to be common across these animals. During the pairwise comparison of human-*D.melanogaster*, human-*C. elegans* genomes, 2758 human-*D. melanogaster* orthologs, 2031 human- *C. elegans* orthologs and 1523 common genes between these sets were identified. These 1523 human proteins are considered as evolutionarily conserved as they have strict orthologs in both *D. melanogaster* and *C. elegans*.

The distribution of the functions of the evolutionarily conserved proteins into several categories such as nucleic acid enzymes, transcriptional machinery comprising of DNA/RNA methyltransferases, DNA/RNA polymerases, helicases, DNA ligases, DNA- and RNA-processing factors, nucleases, and ribosomal proteins. The transcriptional and translational machinery is very well conserved over evolution from bacteria through to the most complex eukaryotes. Many ribonucleoproteins involved in RNA splicing also appear to be conserved among the animals. Enzymes involved in the intermediary metabolism are also conserved, with an exception of hydrolase. Regulatory molecules such as GTPases and cell cycle regulators also well conserved, along with the genes involved in protein transport and trafficking. Chaperones involved in the protein folding and heat-shock response, particularly the DNAJ family, and HSP60, HSP70 and HSP90 families are also conserved. However, this analysis does not provide a complete estimation of conservation across these genomes, as paralogous duplication makes the determination of true orthologs difficult within the members of conserved protein families.

5.10.10 80% of the Human Genome has an Active Function

In 2001, the scientists who completed human genome project thought that 20,000 genes comprising of ~1% of the human genome code for the proteins and the remaining genome are considered as the Junk DNA. ENCODE project (Encyclopedia of DNA Elements), started in 2003, where the Human Genome Project has left has shown far from what have expected in their final results about the junk DNA in the human genome. As initially thought, The junk DNA is not totally a wasteland. The vast desert regions of non-coding DNA in the human genome has been identified to be comprised of a vast amount of features, contributing to the gene regulation and each and every gene in the genome must be using different permutations and combinations of these features for generating their own way of expression. According to the ENCODE project, ~80.4% of the human genome is active with some role or function such as promoters, enhancers, regulatory factors, that recognize the specific DNA sequences in these regulatory regions, thereby creating switches for turning the genes on and off. These switches are the regulatory DNA and the genes are not having any function in the absence of these switches. Interestingly, these switches might be far from the genes making it difficult to determine their relationship. Many studies have revealed that most of the genes in the human genome are controlled by more than one switch. The useless non-coding regions of the genome produce non-coding RNA, which play a role in the activation and silencing of protein-coding genes.

5.11 Conclusions

Nature has a system to get rid of useless body parts or even useless DNA. If the DNA is present in the cell, it must be with a specific function, which our present technology is not permitting us to know it. Finding the genetic basis for the disease resistance has been one of the objective and outputs of the genome sequencing. It has been identified that some genes associated with the disease resistance are present in the non-coding regions of the genome and, this presence is relatively more common in the human population. 76% of the disease-associated variants in the non-coding regions are linked to the regulatory DNA, shedding new light on the mechanisms involved in turning these genes on.

Genome Sequencing, Assembly, and Annotation

6

Abstract

Strategies and the Next Generation Technologies applied for the sequencing and assembly of genomes are discussed in this chapter. Recent methodologies adapted for sequencing the fungal, plant, animal, degraded and ancient fossil genomes are provided with appropriate examples. Information about the transcriptome sequencing and the role of Mass spectrometers in DNA sequencing and their adaptations for increasing the accuracy are included.

6.1 Introduction

The genome comprises the entire genetic information of an organism encoded either in DNA or RNA consisting of both coding and non-coding regions. A genome is composed of double-helical DNA in higher organisms, single circular chains of DNA in bacteria, viruses, and organelle such as chloroplast, mitochondria, linear chains of RNA in some viruses and transposable elements. In eukaryotes, the entire genome is packed into copies of chromosomes and the copy number varies from two in diploids to four in tetraploids. The term sequencing refers to determine the order of the nucleotides in the DNA sequencing and amino acids in protein sequencing. Genome sequencing is a technique that determines the complete DNA sequence of an organism's genome at a time, including the nuclear chromosomal DNA, mitochondrial DNA and in the case of plants, chloroplast DNA also. Genome sequencing has created a revolution in biology research by further opening the doors for studying the molecular processes involved in the complete cellular systems, leading to the concept of systems biology. Genome sequencing has also laid the foundation to the 'omics' technologies such as proteomics and transcriptomics. All this was possible because of the availability of advanced nucleic acid sequencing techniques. Study of a genome will not only yield the information regarding the total number of genes in an organism, but also about the mechanisms that could have led to the production of great variety of genomes that are existing today by comparing with different genomes for their size, codon usage bias,

K. V. Chaitanya, *Genome and Genomics*,
https://doi.org/10.1007/978-981-15-0702-1_6

GC content, repeats (STR), duplication of genes etc. Variations in the genetic information especially regarding the traits related to the diseases require comparisons between individuals, which makes a genome more complex.

The sequential order of nucleotides in the polymer chains of DNA and mRNA ultimately contains the information about the heredity and other important properties belonging to the life processes of an organism. Therefore, it is very important to determine and infer the sequence of such chains for better understanding of the information present inside them. Sequencing of the DNA is determining the order of four nucleotides A, T, G and C that makes up the DNA molecule. This sequential order of nucleotides in DNA contains information for coding, non-coding and regulatory regions, which can be used for the identification of a diseased locus, mutations and regulatory regions that turn the genes on and off. This chapter deals with the techniques developed for sequencing the DNA and its annotation.

6.2 Sequencing Technologies and Genome Sequencing

6.2.1 First-Generation DNA Sequencing

Watson and Crick solved the DNA double helix structure in 1953 based on the DNA X-Ray crystallography data produced by Franklin and Wilkins and has contributed to the framework of DNA replication. But, reading the nucleotides in a DNA sequence was done for the first time by Robert Holley and colleagues in 1965, produced the first whole nucleic acid sequence of *Saccharomyces cerevisiae* alanine tRNA. Parallelly, Fred Sanger and his colleagues developed a technique for the detection of radiolabelled and partially digested fragments after 2-D fractionation, which allowed the reading of ribosomal and transfer RNA sequences.

The next major change, which has made a huge impact on the DNA sequencing, was the replacement of 2-D fractionation by a single separation of polynucleotides using polyacrylamide gel electrophoresis, which has also provided a better resolving power. This technique was applied by combining the protocols of Alan Coulson and Sanger's 'plus and minus' system with Allan Maxam and Walter Gilbert's chemical cleavage method. The plus and minus technique involved the usage of DNA polymerase enzyme for incorporating the radiolabelled nucleotides using a primer, before performing 2-second polymerization reactions. A 'plus' reaction, consisting of only a single type of nucleotide terminating all extensions of that base and a 'minus' reaction, uses three produces of the sequence up to the position before the next missing nucleotide. By running the reaction products on a polyacrylamide gel and comparing between the eight lanes, one is able to infer the position of nucleotides in the covered sequence. Maxam and Gilbert's technique was little different in its approach with utilizing the chemicals for breaking the radiolabelled DNA at the specific base instead of using DNA polymerase and the length of cleaved fragments and the position of specific nucleotides can be determined after separating on a polyacrylamide gel. This technique was the first to be widely adopted and is considered as the real birth of 'first-generation' DNA sequencing.

The major breakthrough work, which has completely revolutionised the DNA sequencing technology was developed in 1977, as Sanger's 'chain-termination' or dideoxy technique, which makes the use of radiolabelled Dideoxynucleotides (ddNTPs) lacking the 3′ hydroxyl group required for the extension of DNA chains and therefore cannot form a bond with the 5′ phosphate of the next dNTP. Such a reaction has resulted in the DNA strands of each possible length being produced, as the dideoxynucleotides get randomly incorporated as the strand extends, halting further progression. By performing four parallel reactions containing each individual ddNTP base and running the results on four lanes of a polyacrylamide gel, one is able to use autoradiography to infer the nucleotide sequence in the original template, as there will be a radioactive band in the corresponding lane at that position of the gel. After the development of chain-termination reaction, a number of improvements have been made to Sanger sequencing in the following years, such as replacement of phosphor or tritium radiolabelling with fluorometric based detection, which has allowed the reaction to occur in one vessel instead of four and improvised the detection through capillary-based electrophoresis. Both of these improvements have significantly contributed to the development of automated DNA sequencing machines and subsequently the first set of commercial DNA sequencing machines used for sequencing the complex genomes.

First generation DNA sequencers are capable of producing the sequence reads only up to the length of 1 Kb. For sequencing, the DNA fragments of bigger size, shotgun sequencing system have been developed, by separately cloning and sequencing the overlapping the DNA fragments and assembling them into a single contiguous sequence with computational aid. The invention of has PCR helped in the generation of pure and high quality DNA samples for accurate sequencing. The invention of polymerases with high significant chemical moieties and modified dNTPs for the sequencing has led to the invention of Dideoxy sequencers such as ABI PRISM of Applied Biosystems. All these inventions have led to the development of simultaneous sequencing of hundreds of samples, which was applied in the sequencing of the mammoth human genome.

6.2.2 Second Generation DNA Sequencing

Development of Dideoxy sequencing machines has triggered the invention of next generation DNA pyrosequencing machines, which are markedly different from the existing DNA sequencers by possessing the luminescence for measuring the pyrophosphate synthesized. This approach is being used to infer the sequence using two enzyme processes, ATP sulfurylase for converting pyrophosphate into ATP, which is used as a substrate for luciferase, producing light proportional to the amount of pyrophosphate. This pyrosequencing technique pioneered by Pål Nyrén et al. (1993) was further improvised by attaching the DNA to paramagnetic beads and enzymatically degrading the unincorporated dNTPs, which can avoid lengthy washing steps. This pyrosequencing was licensed to a biotechnology company named 454 life sciences, where it emerged as a successful commercial next generation sequencing technology. These 454 pyrosequencing machines (which were later purchased by Roche) has permitted a mass parallelization of sequencing reactions, by enhancing the amount of DNA that can be sequenced in any single run. DNA libraries were first attached to beads *via* adapter sequences, undergoes a water-in-oil emulsion PCR to coat each bead in a clonal DNA population, resulting in one DNA molecule on one bead, gets amplifies in its own droplet in the emulsion. These DNA-coated beads were further washed on a picoliter reaction plate that fits one bead per well. Pyrosequencing will be performed using bead-linked enzymes and dNTPs and pyrophosphate release is measured using a charged coupled device (CCD) sensor connected to the picoliter reaction plate beneath the wells. This machine can produce the reads of 400–500 base bp length, for more than a million wells containing DNA coated beads. This parallelization has massively increased the yield of sequencing efforts by orders of magnitudes allowing the researchers to completely sequence a major eukaryotic genome at a far quicker and economically cheaper price. The first second generation 454 high-throughput sequencing (HTS) machine available to the consumers was GS 20, which was later modified and released as 454 GS FLX with more wells in the pico-titer plate, offering a more number of reads with a better quality data and high-resolution imaging.

Following the success of 454 GS FLX, a number of parallel sequencing machines sprung up. Among them, the most important one is Solexa, acquired by Illumina in 2007 for $650 million. This technology has the capability of sequencing a human genome for just $1000 in a day, which was a million fold improvement on the state of art. This technology is being used for population scale human genome sequencing as well as large scale clinical sequencing such as NHS 100,000 genomes project. The DNA sample for the sequencing is prepared into a sequencing library by fragmenting into pieces of ~200 bases length. Custom adapters are added to each end and the library has flowed across a solid surface called flow cell surface, which was designed to provide stability of surface bound DNA as well as access to the enzymes and low non-specific binding of fluorescently labeled nucleotides. The DNA sample will be amplified by a solid phase "bridge amplification" PCR, which generates approximately one million copies of each template in tight physical clusters on the flow cell surface. Solexa sequencing technology can achieve densities of up to ten million single molecule clusters per square centimeter.

The Solexa sequencing is similar to the Sanger's sequencing, with the only difference of using modified dNTPs containing a terminator, which blocks further polymerization. As a result, only a single base can be added by a polymerase enzyme to each growing DNA copy strand. The sequencing reaction is conducted simultaneously on a very large number of different template molecules spread out on a solid surface. The terminator also contains a fluorescent label, which can be detected by a camera. Only a single fluorescent color is used so that each of the four bases must be added in a separate cycle of DNA synthesis and imaging. Following the addition of the four dNTPs to the templates, the images are recorded and the terminators are removed. This chemistry is called "reversible terminators". Finally, another four cycles of dNTP additions are initiated. Since single bases are added to all templates in a uniform fashion, the sequencing process produces a set of DNA sequence reads of uniform length. Although the fluorescent imaging system used in Illumina sequencers is not sensitive enough to detect the signal from a single template molecule, the major innovation of the Illumina method is the amplification of template molecules on a solid surface. The Solexa sequencer comprises millions of short sequence reads, whose deep sampling of more than ten fold coverage is required for generating a consensus and providing high confidence in determining genetic differences. Such differences are identified by comparison of sample sequence reads with a reference. Deep sampling allows the usage of weighted statistical analysis, similar to conventional methods for identifying homozygotes and heterozygotes and also to distinguish the sequencing errors. Each raw read base has an assigned quality score so that the software can apply a weighting factor in calling differences and generating confidence scores.

Similar to Solexa, a number of other sequencing machines have come into the market with variable experimental procedures. The third major sequencing system after 454 and Solexa was Sequencing by Oligonucleotide Ligation and Detection (SOLiD) by Applied Biosystems (became Life Technologies after merging with Invitrogen). SOLiD sequencing is based on the DNA ligase that is widely used in biotechnology for the ligation of DNA ends. A collection of targeted DNA sample fragments are first ligated on to a magnetic bead with universal P1 adapter and the beads containing the single copies of the same DNA molecules are deposited on to a glass slide. A single DNA fragment bound to the adapter is hybridized with a primer of length N and is amplified using an emulsion PCR. Beads are exposed to a library of 8mer probes which have different fluorescent dye at the 5′ end and a hydroxyl group at the 3′ end. First and second bases are complementary to the nucleotides to be sequenced, bases 3–5 are degenerate and bases 6–8 are inosine bases. Only a complementary probe can hybridize to the target sequence, adjacent to the primer. DNA ligase joins the 8mer probe to the primer. A phosphorothioate linkage between bases 5 and 6 allows the fluorescent dye to be cleaved from the fragment, which allows fluorescence to be measured (four different fluorescent dyes with different emission spectra are used) and also generates a 5′-phosphate group, which can undergo further ligation. The sequencing process is continued with another primer and a sequence reading length of about 35 bases. Sequences are determined in parallel for more than 50 million bead clusters, resulting in a very high throughput of the order of Gb per run. The new SOLiD instrument is capable of producing 30 Gb of sequence in an 8-day run that offers 99.94% accuracy. The main disadvantage of this system is its read lengths, which are of short length and depth, making it unsuitable for many applications.

DNA Nano ball sequencing is another high throughput sequencing technology for determining the full length sequence of an organism's genome using ligation. This method is based on rolling circle replication of small genomic DNA fragments to the DNA Nano balls. Fluorescent probes bound to the complementary DNA are ligated to anchor sequences, which are bound to the known sequences on the DNA template. Unchained sequencing by ligation is then used to determine the nucleotide sequence. This method of DNA sequencing allows large numbers of DNA Nano balls to be sequenced per run, at low reagent costs compared to other next generation sequencing platforms. The only drawback with this technology is the difficulty to produce short sequences of DNA from each DNA Nano ball, which makes mapping of the short reads to a reference genome difficult. This technology has been used for multiple genome sequencing projects and is scheduled to be used for more.

Another notable DNA sequencing technology Ion Torrent Semiconductor Sequencing is the first post-light sequencing technology, which uses neither fluorescence nor luminescence by the protons released. In a manner analogous to 454 sequencing, beads bearing clonal populations of DNA fragments produced by emPCR are washed over a pico well plate. The nucleotide incorporation is measured by the difference in pH caused by the release of protons (H^+ ions) during polymerization, using the complementary metal-oxide-semiconductor (CMOS) technology. This technology allows rapid sequencing during the actual detection phase, which was less with other pyrosequencing technologies.

Massively parallel signature sequencing (MPSS) was developed by Sydney Brenner and Sam Eletr at Lynx Therapeutics in the 1990s. MPSS is used for identifying and quantifying of mRNA transcripts similar to the serial analysis of gene expression (SAGE). MPSS is another bead-based technique, used as a complex approach of adapter ligation and adapter decoding, reading the sequence in the increments of four nucleotides. Here, mRNA is reverse transcribed into cDNA and is amplified in an emulsion PCR. A sequence signature of ~16–20 bp is determined from all the beads in parallel. Each signature sequence is cloned onto microbeads and then arrayed in a flow cell for sequencing and quantification. This technology is mainly used for measuring absolute gene expression levels in an organism. The main drawback of this technology is its complexity. MPSS was only performed in-house at Lynx Therapeutics and no DNA sequencing machines were sold to independent laboratories.

Developed by George M Church of Harvard, Polony sequencing was one of the very first second generation sequencing systems used for sequencing a full length genomes in 2005. It is available in a set of in vitro paired-tag library with emulsion PCR, an automated microscope and a ligation-based sequencing. Emulsion PCR isolates individual DNA molecules along with primer-coated beads in aqueous droplets within an oil phase. PCR then coats each bead with the clonal copies of the DNA molecule followed by the immobilization for later sequencing. Using this technique, the cost of sequencing of an *E. coli* genome with an accuracy of >99.9999% was ~1/9 that of Sanger sequencing.

6.2.3 Third Generation DNA Sequencing

The main criteria, which defines the characteristic features for third generation sequencing technologies are single molecule sequencing (SMS), real-time sequencing and simple divergence from previous technologies. Third generation sequencing technologies are those capable of sequencing single molecules, negating the requirement for DNA amplification shared by all previous technologies. The first SMS technology was developed by Stephen Quake, later commercialized by Helicos

BioSciences. DNA templates become attached to a planar surface, and then fluorescent reversible terminator dNTPs, also called as virtual terminators are washed over one base time and imaged before cleavage and cycling the next base over. This was the first technology to allow sequencing of non-amplified DNA, thus avoiding all associated biases and errors.

One of the most widely used third-generation sequencing technology is probably the single molecule real-time (SMRT) sequencing, also referred as PAC Bio sequencing platform from Pacific Biosciences offers longer read lengths than the lengths provided by second-generation sequencing (SGS) technologies, making it highly suitable for unsolved problems in genome, transcriptome, and epigenetics. Here, the DNA polymerization occurs in arrays of microfabricated nanostructures called zero-mode waveguides (ZMWs), which are tiny holes on a metallic film covering a chip. The sequencing is performed with the use of unmodified polymerase attached to the ZMW bottom and fluorescently labeled nucleotides flowing freely in the solution. The wells are constructed in such a way that only the fluorescence occurring by the bottom of the well can be detected. The fluorescent label is detached from the nucleotide at its incorporation into the DNA strand, leaving an unmodified DNA strand. According to Pacific Biosciences, the SMRT technology developer, this methodology allows the detection of nucleotide modifications such as cytosine methylation. This process can sequence single molecules in a very short amount of time. This approach allows reads of 20,000 nucleotides or more, with an average read length of 10 kb useful for the de novo genome assemblies.

Development of a nanopore sequencing technology and utilization of nanopores for the detection and quantification of all biomolecules is perhaps the most anticipated area of third-generation DNA sequencing technology. The potential of nanopore sequencing has been established well before the emergence of second-generation sequencing technology, during the demonstration of a single-stranded RNA or DNA transport across a lipid bilayer through large α-hemolysin ion channels by electrophoresis. The nucleic acid passage through the hemolysin channel blocks ion flow, decreasing the current for a length of time proportional to the length of the nucleic acid. The potential of using solid-state technology for generating suitable nanopores also provides the ability to sequence double stranded DNA molecules.

Oxford Nanopore Technologies is the first company to provide the nanopore sequencers, GridION and small, mobile phone sized USB device MinION, which was first released in 2014. Nanopore platform sequencers are meant to produce incredibly long, non-amplified sequence data faster than previously possible at a far cheaper price. MinION sequencers have been used to generate bacterial genome reference sequences and targeted amplicons. They are also being used for generating scaffolds to map short reads made by Illumina with high read depth and accuracy for combining the ultra-long read length of the nanopore technology. They can even be deployed in the field for sequencing of the pathogens as proved by Joshua Quick and Nicholas Loman earlier this year when they sequenced Ebola viruses in Guinea 2 days after sample collection. Nanopore sequencers are believed to revolutionize not only the composition of the data, but when and where it can be produced, and by whom.

6.2.4 Sequencing of Fungal Genomes

Blumeria graminis is one of the most significant pathogens of cereal crops, which causes the powdery mildew, reducing the crop yields by 40%. In order to support the developments regarding the crop protection and disease resistance, the genome of this pathogen has been completely sequenced. One of the basic experimental challenges posed by *Blumeria graminis* is that it can only be grown on its host. Thus, the supply of biological material was very limited and was contaminated by host tissues. The advent of high-throughput DNA sequencing platforms has revolutionized the depth at which transcriptomes can be analyzed and the development of robust and efficient protocols for generating cDNA that can be introduced directly in the sequencing pipeline is of huge importance. One of the requirements of genome annotation is a collection of full-length cDNA sequences from as many diverse stages of the organism as possible to use in *ab initio* gene discovery by programs such as EuGene and FGENESH. Full length cDNA was synthesized and was sequenced using 454 pyrosequencing (one run on GS-FLX), which yielded 247,306 reads, comprising a total of 50.8 M bases corresponding to an average read length of 205 bases. The data were assembled (using MIRA; www.chevreux.org/projects_mira.html), clustered, and combined with ESTs available in public repositories. The number of unique *B. graminis* genes identified by cDNA sequencing was increased from 4584 to 7727. When the cDNA sequences were compared to genomic DNA, it became evident that there was marked heterogeneity in the RNA populations of *B. graminis*. Some bases were added in the transcript at a considerable distance from the beginning of the mature transcript. The majority of the additions include adenosines, but thymine, cytosine, and guanosine were also found. Similar observations were reported in the sequencing of other related fungi *Magnaporthe oryzae* (syn. grisea). Several high-throughput techniques are available for sequencing of genes in a genome, but their function can be inferred only on the basis of sequence motifs or sequence similarity. The immediate challenge after the collection of genome sequences is to investigate how a genome and an interactome (interactions of protein-protein or protein-DNA) determine the phenome of an organism. To study these interactions, technologies for genome-wide analysis of gene expression such as microarray hybridization were used. However, the complexity of higher eukaryotic genomes makes interactome analysis more complicated and difficult.

One of the approaches to counteract these difficulties is synthetic genetic array (SGA) analysis. Out of the 6000 genes of the yeast genome, 5000 genes have shown to be nonessential in a genome-wide single-gene knockout project, but the double mutants of these non-essential genes produced lethal phenotypes. SGA analysis allows the identification of genetic interactions because if a double mutant has a synthetic lethal phenotype the two corresponding wild-type genes often have a functional relationship. Among the 5000 non-essential genes, the function of 132 genes was tested by making double mutants with each other, it was determined that each gene has an average of 30 synthetic genetic interactions and was hypothesized that there can be 100,000 such interactions in the yeast genetic network. Using cluster analysis of SGA results, the function of an unknown gene can also be predicted on the basis of the genes with which it is connected in the SGA network. SGA analysis gives a much more complex network of the yeast interactome than previously reported.

6.2.5 Sequencing of Plant Genomes

The complexity and size of the plant genomes have limited the ability to obtain their comprehensive genetic information. After the sequence of the *Arabidopsis thaliana* genome, the complete rice genome, poplar genome, and maize have also been completed. The National Center for Biotechnology Information Entrez Genome Projects website reports that sequencing of several more plant genomes is in progress. The first wave of plant genome sequencing has passed, in which all these genomes were sequenced using the traditional approaches, constructing the sequence libraries from individual segments of the genome and are sequenced by Sanger's method. A whole genome shotgun (WGS) strategy, with improved assembly algorithms, has been used for several recent plant genomes, in which the sequencing libraries are made directly from genomic DNA. We are now entering a new era in plant genomics research in which large and complex plant genomes such as sugarcane, wheat, and barley are difficult to decode because of historic duplication events, the amplification of families of transposable elements, and polyploidization conditions. The development of next-generation DNA sequencing technologies has revolutionized plant genome research and stimulated the analysis and sequencing of large plant genomes. These novel DNA sequencing technologies provide ultra-high throughput at a substantially lower cost with huge increases in sequencing throughput and, perhaps more importantly, the ability to avoid the handling of individual clones from shotgun libraries. There are currently four commercially available high throughput technologies, which are used in sequencing the plant genomes; 454 Life Sciences, Solexa, ABI SOLID and Helicos Biosciences. These technologies can be grouped into two classes based on the lengths of the sequence reads produced. Solexa, ABI SOLID, and Helicos all produce very short reads in very large quantities, while the 454 platforms can produce a more moderate amount of sequence, but with much longer read lengths. Development of novel methods for sequencing assemblies of large genomes from short-read sequences have solved this problem and successfully assembled the cucumber genome of size 367 Mb and the genome sequences of Chinese cabbage (500 Mb) potato (830 Mb) are underway. The development of these assembly methods creates new opportunities for building reference sequences and carrying out accurate analyses of unexplored genomes in a cost effective way. Approaches are being made for the identification of genes and promoters by genome analysis of key species and wild crops, which reduces the complexity of genome assembly to establish reference genomes for the major crops. Genes and markers with important agricultural implications can be discovered and further applied in breeding programs for crop improvement.

High throughput DNA sequencing technology is capable of identifying new micro RNAs (miRNAs) in plant species. Several miRNA genes in the plant kingdom are found to be ancient, with the conservation extending mainly between angiosperms and the mosses, whereas many other miRNAs are more recently evolved. Solexa sequencing approach was used to identify eight new micro RNAs in one of the model legume species, *Medicago truncatula*. Using deep sequencing and computational methods, 48 non-conserved miRNA families, nearly all of them representing single genes, were identified in *Arabidopsis thaliana*. With increasing information on plant genomes, comparative analyses of genomes have become an important aspect in the identification of genes and their function. Genome comparisons are more useful in the understanding of plant biology and evolution specifically at the intra kingdom levels, but the plant species for which genome sequences are available span only 200 million years of land plant evolution. Sequencing of the pteridophytes genome like *Selaginella moellendorffii* and moss genome *Physcometriella patens* will add another 200 million years of evolutionary history to comparative plant genomics.

6.2.6 Sequencing of Animal Genomes

High-throughput sequencing (HTS) technologies such as Illumina/Solexa, AB SOLiD, and nanopore have enabled the sequence a full human genome considerably at a lower cost of at least 200-fold less than the regular methods. Selective genome sequencing and re-sequencing of genomes are one of the regularly used approaches for obtaining the contrasting results during biomedical research and clinical practice, for which the underlying cost remains prohibitive. To reduce the cost, in most of the genome experiments specific portions of the genome will be sequenced. Several innovative technologies like Selector, Gene-Collector MegaPlex PCR, Multiplex Exon Capture technique and Sequence-capture Array have recently been developed for selective genomic sequencing. The first three are suitable for selecting genomic loci on the order of a few hundred. In contrast, the Multiplex Exon Capture technology and Sequence-Capture Array strategy are more suitable for applications at the genome scale because tens of thousands of target regions can be simultaneously captured. As far as the performance is concerned, Sequence-Capture Array (65–77%) tend to be less specific than medium-throughput strategies (in general >90%) because of increase in the assay complexity with an exception of Gene-Collector strategy which has a specificity of only 58%. Strikingly, the Sequence-capture Array strategy showed the best, enrichment distribution up to 90% bases within a five fold range due to the simplicity of the overall procedure with minimum amplification steps involved. On the other hand, low uniformity was observed for the Multiplex Exon Capture approach, which is reflected in its high dropout rate of 72%.

6.2.7 Sequencing of Degraded and Ancient Fossil DNA

The ancient DNA isolated from the fossils is always less and is heavily fragmented, chemically modified and contaminated with environmental DNA making the sequencing difficult. Development of HTS technologies has bought solutions to some of these problems, making a significant contribution to the emerging field of paleogenomics, enabling the sequencing of the genomes from extinct organisms. Recently, high throughput sequencing techniques have been applied to sequence the fossil mitochondrial DNA (mtDNA) genome. Using 454 sequencings, mtDNA genome has been recovered from the bone of Neandertal man excavated from Vindija Cave, Croatia.

6.2.8 Structural Variations in the Drosophila Genome

The complete genome sequences of 10 species of Drosophila have been published in 2007, to complement with already published *Drosophila melanogaster* and *D. pseudoobscura*. One common thing in comparative genomes of these 12 drosophila species is structural variation. Copy number variation (CNV), is a type of structural variation includes deletions, duplications, insertions and genomic rearrangements which will affect the number of occurrences of a specific DNA sequence present in the genome. CNV is known to occur extensively in the Drosophila genome with functionally significant consequences. Until recently, comparative genomic hybridization with whole-genome tiling arrays (array-CGH) was the primary method for characterizing CNVs. However, several limitations for this platform reduce its efficacy and efficiency and make this a costly affair. High throughput sequencing provides an ideal and cost-effective platform for CNV characterization by overcoming the inherent limitations of cross-hybridization and provides a digital count of sequence representation without prior knowledge or design work. Using HTS, deletions in three deficiency fly stocks were successfully characterized and the associated breakpoints were accurately determined using the Illumina genome analyzer. Sequence reads obtained were mapped to the Drosophila reference genome release 5.1 using the vendor-provided Eland pipeline.

6.3 Whole Genome Sequencing

6.3.1 Major Strategies for Whole Genome Sequencing

Two different strategies are followed for the genome sequencing: (1) clone-by-clone approach and (2) whole-genome shotgun sequencing approach.

1. Clone by Clone Method

Clone-by-clone also called map-based approach has been successfully used for generating the complete genome sequence of yeast and *C. elegans*. This method is initiated by the preparation of genomic DNA, which were cut into 150,000 bp long fragments with restriction enzymes. Each of these fragments was cloned into a Bacterial Artificial Chromosome (BAC), fingerprinted to give a unique identification tag that determines the order of the fragments. Fingerprinting involves cutting each BAC fragment with a single restriction enzyme and finding common sequence landmarks among the overlapping fragments that determine the location of each BAC along the chromosome, followed by overlapping every 100,000 bp of BACs with markers to form a map of each chromosome. Each BAC is broken randomly into 1500 bp pieces placed randomly into M-13 phage to make a collection known as M-13 library. The whole library is sequenced and the overlapping sequences are assembled by computer into a contiguous sequence. The complete genome sequence is obtained by integrating the sequences of the individual clone inserts into the contiguous sequence.

2. Whole Genome Shotgun Sequencing (WGS)
Whole genome shotgun (WGS) was proposed for human genome sequencing, as an alternative to the clone-by-clone strategy. This approach was successfully applied for sequencing λ phage DNA (~49 Kbp), which was low cost efficient and takes less time, compared to that of the clone-by-clone method. This approach of genome sequencing was later applied to other small genomes such as pox virus, *Haemophilus influenzae,* mainly due to the fact that smaller genomes do not have repeated sequences and less amount of non-coding regions, whose assembling will not cause problems. But, WGS was successfully applied for sequencing the genome of *D. melanogaster* to test its adaptability for sequencing the huge human genome, which was initially doubted by many scientists.

The first step in whole genome shotgun sequencing is the generation of small overlapping random fragments, which covers the genome of an organism. The highly purified genomic DNA was randomly fragmented two different sizes; long insert (2–25 kbp) and short insert (0.5–1.2 kbp). The long inserts are cloned in the phage or cosmid vectors and the short inserts are cloned into plasmid vectors. Clones from the short insert library are sequenced from both the ends. Since the size of the short insert is small ($<$1.2 kbp), sequencing of this insert from both ends gives read lengths, which overlap. This is known as a paired-end sequence or mate pair sequencing. Large numbers of clones from the short insert library are sequenced, which results in huge nucleotide sequence data. Each of the genomes will be covered more than 10 times. Use of both short insert and long insert clones increases the efficiency of sequence assembly.

6.4 Genome Sequencing by Mass Spectrometry

One of the primary goals of a genome project is to minimize the time and cost of the sequencing by taking the support of the technical advances. Mass spectrometry is one of the promising technologies, which increases the speed and minimizes the cost. It is an important tool for correlating proteins to their respective genes. Mass spectrometry plays a pivotal and long-standing role in the protein sequence analysis and its structure prediction. Mass spectrometer generates multiple ions from the sample, separates them according to their specific mass by charge ratio (m/z) and records the relative abundance of each generated ion in the form of a mass spectrum, displayed as a plot of ion abundance versus mass-to-charge ratio. Recent advances in the mass spectrometry have led to the emergence of two mass spectrometers; MALDI-TOF and ESI-TANDEM.

DNA is a biopolymer consisting of deoxyribonucleotides, each containing a sugar, phosphate group and one of the four bases, two purine derivatives A & G, two pyrimidine derivatives T & C. For the mass spectrometry to be more suitable for sequencing DNA, the ionization must be capable of generating the DNA ions containing several hundred bases with a mass resolution and mass accuracy adequate to identify the peaks corresponding to a difference of one base. Before the invention of MALDI and ESI, this was far from being achievable, with mass spectrometry of nucleotides was limited to the techniques such as Fast Atom Bombardment (FAB, a ionization technique, where the sample material is mixed with a non-volatile chemical protection environment called matrix and is bombarded in vacuum with a high energy beams of 4000–10,000 Electron Volts. Bombardment produces protonated and deprotonated molecules) and Plasma Desorption Mass Spectrometry (PDMS, A technique used for the analysis of non-volatile molecules, which are exposed to a 100,000 Electron Volts beam of californium, resulting in the ejection of ions and shockwaves from the matrix surface). Present day, three strategies are being followed for sequencing.

1. Mass analysis of Sanger Sequencing Reaction Products
2. DNA Ladder Sequencing
3. Gas-phase Fragmentation

6.4.1 Mass Analysis of Sanger Sequencing Reaction Products

This is one of the most simple and straight forward approaches, involves the replacement of gel electrophoresis determination with Mass spectrometry. Mass spectrometers have less time and low cost advantages over the gels. A conventional gel may take several hours, whereas the mass spectrum may be obtained in a span of less than a minute. The sequence readout of the mass spectrum is very much similar to that of readout from the gel. Instead of one lane for each of the four dideoxy termination reaction solutions, four mass spectra are obtained for each of the A, G, T, and C bases. Although it is simple to envision, it is less simple to put into the operation with current mass spectrometer capabilities. Each sequencing solution contains a mixture of DNA strands, several hundred bases in length. The mass spectrometer must be able to generate ions from this mixture at a resolution sufficient to distinguish between two bases in strands of at least 500 bp in length, which corresponds to a mass resolution of 500 at a mass range of 100,000 Da.

An excellent illustration of a recent work done on the DNA sequencing of the mass spectrometer is illustrated here. Four mixtures of oligonucleotides between 17–40 bases in length were synthesized to stimulate the products of Sanger's sequencing reaction. The oligos are free from impurities such as buffers or Sanger's reagents. Calibration was performed using their oligonucleotides as internal standards. Picomole concentration of each sample was used, which was at least 100 times lesser than the amount typically used for the Sanger sequencing. If it was found that the signal for the high mass molecules was weaker in the mixture, their components were analyzed separately. It was found that a mass resolution equal to the size of 100–200 bp was sufficient for the separation of the mixture component. The sequence information was read to 19 bases past the primer, with a picomole concentration of the sample used for the reaction. These results illustrate that MALDI must be improved in detection limit and mass range to be a legitimate candidate for the replacement of gels. An important application, which was possible using the current technology is the detection of a mutant form of the gene responsible for cystic fibrosis, an autosomal recessive inherent disorder passed through the generations by causing thick sticky mucus to build up in the lungs, digestive tract, pancreas and other areas of the body. It is one of the most common lung diseases in young children and adults, causing male infertility and death due to the mutation leading to a deletion of three bases that results in loss of the amino acid phenylalanine in the cystic fibrosis transmembrane conductance regulatory protein. A method was developed using Mass spectrometry that produces 59 bp fragment from a normal gene and a 56 bp fragment from the gene containing the deletion of three bases. Compared with gel electrophoresis, mass spectrometry gave accurately same results for 30 samples obtained from the patients.

6.4.2 DNA Ladder Sequencing

Ladder sequencing of DNA by a Mass spectrometer is similar to protein ladder sequencing, which determines the amino acid sequence of a peptide by measuring the molecular mass difference between the members of a family of fragments produced from that peptide using Mass spectrometer. Here, the oligonucleotide is cleaved using an enzyme that sequentially removes the nucleotide from either 3′ or 5′ end of the molecule. The sequence was determined by the mass change after each cleavage of 313.2 Da for Adenine loss, 304.2 Da of Thymine loss, 329.2 Da for Guanine loss and 289.2 for Cytosine loss. Accurate mass measurement is the prime factor of utmost importance in the success of DNA ladder sequencing, which can be performed with MALDI-TOF if the length of the nucleotides is 24 or with ESI-TANDEM Mass spectrometers if the length is 10 nucleotides. The nucleotides from the 5′ end can be cleaved using calf spleen phosphodiesterase and from the 3′ end using snake venom phosphodiesterase.

6.4.3 Gas-Phase Fragmentation

Fragmentation of DNA is regarded as a major hurdle for its sequence by using Mass spectrometry, which can be overcomed by Fourier Transform Ion Cyclotron Resonance Mass spectrometer (FTMS). It is a high resolution instrument that traps ions in a magnetic field. The ion cyclotron frequency, which is a function of a specific ion's mass and charge can be measured with high accuracy. With MALDI, fragmentation can be done by using reflectron mass spectrometer using post source decay.

6.4.4 Mass Spectrometry and SNP Genotyping

Mass spectrometers have predominantly been applied in the DNA analysis, mainly for the SNP genotyping. Among the available mass spectrometers, MALDI-TOF has been the main choice due to its high reliability, accuracy, and speed. The most evolved methods for DNA sequencing by MALDI mass spectrometer use PCR amplification with primers containing the ORF start sites at their 5′ ends. After the transcription of the PCR product, sequence-specific RNA endonucleases are used to generate the fragments terminating on a specific base. Sizing the fragments by MALDI allows the identification of the sequence differences in a genome. With a little optimism about the contribution of a mass spectrometer in the DNA sequencing, there is a lot of scope about their role in the genome sequencing in the coming years with significant progress along with the development of high throughput, parallel processing, simplified handling, and low-cost reagents. Recently, novel methods for determination of more complex sequence variability in the defined loci have been developed along with the complete genotyping of the hypervariable regions in the human mitochondrial DNA using MALDI mass spectrometry for detection.

6.5 Mapping of Genomes

A genome map defines the relative positions of the salient features in the genome, that are of specific interest or can serve as reference points for the navigation. Genome mapping consists of two broad categories; (1) Genetic mapping and (2) Physical mapping. Both these mapping methodologies will provide the probable order of feature arrangement across the chromosome. The genetic map provides an indirect estimate of the distance between two features. Physical maps predict the true distance between two genes of interest in base pairs.

6.5.1 Genetic Map

Genetic maps are generated by using genetic markers. A genetic marker is an observable variation in a locus due to alteration, or a mutation. Genetic markers can be considered as a landmark in a genome if the same DNA fragment consisting of that genetic marker gets inherited from a parent to offspring in accordance with Mendel's laws of inheritance. A marker can also be the segment of a gene that encodes for noticeable physical characteristics such as eye color or an unnoticeable trait such as disease. DNA markers are highly resourceful for generating genetic maps. These markers are of great use during the occurrence of occasional and predictable mutations during meiosis, leading to a high degree of variability in the DNA content of the marker from one individual to another. Most commonly used genetic markers are:

1. Restriction Fragment Length Polymorphisms (RFLPs)

Restriction Fragment Length Polymorphisms is one among very first developed DNA markers used for a variety of genetic and molecular studies. RFLPs are generated by recognizing the presence or absence of a restriction site in the genomic DNA, for a specific restriction endonuclease (*HinD* III). This enzyme digests the DNA by recognizing the sequence site. As a marker, RFLP is highly specific for a restriction enzyme combination. Most of the RFLP markers are co-dominant and are highly specific to a locus. An RFLP probe is a labeled DNA sequence that hybridizes with one or more fragments of the digested DNA sample after they were separated on an agarose gel by electrophoresis. Probe-target hybridization reveals a unique blotting pattern, characteristic of a specific genotype at a specific locus. Short, single or low-copy genomic DNA is typically used as RFLP probes.

2. Variable Number of Tandem Repeat Polymorphisms (VNTRs)

Variable Number Tandem Repeat (VNTR) is a DNA sequence motif that is repeated several times continuously in a genome and is inherited in a Mendelian fashion. VNTRs occurs in the coding (human X-fragile syndrome) and non-coding regions of the DNA. A VNTR marker is defined by the presence of a di, tri, tetra or pentanucleotide sequence that is repeated several times. The number of times the VNTR sequence repeated will vary between individuals, making them a powerful entity for studying polymorphisms. The VNTR polymorphisms are of two types: (1) Microsatellite polymorphisms, also known as simple sequence repeat polymorphisms, short tandem repeats, and simple sequence length polymorphisms are the most predominant and interspersed repeats of 1–6 bp that are found in the euchromatin regions of prokaryote and eukaryote genomes. (2) minisatellite polymorphisms are long continuous motif sequence blocks, which can span hundreds of nucleotides, within a DNA sequence.

3. Single Nucleotide Polymorphisms (SNPs)

Single nucleotide polymorphisms are considered as the most common type of genetic variation among humans. Each SNP represents a difference in the nucleotide of the DNA. For example, an SNP may replace the nucleotide C of one human DNA with the nucleotide thymine T in another human. SNPs occur throughout the DNA of a person at an average of one in every 300 bp. There are ~10 million SNPs in the human genome, with most of the variations found in the DNA sequences corresponding to the genes of two individuals. SNPs act as biological markers for targeting the genes associated with a specific disease. When SNPs occur within a gene or in its regulatory regions, their role is more prominent in affecting that gene function. Most of the SNPs have no or little effect on human health or development, but a few of them are. SNPs help in predicting an individual's response to certain drugs, susceptibility to environmental factors and toxins, risk of developing a particular disease. They can also be used to track the inheritance of a disease within the families.

6.5.2 Physical Mapping

Physical mapping involves finding a contiguous series of cloned DNA fragments containing the overlapping regions of the genome. These overlaps define the positions of the clones relative to one another. Mapping of a portion of the clones containing specific markers will lead to identifying the position of the entire contig in the genome. Physical mapping starts with a library of cloned genome DNA fragments generated either by subjecting the genome to random mechanical breakage or through partial restriction digestion. These fragments are cloned into bacterial hosts such as *E.coli*, using bacteriophage, cosmid, plasmid, etc. For physical mapping of large genome fragments, P1 Artificial Chromosomes (PACs), Bacterial Artificial Chromosomes (BACs), capable of carrying the inserts larger than 100 kb are used. Larger fragments of the size over 1 Mb can be cloned into Yeast Artificial Chromosome (YAC) vectors. Once the library is constructed, overlapping clones can be identified using the following approaches.

6.5.2.1 Sequence Tagged Sites (STS) Mapping

Sequence Tagged Sites are the short sequences of genomic DNA, which can be specifically amplified by the PCR. As each STS is unique, they are often used in genome mapping. Here, libraries are screened for identifying the clones containing specific STS marker by PCR using the appropriate primers. Two clones with the same STS indicates the overlapping parts of the genome.

6.5.2.2 Restriction Fingerprinting

Each member of the clone library is digested with one or more restriction enzymes and the sizes of the resulting fragments are measured by gel electrophoresis. If two clones are found to have several fragment sizes in common, then they must represent overlapping parts of the genome, with the shared fragments coming from the region of overlap. Clones are fingerprinted to find a contiguous series of overlapping clones.

6.5.2.3 Fluorescence In Situ Hybridization (FISH)

FISH is an elegant way of observing the physical arrangement of markers along the chromosomes. Fluorescently labeled probes are hybridized to metaphase chromosomes on a glass slide, and their position is observed using a fluorescence microscope. Lymphocytes or fibroblast cells are cultured in the presence of a mitogen such as a phytohaemagglutinin and are treated with colcemid to block their cell cycle at metaphase. Cells are swollen in a hypotonic salt solution and stabilized with a fixative. Drops of the cell suspension are placed on glass slides, to which the chromosomes adhere. Their DNA is denatured and are ready for the hybridization. Probes used for the FISH are cloned fragments of genomic DNA or cDNA, which have been labeled using nick-translation or by the incorporation of digoxigenin. The probes are denatured for allowing the hybridization with their corresponding sequences of the metaphase chromosomes. Nonspecific hybridization is suppressed by using an unlabeled competitor such as total genomic DNA. For probe detection, the slide is incubated with fluorescently labeled proteins such as anti-digoxigenin antibodies, whose binding to the hapten in the probe can be visualized using the fluorescence microscope.

6.5.2.4 Cytogenetic or Chromosomal Mapping

Cytogenetic mapping is the lowest resolution physical map available, which is based on the distinguishing banding patterns of stained chromosomes detected under a light microscope. Chromosomal mapping is used for locating the genetic markers that are defined by different traits observed only in whole organisms. As the chromosomal maps are completely based on the estimates made by the physical distance between the markers, they are also known as physical maps.

6.5.2.5 Radiation Hybrid (RH) Mapping

Radiation hybrid mapping is a genetic technique that was originally developed for constructing long-range maps of mammalian chromosomes. It is based on a statistical method to determine not only the distances between deoxyribonucleic acid (DNA) markers but also their order on the chromosomes. In radiation hybrid mapping, human chromosomes are separated from one another and are broken into several fragments using high doses of X rays. Similar to the underlying principle of mapping genes by linkage analysis based on recombination events, the farther apart two DNA markers are on a chromosome, the more likely a given dose of X rays will break the chromosome between them and thus place the two markers on two different chromosomal fragments. The order of markers on a chromosome can be determined by estimating the frequency of breakage that, in turn, depends on the distance between the markers. This technique has been used to construct whole-genome radiation hybrid maps.

6.5.2.6 HAPPY Mapping

In HAPPY mapping, DNA is isolated by embedding the living cells in agarose gel and then treating them with a solution of detergents and proteases. As a result, the cellular debris containing proteins will diffuse out of the agarose and the chromosomal DNA will remain trapped inside without any shearing or damage. The DNA is randomly broken by gamma radiation into a pool of random DNA fragments, whose average size depends upon the intensity of the radiation. Broken DNA is diluted into a very low concentration and about one hundred samples are dispensed into separate tubes, referred to as mapping panel. These samples are screened by PCR for the presence of markers such as STS, ESTs, etc. in each one. The amount of the sample in each panel tube is so small such that it will contain only a randomly sampled subset of the markers, rather than the complete genome. Hence, a particular marker will be present in some, but not in all tubes of the mapping panel. If two markers are close together in the genome, then they co-segregate. As the distance between markers increases, it is more likely that the random breakage will separate them so that they lie on separate DNA fragments and are hence less likely to co-segregate. If the markers are far, they are subjected to the random breakage, hence will not co-segregate.

6.5.2.7 Optical Mapping

Optical mapping was developed by David Schwartz of New York University in 1998. This method uses fluorescence microscopy to image individual DNA molecules that have been fragmented by restriction endonucleases. By imaging, a large number of DNA fragments of an organism, an optical map of a genome can be produced at a very low cost. Optical mapping can be used for measuring the absolute length of a genome and for quick detection of the differences in length and structure between the two genomes. Thus, comparing the optical maps of healthy and diseased genomes can help in the identification of crucial changes such as mutations.

6.6 Genome Assembly

Assembling a genome comprises the entire process of arranging a large number of short DNA sequences as a represented in the original chromosomes from which the DNA has been isolated. During whole genome shotgun sequencing, entire DNA of a single organism is first fragmented into small pieces, which are read by automated sequencing machines, capable of reading up to 1000 nucleotides or bases at a time. These fragments are aligned with one another and the places of overlapping sequences are detected so that they can be merged.

The assembler algorithm relies on the basic assumption that two sequences read sharing the same string of letters originated from the same place in the genome. Using such overlaps between the sequences, an assembler can join the sequences together in a manner similar to solving a jigsaw puzzle. The assembly process is more complicated if shotgun sequencing is used. Due to the randomness of the shearing process, genome assembly is possible only if a sufficient amount of sequences are generated to cover the genome 8–10 times. This phenomenon can be well understood by looking at a sidewalk when it begins to rain. As the raindrops randomly fall across the sidewalk, a few dry spots persist for a while, corresponding to regions of the genome that are not represented in the set of shotgun reads. Mathematically, this phenomenon was modeled by Eric Lander and Michael Waterman in 1988. They examined the correlation between the oversampling of the genome called coverage and the number of contiguous pieces of DNA commonly called contigs that can be reconstructed by using an ideal assembly program.

The genome sequenced using whole genome sequencing is assembled by a three phase overlap, layout and consensus strategy. During the overlap, all read-to-read sequence overlaps are detected. During layout, subsets of the overlaps are selected in order to merge reads into longer scaffolds of ordered and oriented reads. Those scaffolds are converted into sequence through multiple alignments during the consensus phase.

6.6.1 Overlap Phase

In the overlap phase, read-to-read sequence overlaps are detected. As the read orientation is not known, all possible overlaps between reads in forward or reverse complement direction are detected. The nature of the data requires several pre-processing tasks to take place prior to overlap detection, such as trimming of reads to remove bases with quality scores that are too low, and filtering of contamination from sequencing/cloning vector or mitochondrial sequences. In the case of large-scale WGS, a significant algorithmic challenge is to find the overlaps efficiently. Strategies that have been employed rely on detecting exact word matches between the reads. Alignment is done with fast procedures that rely on the fact that only almost exact sequence identity between two reads can induce an overlap. Celera, Phusion and Arachne's are the three popular assemblers available for assembling the genomes.

6.6.2 Layout Phase

This phase is also called as contig phase. One of the most prevailing ways of representing the overlaps is the *overlap graph* containing the reads as nodes and overlaps are labeled as edges that contain information on the location and quality of overlap. Overlap graph consists of only full overlaps between reads x and y. Repeat boundaries can be detected in an overlap graph by finding a read *x* that extends to two reads *y* and *z* to the left (or right), where *y* and *z* do overlap one another.

Layout phase is the most complex phase in the genome assembly divided into two sub-phases: (1) *contig layout*, during which the assembly systems identifies the repeat boundaries and connect the reads into contigs that are as long as possible without crossing those boundaries (if the boundaries are crossed, they are considered as misassemblies) The resulting contigs are classified as unique, representing a sequence that occurs once in the genome or repetitive, representing a high-fidelity repeat sequence whose reads were merged together. Unique and repetitive contigs vary in the density of reads within a contig with repetitive contigs are highly dense than that of unique contigs. (2) *supercontig layout* is the toughest phase where contigs are chained into longer structures by using the paired reads crossing contig boundaries. After the classification of contigs as unique or repetitive, assemblers link them into larger structures, called *scaffolds* or *supercontigs*. A *supercontig* is a structure of ordered and oriented contigs derived by linking contigs according to paired reads. A series of supercontigs are linked to form *ultracontigs*. Most of the assemblers first join the unique contigs while excluding the likely repeated ones, to produce supercontigs that represent correctly ordered and oriented unique sequences with gaps. These gaps are filled by recruiting the repeat contigs.

6.6.3 Derivation of a Consensus Sequence

After obtaining layout, a consensus sequence is derived by starting from the leftmost read of each contig, by the multiple alignments of reads. Conservative quality scores are given to each consensus base according to the quality and agreement of bases in the multiple alignment column. Additionally, the Celera Assembler report the positions that appear polymorphic and give an estimate of the likelihood that the polymorphism is real versus repeat-induced.

6.6.4 Repeats and Sequencing Errors in the Genome Assembly

Assembling any genome would be easy without any repeats. But eukaryote genomes are composed of highly repetitive sequences. Another problem in assembling the genomes is to distinguish between a true overlap and a repeat-induced overlap. More than half of the mammalian genomes consist of repeats, whose true overlaps can be confused with repeat-induced overlaps. The main challenge of assemblers is to merge the reads of true overlaps and avoid the merging of repeat-induced overlaps to prevent misassembling of the genome.

Two broad methods are being practiced for assembling the genomic regions that are rich in repeats;

(i) Clone-based sequencing, where the reads will be clustered by performing shotgun sequencing on Bacterial Artificial Chromosome (BAC) clones of length ~200 Kb. With this approach, the effective repeat content can be drastically reduced as most of the repetitive segments in a genome will be unique. In such a short region, the number of reads to be assembled is reduced by a factor of 10^4. Therefore, the computational assembly will be much easier.

(ii) Double-barreled whole-genome shotgun sequencing, whereby paired reads that are obtained from both ends of inserts of various sizes. These paired reads can resolve the repeats by jumping across them. Paired reads from longer inserts such as cosmids and BACs can resolve duplications, help build longer scaffolds and verify the large-scale accuracy of an assembly.

6.6.5 Assembly Algorithms and Notable Assembly Programs

6.6.5.1 Assembly Algorithms

1. Greedy Assemblers

The first assembly programs followed a simple but effective strategy in which the assembler greedily joins together the reads that are most similar to each other. Here, the assembler joins reads in the order of 1 and 2 with an overlap of ~200 bp, then reads 3 and 4 with an overlap of 150 bp, then reads 2 and 3 with an overlap of 50 bp, thereby creating a single contig from the four reads that are provided in the input. The major disadvantage of the simple greedy approach is the consideration of local information at each step; the assembler can be easily confused by complex repeats, leading to misassemblies.

2. Overlap-Layout-Consensus

The relationships between the reads provided to an assembler are represented in the form of a graph, with the nodes representing the reads and an edge connects two nodes only if the corresponding reads overlap. Assembling thus becomes the problem of identifying a path through the graph that contains all the nodes—a Hamiltonian path. The overlap-layout-consensus formulation allows the researchers to use techniques developed in the field of graph theory for resolving the assembly problem. An assembler following this paradigm starts with an overlap stage during which all overlaps between the reads are computed and the graph structure will be computed. In a layout stage, the graph is simplified by removing redundant information. Graph algorithms are then used to determine a layout of the reads along the genome. In a final consensus stage, the assembler builds an alignment of all the reads covering the genome and infers, as a consensus of the aligned reads, the original sequence of the genome being assembled.

3. Eulerian Path

Eulerian path approaches are based on early attempts to sequence genomes through a technique sequencing by hybridization, which helps in identifying all strings of length k (k-mers), instead of generating reads. This approach is also based on a graph-theoretic model, breaks up each read into a collection of overlapping k-mers. Each k-mer is represented in a graph as an edge connecting two nodes, corresponding to its k-1 bp prefix and suffix respectively. The graph contains the comprehensive information obtained from all the reads, with a solution to the assembly problem corresponds to a specific path in the graph that uses all the edges (Eulerian path). A major advantage of this approach is the repeats can be recognized immediately compared with that of an overlap graph, which makes more difficult for identification.

4. Align-Layout-Consensus

With more number of genomes are available in the public databases, there are more chances that a closely related genome exists in the databases, which makes the assembly more easy by inferring the position of the reads from their alignment with reference genome in a process called a comparative assembly. Thus, the overlapping stage of assembly is replaced by alignment and the layout stage is also much simplified due to the additional constraints provided by the alignment to the reference.

6.6.5.2 Notable Assembly Programs

1. TIGR Assembler

Developed at The Institute for Genomic Research (TIGR), this assembler was mainly used for generating the first genome sequence of a free living organism *Haemophilus influenzae* (refer to journal *Science* 1995).

2. Celera Assembler

Developed at Celera Genomics, this assembler has demonstrated the applicability of a shotgun method for assembling a eukaryotic genome through the genome of *Drosophila melanogaster*. Celera Assembler was a key element in the successful assembly of the human genome by Celera Genomics and is currently being used in numerous bacterial and eukaryotic genome projects.

6.7 Scaffolding

Contigs that are produced during assembly are arranged in an oriented order along the chromosome. This is more important while using a shotgun approach, thus providing a link between the sequence reads that are generated by the ends of the same fragment. During whole genome shotgun sequencing, multiple libraries comprising the collection of fragments with similar sizes are generated, providing additional constraints to the assembler. While placing the paired end reads of a library at a distance consistent with its size, each fragment in the library must be assigned with an orientation similar to the corresponding DNA strand from which the fragment has been generated. The process by which the sequence information of the fragments is used for arranging the contigs in a right orientation is called scaffolding.

6.8 Finishing

The ultimate goal of any genome sequencing project is to determine every single base-pair of the original set of chromosomes. During genome assembly, position or orientation of a few sequenced fragments cannot be determined, leading to the generation of gaps during the reconstruction of the genome. These gaps are filled by conducting additional laboratory experiments followed by extensive manual curation for validating the final assembly with mere perfection. This process of gap filling leads to a high-quality reconstruction of the original genome is called finishing or gap closure.

6.9 Genome Annotation

Genome annotation is the process of extraction, definition, interpretation, and identification of the functions to the features on the genome sequence such as loci by integrating the computational tools and biological knowledge after genome assembly. Annotation gives meaning to a given sequence by identifying the coding and non-coding regions, which makes it much easier for the researchers to view and analyze its contents. When a group of researchers assembles a genome, the annotation also takes place simultaneously by the same group and this process of assembly with annotation together is known as a **build**. Nowadays, the genome assembly process and its annotation process are often completely uncoupled. Genome annotation consists of three major goals; (1) Identification of portions in the genome that do not code for proteins, (2) Identification of the coding region in the genome and (3) Attachment of biological information to these elements. Genome annotation is performed under three main steps;

6.9.1 Nucleotide Annotation

Is performed under following heads.

1. Mapping

The first step in genome annotation is the identification of genes, genetic markers and other landmarks previously identified by genetic, cytogenetic or radiation hybrid mapping. Identification of tRNAs, rRNAs, other non-translated RNAs, repetitive elements is done by physical mapping. Along with the gene finding, the principal activity of this phase of an annotation is identification and positioning of all known landmarks into the genome, which provides a path to connect the pre-genomic literature with post-genomic research.

2. Finding Landmarks

Finding landmarks on the genome sequence is a relatively straightforward task, which can be achieved by using short sequences such as PCR-based genetic markers, can be identified rapidly using the e-PCR program. Long sequences, such as RFLP markers, is found using BLASTN, SSAHA or other rapid sequence-similarity searching algorithms. During annotation of the human genome, integration of the sequence with the cytogenetic map was done by systematic fluorescence in situ hybridization mapping of BAC clones against metaphase chromosomes. The assembly of the Drosophila genome was aided by in situ hybridizations of each BAC clones in the physical map to polytene chromosomes, providing bridges between the cytogenetic and physical maps with an accuracy of a few 100 Kb.

3. Gene Finding

Gene finding is the most vital part of the nucleotide annotation. In prokaryotic genomes, gene finding is mostly involved with the identification of long overlapping open reading frames (ORFs) and true coding regions. As genomes get larger in eukaryotes, gene finding becomes increasingly tricky mostly due to an enormous increase in the non-coding regions. In prokaryotic genomes, such as *Haemophilus influenzae*, 85% of its genome falls in the coding region, whereas the number gets decreased up to 70% in yeast, 25% in *Drosophila* and *C. elegans*. The percentage further goes down with the human genome. Further, the process of finding genes is more complicated by the presence of splicing and alternative splicing. In the human genome, a typical exon is 150 bp in length and a typical intron of several kilobases. There is no clear delineation between the intergenic regions that separate adjacent genes and the intragenic regions that separate exons. In these genomes, identification of genes is a major exercise achieved by running a computer program that carries out a six-frame translation and identifies all ORFs that are longer than a chosen threshold. Even in the small genomes, finding genes is not so easy as in the case of the yeast genome, which took several years to get a proper number for the active genes and still there are several short ORFs whose status still remains uncertain. Defining the precise start and stop position of a gene and the splicing pattern of its exons among all the non-coding sequences is the fundamental step in the identification of a coding region in the genome, which is being done by several sophisticated software algorithms such as GENSCAN, Genie, GeneMark.hmm, Grail, HEXON, MZEF, Fgenes and Gene Finder. All these algorithms consist of one or more sensors, such as transcriptional start site sensors, splice site sensors that attempt to adduce the presence of a gene feature from motifs or statistical properties of the DNA. Despite great progress, still, gene prediction entirely based on DNA analysis is still far from perfection.

4. Non-Coding RNAs and Regulatory Regions

Genome is the composition of coding, non-coding, and regulatory regions. Non-coding RNAs include tRNAs, rRNAs, small nucleolar RNAs, and small nuclear RNAs. rRNAs can be found in the genome sequence by similarity searching, but identification of other RNAs is little complicated mainly because of their short length and their high nucleotide diversity. tRNAs are amenable to the de novo prediction through algorithms that search for characteristic structural signatures, such as hairpin formation. The most widely used tRNA prediction program is tRNAScanSE32, which identify tRNAs with high accuracy. It also distinguishes the active tRNAs from tRNA pseudogenes. Using tRNAScanSE32, 497 tRNAs and 324 putative pseudogenes have been identified during the annotation of the public human sequence. Other non-coding RNAs, such as telomerase RNA and the U1–12 series of spliceosome RNAs, have been identified by sequence similarity, to a very small extent. Still many non-coding RNAs have not yet been identified.

The situation is still worse in the case of regulatory regions, with a relatively small number of transcriptional-factor-binding sites have been identified by classical experiments, curated TRANSFAC and PROSITE databases, obtained by applying similarity search methods. However, the number of known regulatory regions is almost certainly a small fraction of what is out there.

6.9.2 Protein-Level Annotation

Identification of genes in a genome should lead to its nomenclature and assign them with putative functions. *H. influenzae* is reported to have 1709 genes, yeast has ~5600, fly, worm, mustard weed, and human have ~13,000, ~19,000, ~24,000 and >30,000 genes respectively. Among these huge number of the gene sets, only a small fraction corresponds to known and well-characterized proteins. With numerous proteins of unknown function, annotators generally begin by classifying them into more manageable groups or protein families, and by searching for similarities with the better-characterized proteins of other species. This searching for similarity involves an intrinsic problem from the evolutionary process of that protein family. During the evolution of a protein family, an ancestral protein is duplicated into single or multiple copies, which diverges to form a family of related proteins known as paralogues. However, these proteins will not adopt a similar protein as that of their ancestor but will have strikingly divergent functions as identified in the case of lens crystallins, which are the derivatives of a family of proteins which functions as enzymes and chaperonins. Another significant feature is the proclivity of genes to pick up or lose functional domains during the course of evolution. Such a process leads to the creation of chimeric proteins that share two or more unrelated ancestors. Most of the protein functions are identified by bibliographic references, descriptions about their function, the biological role of the protein, protein family assignments and pointers to structural data if available. Because of its well-curated nature, the protein sequences of the SWISS-PROT database and SWISS-PROT TrEMBL, an automated translation of coding DNA sequence (CDS) entries submitted to the nucleotide databases are the sources of high reliability.

6.9.3 Process-Level Annotation

Final and the most challenging part of genome annotation is relating a genome to the biological processes. How do the genes and proteins in a genome are related to a specific function such as cell cycle, programmed cell death, embryogenesis, metabolism is the question which is puzzling the genomics analysts since a long time. One of the answers to the questions is the functional annotation. Publication of every new genome is accomplished by a pie-chart showing the distribution of proteins classified by function, for example, 'metabolism' and 'cytoskeleton'. This part of the analysis was lacking a common and uniform classification and the depth required to describe a specific biological function as well as to distinguish a particular protein from other members of its family. The lack of such a standard annotation has also hampered the ability to relate the genes that were previously annotated by different research groups, particularly while crossing species borders. In order to solve this issue, databases of three model organisms namely *Saccharomyces* Genome Database, FlyBase and the Mouse Genome Database, formed a consortium to create Gene Ontology, a standard vocabulary for describing the function of eukaryotic genes. Gene ontology consists of three subparts: molecular function, biological process, and cellular component. Molecular function describes the tasks carried out by individual gene products, such as its enzymatic activity. Biological processes are applied for broader biological goals, such as meiosis. Cellular component tends to describe the genes in terms of the subcellular structures such as organelles, as well as the macromolecular complexes such as the ribosome.

Genomes are actively annotated by several groups such as Celera, Ensembl, and the National Center for Biotechnology Information, the computational biology group at Oak Ridge National Laboratory. The strength of having several groups involved in this process is that researchers benefit from their diverse approaches. The main weakness is that a diversity of sources for the annotations tends to fragment the information. A researcher seeking to compare the annotations made by one group to those of another must visit several different web sites and surmount various obstacles, including incompatible user interfaces, coordinate systems, file formats, and naming conventions. Without extensive bioinformatics support, this task can be nearly impossible to carry out on more than a few genes at a time.

6.10 Applications of Next Generation Sequencing Systems

Next generation sequencing system can be applied for sequencing as well as re-sequencing of the genomes, their transcriptome profiling, DNA-protein interactions, epigenome characterization and resurrection of the ancient genome.

6.10.1 Transcriptome Sequencing

Genome wide gene expression levels are studied using qPCR, SAGE, and microarray with a few limitations. Next generation sequencing technologies, along with SAGE tags to sequence the RNA populations from the cells expressed specifically noncoding RNA (ncRNA). ncRNAs are any RNA's that are transcribed but are not translated into a protein. ncRNAs play a significant role in the regulation of gene expression of plants and animals. Next-generation sequencing technology has discovered many novel ncRNAs, which are unique, diverse and regulate genes by a variety of mechanisms. Readouts obtained from the next generation sequencing technologies are quantitative, which allows the detection of changes in their expression levels either due to changes in environment or onset of a disease. Studying the roles of these ncRNAs may help in uncovering certain aspects of diseases such as cancer, which can be very well accomplished by using the next generation sequencing technologies. Discovering these new specific RNAs and transcriptome sequencing using next generation sequencing technologies would provide us new insights on the genome wide expression patterns of an organism.

6.10.2 The Resurrection of Ancient Genomes

Samples from fossils and ancient remnants are inevitably found in a state of degradation and sequencing the fossil DNA has several non-trivial technical complications, most notably DNA contamination. With the availability of molecular techniques such as PCR and the next generation DNA sequencing technology, deciphering of mitochondrial genomes from fossils and ancient remnants is possible. Next generation sequencing can be implemented to obtain sequence information from the degraded nature of the ancient genome. Sequence information from single fossil bone of the Neanderthal genome was obtained using next generation sequencing technologies. In addition to the Neanderthal sequence, information was also obtained directly from the nuclear genomes of ancient remains of the cave bear and mammoth. In this way, next generation sequencing technology can be used in the resurrection of ancient genomes.

6.10.3 Analysis of Epigenetic Modifications of Histones and DNA

With the predominant improvement in the capability of next generation sequencing technologies, DNA methylation profiling is made available by bisulfite DNA sequencing, mapping histone modifications, mapping the locations of DNA-binding proteins, DNA accessibility, and chromatin structure. The interaction between DNA and proteins is vital in regulating gene expression and makes the availability of DNA for transcription, replication, and many important processes. Genome-wide chromatin immunoprecipitation (ChIP)-based studies is a useful technique for understanding the DNA-protein interactions. ChIP-based approach along with the Illumina platform has provided valuable insights regarding the transcription factor binding sites in the human genome including neuron-restrictive silencer factor (NRSF) and signal transducer and activator of transcription 1 (STAT1). Next generation sequencing technologies can also be used for the understanding of gene expression–based cellular responses.

6.10.4 Sequencing of Cancer Genome

Cancer is a major concern to the medical and scientific community and is an important arena where high throughput sequencing technologies have been applied to a greater extent. Using the high throughput sequencing, The Cancer Genome Atlas (TCGA) and International Cancer Genome Consortium (ICGC) have performed genome and exome sequencing on thousands of tumor-normal pairs, which has identified and described the locations of the mutations responsible for more than 20 types of cancer. These descriptions on the mutation rates are necessary for the detection of cancer driver genes. Projects concerned with TCGA have identified several novel cancer drivers and whole genome sequencing of cancer samples has also identified high-frequency, non-coding mutations.

The scale and sensitivity of high throughput sequencing technology have also enabled the global descriptions of tumor heterogeneity, clonal evolution, and the mechanisms underlying drug resistance. Using single cell sequencing technology, aberrations in the copy number of primary breast cancer cells such as copy number rearrangements has been tracked to occur in bursts, followed by persistent clonal expansion. High throughput sequencing has also been used to compare primary tumors with relapse lesions, allowing characterization of the effects of chemotherapy and identifying the molecular mechanisms involved in the resistance to therapy. Molecular portraits of cancer and high throughput sequencing technologies together are laying the foundation of new paradigms for the early diagnosis and treatment of cancer.

6.10.5 Diagnosis of Rare Diseases and Exome Sequencing

The capacity to rapidly sequence the genomes, exomes and transcriptomes have profoundly increased our understanding of the genetics of human disease, especially rare disorders (also called Mendelian disorders). Online Mendelian Inheritance in Man database has identified more than 7800 Mendelian disorders, but the causative genes for more than one half of them are not yet determined. An exome is defined as the collection of all exons of protein coding genes in a genome. An exome covers between 1% and 2% of the genome, depending on the species. Functional non-protein coding elements such as microRNA, long intergenic noncoding RNA, etc. are also included in the exome. By sequencing the exomes of unrelated patients, affected and unaffected family members, causal alleles for a variety of inherited diseases have been identified. In rare cases, sequencing of patient samples has suggested specific clinical interventions. Exome sequencing of a child suffering from severe inflammatory bowel disease using high throughput sequencing has identified a mutation in X-lined inhibitor of apoptosis (*XIAP*), an important regulator of inflammation, suggesting the patient to undergo bone marrow transplantation. Despite the availability of high throughput sequencing technology for identification of diseased genes, exome sequencing currently identifies the genetic defect up to only 25% of patients.

6.11 Conclusions

DNA sequencing is a vital source of biological research, applied for defining and differentiating the properties of various life forms on this planet. Many researchers have invested most of their valuable time and resources in developing and improving the technologies for a better understanding of the components involved in the nuclear, mitochondrial and chloroplast DNA. As a result, there is a tremendous increase in the capabilities of sequencing with decreased cost, allowing reading hundreds and thousands of base pairs in length, massively parallelized to produce gigabases of data in one run. This technology has permitted to decode the genomes, development of databases, which has revolutionized modern biology. Altogether, DNA sequencing technology has a rich history in many respects. A better understanding of this history will provide new insights for the better development of future ones.

Other 'Omics' Integrated into Biosciences 7

Abstract

The role of other omics such as transcriptomics, proteomics, metabolomics, exposomics, connectomics, microbiomics that are integrated into the genomics have been emphasized, which will provide a better understanding of the cell function or an organism. Concepts related to the proteomics such as codon biasedness that affect the protein levels, metabolic profiling, metabolomics, and its applications were accentuated. Measurement of regional connections in the living Human Brain through connectomics and the entire human microbiome has been discussed in detail.

7.1 Introduction

'Omics' studies provide a comprehensive view of the biomolecules that builds a cell of an organism. Primarily, these omics are involved in the detection of genes (genomics), mRNA (transcriptomics), proteins (proteomics) and metabolites (metabolomics) in a specific biological sample. Integration of these high dimensional omics will provide a complete understanding of the complexities involved in the cell function. Continuously developing Omics technology is being applied not only for greater understanding of normal physiological processes but also in screening, diagnosis, and prognosis as well as aiding our understanding of the etiology of the diseases. Biomarkers developed by using Omics strategies have enabled the investigation on multiple molecules simultaneously. Omics is also playing a pivotal role in the drug discovery, assessment of toxicity and efficacy. For example, the intersection of genomics and pharmacology has evolved Pharmacogenomics, to study the role of inheritance in individual's variation in drug response which can be potentially used for optimizing the drug therapy. Pharmacogenomics is especially important in selecting the novel targets for treatments and drug development. The role of other omics whose intersection helped in better understanding of the complex processes involved in the cell function is discussed in this chapter.

K. V. Chaitanya, *Genome and Genomics*,
https://doi.org/10.1007/978-981-15-0702-1_7

7.2 Transcriptomics

The complete collection of all transcription products (mRNAs) of a cell, aspecific tissue or an organism along with their quantity at a specific developmental stage or a physiological condition is denoted as the transcriptome. The main objective of transcriptomics is:

1. To collect all RNAs (mRNAs, ncRNAs, and snRNAs) of a cell, tissue or an organism.
2. To determine the transcriptional organization of genes, in terms of their arrangement regarding start and stop codons, introns, alternative splicing patterns and other post-transcriptional modifications such as 5′ capping and 3′ tailing.
3. To develop the technology for quantifying the changes in the expression levels of each transcript under different physiological and development conditions of organisms.

7.2.1 Categories of RNA

A great variety of RNAs play structural roles, functional and regulatory roles in many cellular processes of an organism. Based on their products, RNAs are broadly divided into two classes: protein coding RNAs (mRNAs) and non-protein coding RNAs (ncRNAs). ncRNAs are further divided into house-keeping and regulatory ncRNAs. House-keeping ncRNAs such as tRNAs and rRNAs are involved in translation, small nuclear RNAs (snRNAs) are involved in mRNA splicing, small nucleolar RNAs (snoRNAs) involved in rRNA splicing, guide RNAs (gRNAs) involved in RNA editing. Many ncRNAs also play some regulatory roles in a diverse variety of biological processes. Regulatory ncRNAs are divided into small (sncRNAs) and long (lncRNAs). sncRNAs are 17–35 nucleotides in length, including microRNAs (miRNAs), siRNA and Piwi-interacting RNA (piRNA). miRNAs play certain crucial regulatory roles in many biological processes, such as development, abiotic and biotic stress response, and cell behavior. miRNAs are ~22 nucleotides in length, derived from pri-miRNA containing hairpin structures, which are subsequently processed by RNase III Drosha and Dicer to form mature miRNAs. Pairing with their target mRNAs, miRNAs inhibit gene expression by translational repression or by promoting mRNA degradation. siRNAs and piRNAs are small RNAs, which mainly act in the gene silencing of transposons and repetitive sequences for maintaining genomic stability. The length of lncRNAs is more than 200 nucleotides, which lack any ORF and RNA polymerase II transcripts. lncRNAs play essential roles in many biological processes. They can be antisense, sense, intergenic, bidirectional and intronic transcripts, involved in the regulation of protein-coding gene expression in different ways: Transcriptional regulation occurs directly through the lncRNA-protein interaction that inhibits the activity of transcription factors or RNA polymerase II, or indirectly with lncRNA helping to recruit regulatory protein factors of chromatin structure, which influences the transcription indirectly. lncRNA may also affect mRNA stability at the posttranscriptional level. The present focus of lncRNA research is to identify and quantify different lncRNAs in different tissues or under various physiological conditions and then to determine their biological functions.

7.2.2 Transcriptome Sequencing and Analysis

Transcriptional level in an organism varies among different tissues under different physiological conditions or environmental stimuli. Identification of differentially expressed genes in different tissues is vital during transcriptome analysis. Expressed Sequence Tag-based methods such as Serial Analysis of Gene Expression (SAGE), hybridization based methods such as microarray and Next generation sequencing based RNA sequencing (RNA-seq) technology have been developed for quick scanning of transcriptome and obtain the differentially expressed genes. Many key genes from various developmental, physiological, or pathological processes were identified by these means.

To understand the structure of a transcript and its promoter, it is essential to map the transcription initiation site. Cap Analysis of Gene Expression (CAGE) method was developed for sequencing the 5′ end of mRNA. Paired-End Analysis of Transcription initiations sites (PEAT), deep CAGE, nano CAGE, and CAGE scan have revealed the Transcription Initiation Site of each gene. Precision nuclear run-on and sequencing (PRO-cap) method allow the detection of transcription initiation site of nascent RNAs. Genome-wide nuclear run-on and sequencing (GRO-seq) and the improved PRO-seq has not only succeeded in monitoring the nascent mRNA globally at a very high resolution of a single nucleotide but also in determining the biogenesis rate of RNA. These two methods provide the rates of transcription initiation and elongation, along with the RNA polymerase pausing positions.

Post-transcriptional processing of the mRNA leading to its maturity consists of a series of steps such as 5′ capping, splicing, 3′ cleavage and addition of polyA tail. Through pair-end sequencing and improvising the read length and depth, mRNAs of the genes comprising introns were identified. RNA editing such as uridine insertion and deletion, A-to-I shift is post-transcriptional processing, obtained by comparing the RNA-seq results to the reference genome sequence. Processing of mRNA 3′ end involves endonucleolytic cleavage and adding multiple adenosines. The mRNAs of many genes have more than one 3′ cleavage sites, known as alternative polyadenylation (APA), affecting either the length of protein-coding sequence or it's 3′ untranslated (UTR) region, thereby regulating mRNA translation efficiency and/or its half-life. Modified RNA sequencing techniques PAS-seq and 3P-seq, together with specific bioinformatics analysis can be applied for studying the alternative polyadenylation mechanisms in mRNAs. mRNA degradation is an important step in the metabolism of RNA. Half-life of RNA can be determined through 5′-Bromo-uridine immunoprecipitation and chase-deep sequencing analysis (BRIC-seq) method.

7.2.2.1 Sequencing and Characterization of Non-coding RNA

Relatively, a lot of progress has been made in the isolation and characterization of non-coding RNAs that are mainly involved in gene regulation. Short non-coding RNAs including miRNA, siRNA, and piRNA of an organism were isolated, pooled and were made into a cDNA library, purified on a denaturing polyacrylamide gel, ligated with a 3′ and 5′ adapter and finished by reverse transcription. Using the library based method for screening short nc RNAs, nearly 20,000 miRNA genes have been cloned from more than 200 species. Identification of the miRNA's target mRNA for elucidating their function was achieved by using Argonaute cross-linking immunoprecipitation and sequencing (CLIP-seq), degradome-seq, and parallel analysis of RNA end (PARE). These techniques are based on the interaction of miRNA with its target mRNA through Argonaute protein. CLIP-seq was designed to immunoprecipitate the Argonaute-RNA complex, allows to sequence the Argonaute associated RNA, While, degradome-seq or parallel analysis of RNA end (PARE) method sequence the 5′ ends of the target mRNA cleavage products by miRNA. Long non-coding RNAs, such as antisense transcripts, were identified by the strand-specific RNA-seq, which provides the direction information of transcripts sequenced, allowing the distinguishing antisense ncRNA from sense coding transcripts. Strand-specific sequencing of polyA RNA has identified the expression of more than10,000 lncRNAs in the human genome.

7.2.2.2 EST Libraries and Microarrays

Sanger sequencing of EST or cDNA library has provided the required information necessary for the genome annotation during the early days of genome research. Due to the limitations on cost, it was not possible to achieve the transcriptome quantitative analysis using these library methods. With the availability of SAGE and CAGE, multiple 3′ and 5′ cDNA ends were concatenated into a single clone, facilitating the recovery of multiple sequence tags from a single Sanger sequencing reaction overcoming the limitations of quantitative analysis. However, due to the high cost of Sanger sequencing and the difficulty in mapping the short sequence (~20 bp) tags, CAGE and SAGE were replaced by cDNA microarray (Fig. 7.1).

Microarray is based on nucleic acid hybridization. High-density gene chips have allowed relatively low-cost gene expression profiling. Specific microarrays were designed to detect different isoforms from alternative splicing. Genome tiling array is considered as an unbiased design for the detection of whole-genome expression with the resolution up to a few nucleotides, without prior knowledge of genome transcription information. However, titling array is cost-effective for large genomes. Another limiting factor is high background, which will not permit to distinguish RNA molecules with high sequence similarity.

7.2.2.3 RNA Sequencing (RNAseq)

Recent advances in the development of next generation sequencing technologies for DNA sequencing have facilitated a novel approach to map and quantify the entire transcriptome of an organism. This technology termed as RNA sequencing (RNAseq) has been successfully applied for elucidating the transcriptomes of

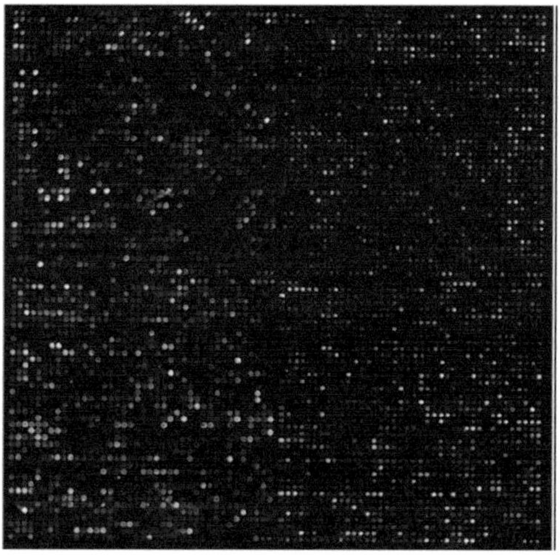

Fig. 7.1 cDNA microarray

Saccharomyces cerevisiae, *Schizosaccharomyces pombe*, *Arabidopsis thaliana*, mouse, and human cells so far and is expected to revolutionize the entire eukaryotic transcriptome analysis.

This technique uses deep-sequencing technologies for determining the transcriptome sequence. A total or fractionated population of RNA is made into a library of cDNA fragments with adaptors ligated to either 5′ or both 5′ and 3′ ends. Each cDNA molecule after PCR amplification will be sequenced using a high-throughput technology as Illumina IG18, Applied Biosystems SOLiD, and Roche 454 Life Science systems for obtaining the short sequences from one end called single-end sequencing or both the ends, which is also known as pair-end sequencing. Following sequencing, the reads (30–400 bp) are either aligned to a reference genome or reference transcripts or assembled *de novo* without the genomic sequence to produce a genome-scale transcription map that consists of both the transcriptional structure and the level of expression for each gene.

Applications of RNAseq

RNAseq offers several advantages over the existing technologies.

(i) RNAseq is not limited to detecting the transcripts corresponding to the sequenced genomes, but can also be used for sequencing the transcriptomes of non-model organisms, whose genomes are yet to be determined. For example, the transcriptome of *Glanville fritillary* butterfly has been sequenced by using 454-based RNA-Seq technology.
(ii) Precisely, the location of transcription boundaries can be identified to a single-base resolution. A short RNAseq read of 30 bp provides the information regarding the connectivity of two exons, whereas the connectivity between multiple exons of a transcript can be identified by the long reads of RNAseq. RNAseq can also provide sequence variations such as SNPs in the transcribed regions.

(iii) RNAseq has a very low or no background signal as in the case of microarrays. There is no upper limit for determining the quantity of a transcript, which permits the detection of a large dynamic range of expression levels. In *Saccharomyces cerevisiae*, 16 million mapped reads were analyzed with a fold range of >9000. RNAseq is extremely accurate in the quantifying the expression levels of a transcript, as determined using quantitative PCR (qPCR).

(iv) Data of RNAseq also exhibits a high level of reproducibility, both for the technical and biological replicates.

(v) RNAseq requires less RNA sample for its analysis without any cloning steps. With the upcoming Helicos technology, there is no need even for its amplification.

(vi) RNAseq is the first technology that facilitates a very high-throughput sequence of the entire transcriptome in a quantitative manner. This method offers both single-base resolution for annotation and 'digital' gene expression levels at the genome scale at a much lower cost than either tiling arrays or large-scale Sanger EST sequencing.

Challenges of RNAseq

(i) Unlike small RNAs, which are directly sequenced after ligation to an adaptor, large RNA molecules of size 200–500 bp must be fragmented into small pieces in order to be more compatible with the current deep sequencing technologies. Common fragmentation methods such as RNA fragmentation and cDNA fragmentation have their own biasness. RNA fragmentation has a little biasedness over the transcript body, but is depleted for the transcript ends when compared with other methods. cDNA fragmentation is highly biased towards the sequence identification from its 3′ ends, providing valuable information regarding the identity of these precise ends.

(ii) Many short reads with a similar identity can be obtained by the construction and amplification of cDNA libraries. These similarly identical short reads could be a genuine reflection of abundant RNA species, or they could be the PCR artifacts. One way of discriminating these possibilities is by determining the occurrence of these similarly identical sequences in the replicates of different biological samples.

(iii) Another key consideration is the preparation of strand specific libraries, which provides the information regarding the orientation of transcripts, valuable for their annotation, especially in the regions with overlapping transcription from opposite directions. However, preparation of strand-specific libraries is laborious.

(iv) It is essential to ensure that the antisense transcripts are not the main artifacts of reverse transcription. Because of these complications, most studies thus far have analyzed cDNAs without strand information.

(v) Another crucial issue is the depth of the sequence coverage, which has cost implications. Greater coverage requires more sequencing depth as evidenced in yeast transcriptome. 30 million 35-nucleotide reads generated from the poly (A) mRNA libraries are enough to observe the transcription of >90% genes in the cells that are grown under a single condition. Analyzing these genes under different physiological conditions can further increase their coverage, which also significantly increases the cost.

Despite these challenges, RNAseq has emerged to be one of the vital sources in transcriptome analysis, which has engendered an exceptional and comprehensive study of the transcriptome and its organization in a number of species and cell types. With its high resolution and sensitivity, RNAseq has revealed many novel transcribed regions and splicing isoforms of known genes and has mapped 5′ and 3′ boundaries for many genes. Still, RNAseq is required to target complex transcriptomes for the identification and tracking the changes in expression of rare RNA isoforms of genes. As the cost of next generation sequencing continues to fall RNAseq is expected to be more vital in determining the structure and dynamics of a transcriptome.

7.3 Proteomics

Proteomics is the study of a proteome, which is the collection of all proteins synthesized by an organism at a specific age and stage. The terms "proteomics" and "proteome" were coined by Marc Wilkins et al. in the 1990s. Protein is a highly complex biomolecule made up of a polymer of amino acids, plays a pivotal role in the metabolic and signaling activities of a cell, which are vital to all living organisms on this planet. Proteins were first reported by Emil Fischer and Franz Hofmeister in 1902.

7.3.1 Life and Death of a Protein

A protein is born as a product of mRNA translation, performed on a ribosome. After emerging from the ribosome, the nascent polypeptide chain will undergo some posttranslational modifications before it assumes its functional role in a living system. There are more than 200 different types of posttranslational modifications reported to date. One of the most prominent and best-understood post translational modifications is the phosphorylation of serine, threonine and tyrosine residues. Phosphorylation is considered as a key on-off switch for rapid control of signaling cascades, cell-cycle regulation and several other important cellular functions, which activates or deactivate enzymes, alter protein-protein interactions and their associations, change the protein structures and targets proteins for their degradation. Polypeptide chain will also undergo a number of permanent modifications such as deletion of N-terminal methionine and carboxylation of glutamate, etc.

The primary structure of a protein is determined by the sequence of specific amino acids, encoded by the mRNA, which directs the proper folding of the polypeptide chain into the secondary structure. Folding and oxidation of cysteine thiols into disulfides confers secondary structure on the random-coil polypeptide chain. There are two major components of a protein secondary structure; a region of the polypeptide that folds into a corkscrew shape denoted as the alpha helix and are linear structures of polypeptides, bonding together to form a flat Beta strand. Turns and coils interact chemically with each other to create a unique three dimensional (3-D) structure of the protein. Many proteins have several different polypeptide subunits whose interactions make the final active protein. These interactions among the subunits form quaternary structure. Proteomics is the study of complex proteins, their composition, structures, functions and their interactions with other proteins determining the cellular activities. Proteomics provides a better understanding of an organism and is considered as the key tool in understanding the biological systems. Transportation of a protein to the specific cellular location is often achieved by a proteolytically cleaved signal sequence. After reaching their cellular destinations, proteins perform their functions, the majority are controlled by their posttranslational modifications. Phosphorylation of some proteins will be followed by their ubiquitination (conjugation with ubiquitin), leading to their degradation by the 26S proteasomal complex. Protein ubiquitination and turnover are also stimulated other factors such as oxidative damage and other protein modifications. Proteins also undergo degradation by lysosomal enzymes.

7.3.2 Types of Proteomics

Proteomics is broadly classified into three categories.

1. **Expression Proteomics** is involved in the differential understanding of total protein expression of the cells isolated under normal and diseased or treated conditions, by using 2-D gel electrophoresis and mass spectrometry techniques for comparing and understanding the differentially expressed proteins under the specific conditions. These differentially expressed proteins can be used as the diagnostic markers for understanding the disease-specific manner or as therapeutic targets.
2. **Structural Proteomics** is concerned with the structural complement of the functional proteins. Prediction of a protein structure is performed by homology modeling and confirmed by the X-ray crystallography, electron microscopy (in the case of receptor proteins) and NMR spectroscopy. Structural proteomics provides detailed information about the structure and function of protein complexes located in specific cell organelle. Using structural proteomics, it is possible to identify all proteins present in membranes, ribosomes, and other cell organelles, to characterize all their interactions that can be possible between these proteins and protein complexes.
3. **Functional Proteomics** will aid in understanding the specific function(s) of cellular proteins or protein complexes through the identification of their interacting protein partners. The interaction of an unknown protein with the ones belonging to a specific protein or a complex involved in a particular function would strongly suggest its biological function. Furthermore, a detailed description of the cellular signaling pathways might greatly benefit from the elucidation of protein-protein interactions in vivo.

7.3.3 Gene Expression and Codon Bias Affecting the Protein Levels

One of the primary questions in the proteomics is the maximum amount of a protein expressed in a cell. Levels of protein expression in a cell vary tremendously, from a few copies to more than a million depending on its importance to the cell and the organism. Essential enzymes involved in the intermediary metabolism are often present in the thousands of copies per cell or even more, whereas protein kinases involved in cell-cycle regulation are found only tens of copies. The level of a protein in a cell at a given time is controlled by three factors,

1. Rate of DNA transcription into mRNA,
2. The efficiency of translating the mRNA into protein, and
3. The rate of protein degradation in the cell.

One of the intrinsic determinants of the expression levels of many genes is "codon bias," defined as the tendency of an organism to prefer certain codons over others that code for the same amino acid in a gene sequence. Calculated codon bias values for the genes in the yeast genome ranges from −0.2 to 1.0 (1.0 is considered as the highest level of gene expression). Most yeast genes display codon bias values of <0.25 are expected to be expressed at relatively low levels. Genes with low codon bias values tend to be expressed at low levels, whether analyzed on the basis of mRNA expression or protein levels. mRNA levels correlate poorly ($r < 0.4$) with protein levels in genes with the codon bias values of 0.25 or less. The correlation between mRNA levels and protein levels is much higher ($r > 0.85$) for the most highly expressed genes with codon bias values of above 0.5.

7.3.4 Techniques that are Involved in the Proteomics Studies

Extraction of Proteins from Biological Samples

The study starts with a biological sample comprising a piece of tissue, a flask of bacterial cell culture, a leaf, etc. The sample is pulverized or homogenized or sonicated to yield a soup of suspension containing cells, subcellular components, and other biological debris. Proteins are extracted from this soup through an appropriate technique. The objective of any technique is to obtain as many proteins as possible from the soup with minimum contamination of other biomolecules such as lipids, cellulose, nucleic acids for the proteomics analysis. This can be achieved by using the buffer consisting of;

1. Detergents such as SDS, Tween for solubilizing and separating the membrane proteins
2. Reductants, such as dithiothreitol [DTT], β-mercaptoethanol, which reduce disulfide bonds and prevents the protein oxidation.
3. Denaturing agents like urea, which disrupts the protein-protein interactions, secondary and tertiary structures by altering solution ionic strength and pH

4. Enzymes such as DNAse, RNAse, which digests the contaminating nucleic acids, carbohydrates, and lipids. For extraction of proteins from plant samples, protease inhibitors such as phenylmethylsulfonyl fluoride (PMSF), a serine protease inhibitor, is frequently used to prevent protein degradation during tissue processing.

Methods of Protein Separation
Separation of proteins for the digestion can be achieved by either of the two major approaches.

1. Gel based approach: uses 1-D and 2-D gels for the separation of proteins along with isoelectric focusing (IEF).
2. Column based high throughput shotgun approach: applies HPLC such as reverse phase, size exclusion, ion exchange, or affinity chromatography. Both of these approaches are discussed below.

1. Gel based approach:

Purpose of applying the gel based approach is to separate intact proteins by taking advantage of their diversity in physical properties, especially isoelectric point, and molecular weight. The mixture is separated into a small number of fractions using 1D-SDS-PAGE and IEF or into a large number of fractions (spots) using 2D-gel. These fractions are subjected to protease digestion for their mass spectrometric analysis.

7.3.4.1 SDS-PAGE

SDS-PAGE is the single most widely used analytical separation technique in protein chemistry. This method is particularly useful for protein purification as the proteins are separated by their sizes. In 1D SDS-PAGE, the protein sample is boiled in a sample buffer (also called loading buffer) that contains SDS (anionic detergent), a thiol reductant such as β-mercaptoethanol and bromophenol blue (a low molecular weight blue coloured dye that locates the sample in the gel). Each protein in the sample is fully denatured by the SDS, which opens up into a rod-shaped structure with a series of negatively charged SDS molecules along the polypeptide chain. The sample also contains a low molecular weight dye bromophenol blue for monitoring the electrophoretic run. Under the influence of electricity, the protein-SDS complexes migrate through the cross-linked polyacrylamide gel at rates based on their ability to penetrate the pore matrix of the gel. The main separating gel (resolving gel) is poured between the glass plates along with a short stacking gel of large pore size (4% acrylamide) poured on top of it. The purpose of the stacking gel is to concentrate the protein into a sharp band before it enters main separate gel, which is achieved by utilizing the differences in the ionic strength and the pH between the electrophoresis buffer and the stacking gel. This phenomenon is known as isotachophoresis. The pH of the stacking gel is 6.8 and the band concentrating effect relies on the fact that the glycinate ions of the electrophoresis buffer are negatively charged and have low electrophoretic mobility than chloride ions of loading buffer and stacking gel. Under the electricity, all the ions have to move at an equal velocity to prevent the circuit breakage. This could be possible only at high field strength, which is inversely proportional to the concentration. Here sample and the two charged ions will adjust to the concentration as $Cl^- >$ protein sample $>$ glycinate and will migrate into the separating gel. The pH of the resolving gel is 8.8, which makes the gel more ionized results in the increased interface between the Cl^- and the SDS-protein complexes, which continue to move towards anode with uniform mobility are resolved into bands according to their molecular weight. 1D-SDS-PAGE is prepared with the polyacrylamide varying from 5–15%, with the low percentage of acrylamide, permits the passage of large proteins through the gel. When the bromophenol blue reaches the bottom of the gel, power is turned off and the gel removed between the plates and is stained in a solution containing Coomassie brilliant blue, methanol, water, and glacial acetic acid for a few hours and is destained in a solution containing methanol, water, and glacial acetic acid overnight, which removes the unstained background from the gel. Stained proteins are visible as blue coloured bands on a clear background. A typical 15% gel gives a separation of proteins with molecular weight 10,000 KDa, which can be identified by loading standard molecular weight marker of known molecular weight run on the same gel (Fig. 7.2).

7.3.4.2 Isoelectric Focussing

This method of separation is ideal for amphoteric substances such as proteins or peptides according to their isoelectric point (pH at which the molecule is without any net charge). This technique enables the separation of proteins with a difference in their isoelectric points as little as 0.01 of a pH unit. The most widely used system for the IEF uses horizontal gels in a glass plate or between plastic sheets containing ampholytes (synthetic mixtures of polyamine polycarboxylic acids), with a wide or a narrow pH range chosen according to the sample.

IEF gel is prepared by mixing the ampholytes and acrylamide solution in riboflavin and is poured between a set of glass plates containing spacer. The gel is

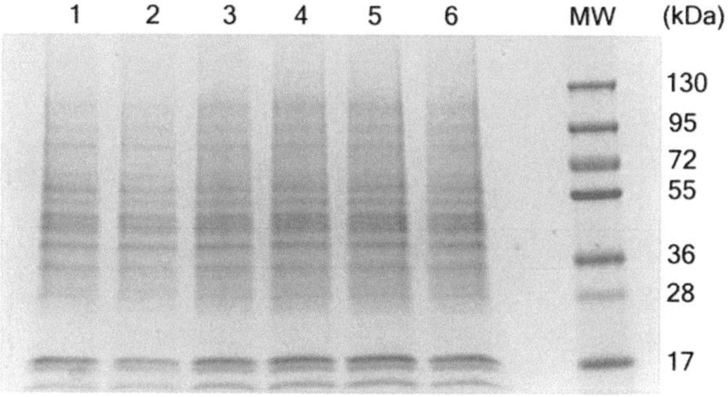

Fig. 7.2 Separation of proteins on an SDS-PAGE gel

photopolymerized by placing in front of a light bank, which generates free radicals for the photodecomposition of riboflavin, which initiates polymerization. After polymerization, the potential difference between the electrodes is applied by placing pre-wetted filter paper comprising of phosphoric acid near the anode and NaOH near the cathode. Under the influence of potential difference, the ampholytes form gradience between cathode and anode. The protein sample is loaded into the wells and the power supply is initiated to the gels. The protein at the pH below their low isoelectric point will be positively charged and will migrate towards the cathode, with a steadily increasing pH. As the sample moves, the positive charge will be decreased correspondingly and will arrive at a point where its pH is equal to its isoelectric point. The protein will now be in the zwitterionic form without any net charge. Hence, no further movement is possible. Similarly, the proteins at the pH above their isoelectric points are negatively charged and will migrate towards the anode until they reach their isoelectric points and will be stationary.

Following electrophoresis, the gel is incubated in 10% trichloroacetic acid solution, which washes ampholytes and precipitates the proteins, which can be stained in a solution containing Coomassie brilliant blue, methanol, water, and glacial acetic acid and is destained in a solution containing methanol, water, and glacial acetic acid (Fig. 7.3).

7.3.4.3 2D-Gel Electrophoresis

2D gel electrophoretic separation of proteins is the best available method resolving the highly complex proteins and has become synonymous with proteomics. 2D gel electrophoresis combines isoelectric focusing, which separates proteins in a mixture on the basis of their isoelectric point with the polyacrylamide gel, which further resolves the proteins according to their molecular weight. Thus, 2D-gel

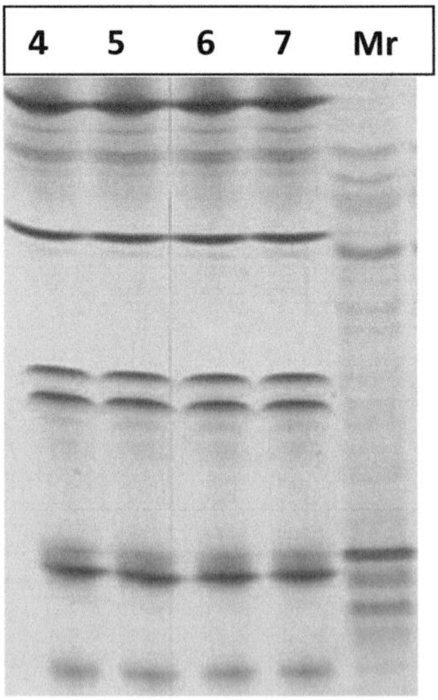

Fig. 7.3 Isoelectric focusing of the proteins with pH 4–7 and an IEF marker

electrophoresis resolves proteins in the first dimension by their isoelectric point and in the second dimension by their molecular weight. Even though the 2D-gel was introduced in the early 1970s, it was not widely used for many years due to the technical difficulties in getting the proteins separated by their isoelectric points into the polyacrylamide gels. This situation was much changed after the introduction of specific 2D-SDS-PAGE systems that use immobilized pH gradient (IPG) strips, which facilitates the transfer of proteins from the IPG strip into the SDS-PAGE slab gel. IPG strip consists of polyamino polycarboxylic acid ampholytes, immobilized on supports which can reproducibly create stable pH gradients. IPG strips with different pH range are commercially available as narrow and broad pH range strips.

IPG strip is hydrated with appropriate buffer and the sample protein, loaded into the strip and is connected to the power supply for achieving the dimension 1 isoelectric focusing. After the dimension 1, the strip is treated with a buffer containing β-mercapto ethanol, SDS and is joined to the SDS-PAGE slab gel for the dimension 2 separation. In this context, the IPG strip containing the focused proteins acts as the stacking gel in 1D-SDS-PAGE, which resolves the focused proteins on the SDS-PAGE slab gel in the manner similar to that of 1D-SDS-PAGE. Proteins separated by the 2D gels are visualized by staining either with silver or Coomassie brilliant blue stains (Fig. 7.4).

The proteins with a specific interest are eluted from the gel and are subjected to the protease digestion for their cleaving at specific amino acid residues to yield the fragments for their identification through Mass spectrometry. Peptide fragments of 6–20 amino acids are ideal for the mass spectrometry analysis and database comparisons, which can be achieved by the digestion with proteases listed in Table 7.1.

7.3.4.4 High Throughput Shotgun Proteomics

With the gel-based analysis for the proteomics, new global approaches such as shotgun proteomics have been gained wide acceptance. Shotgun proteomics approach is developed based on high-resolution liquid chromatography coupled

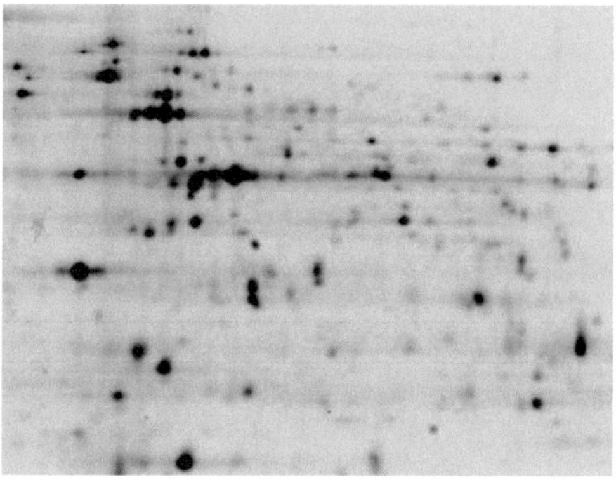

Fig. 7.4 2D electrophoresis gel

Table 7.1 List of proteases that are used in the peptide digestions with their target sites

S. No.	Protease	Target amino acids
1	Trypsin	Lysine and arginine residues without Proline at the C-terminal end
2	Glu-C	C-terminal end of glutamate residues
3	Lys-C	C-terminal end of lysine residues
4	Chymotrypsin	Tyrosine, tryptophan and phenylalanine
5	Cyanogen bromide	Methionine
6	Acid hydrolysis	Cleavage at aspartyl residues
7	Hydroxylamine	Cleavage of aspargine and glycine residues
8	BNPS skatole	Cleavage at tryptophan

with high-speed tandem mass spectrometry. In shotgun proteomics, the proteomes existing in the complex form are analyzed directly in a quasi-random nature. High-resolution liquid chromatography such as HPLC and mass spectrometry are employed for separating the proteins and analyze them through a mass spectrometer. Utilization of HPLC to separate proteins is in practice since the 1970s, where Chang used steric exclusion and ion exchange columns to separate three isoforms of creatine phosphokinase and lactic dehydrogenase. However, its first global application in proteomics has come into existence by the introduction of Multidimensional Protein Identification Technology (MuD-PIT) in 1991.

Utilization of MuD-PIT involved the application of two liquid chromatographic separation processes to resolve protein complexes. Here, samples in their crude form such as whole cell lysate are digested by proteases, such as trypsin. The peptide fragments are subjected to a first dimension separation, comprising of strong cation exchange chromatography, to separate the peptides based on their net positive charge. The samples eluted from the cation exchange separation are subjected to a second separation stage, reverse phase chromatography, where the peptide fractions are further fractionated based on their hydrophobicity. The samples are further transferred on to a tandem mass spectrometer where peptides are subjected to tandem mass sequencing. MUDPIT approach of proteomics study has provided with the identification of 1500 new proteins in *Saccharomyces cerevisiae* with no biasness to any type of protein in a span of 24 h.

With HPLC based shotgun proteomics approach is becoming an alternative to the gel based approach, other high throughput methods involving the mass spectrometers have been developed. Among them, LC-MALDI is one of the vital high throughput methods, which effectively replaces RP-LC/MS with spotted matrices of fractionated samples. Integration of capillary electrophoresis with mass spectrometer has gained a lot of popularity due to the decreased cost and chemicals. Usage of affinity chromatography through a specific pull down methods for proteomics such as IMAC (immobilized metal ion affinity chromatography) phosphoproteomics and glyco-proteomics shotgun profiling are gaining popularity. Development of automated nanoscale-direct ESI infusion for high throughput electrospray ionization (ESI) has enabled the high speed of samples.

Rapid improvements in the mass spectrometer technologies, such as new generation quadrupole ion traps and quadrupole time-of-flight (Q-TOF) have facilitated the development of high resolution LC methods including the 'ultra-performance' [UPLC] technology, coupled with increasingly matured mass spectrometry proteome database search engines, Phenyx, Sequest, Paragon/ProteinPilot and Mascot, which are capable of sequencing and identifying thousands of proteins without much optimization.

7.3.4.5 Mass Spectrometry for Peptide Sequence Analysis

Two completely different types of mass spectrometers are used for the proteomics studies; (1) MALDI-TOF instruments (2) ESI-tandem instruments. These two mass spectrometers operate in entirely different ways and generate differently, but complementary information. Both these mass spectrometers have three essential parts; (1) Source produces ions from the sample. (2) Mass analyzer resolves ions based on their mass/charge (m/z) ratio. (3) A detector detects the ions resolved by the mass analyzer. Mass spectrometer converts the components of a sample containing the protein mixture to ions and analyzes them based on their mass/charge. The data generated by the detector is automatically recorded by the data system, which can be retrieved for manual or computer-assisted interpretation. Modern mass spectrometers are controlled by sophisticated computers and software programming.

1. MALDI-TOF

MALDI-TOF is the standard acronym for Matrix Assisted Laser Desorption Ionization-Time Of Flight. MALDI refers to the source and the TOF refers to the mass analyzer. Protein sample to be analyzed is mixed with a chemical matrix, containing a small organic molecule with a desirable chromophore that absorbs light at a specific wavelength. A mixture of sample and the matrix is spotted onto a watch glass or slide and then allowed to evaporate in air. The evaporation of residual water and other solvents from the sample leads to the formation of a crystal lattice into which the peptide sample is integrated. The sample containing crystal is placed into the source, equipped with a laser, which fires a beam of light. The crystal absorbs photons from the beam and becomes excited. This excess energy is then transferred to the peptides or proteins in the sample, leading to the ionization, producing both positive and negative ions, depending on the nature of the sample. Positive ions are formed by accepting a proton as they are ejected from the matrix. TOF mass analyzer measures the time taken by the ions to fly from one end of the analyzer to the other and strike the detector, which is proportional to their mass/charge values. Greater is the mass/charge, faster their flight.

2. ESI-TANDEM Mass Instruments

ESI tandem is the standard acronym for Electrospray Ionization Tandem Mass Spectrometry. ESI refers to the source for the production of ions and Tandem refers to a series of mass analyzers that are capable of performing two-stage or even multistage mass analyses of ions used in ESI-2D-MS instruments, comprising of quadrupole and ion trap analyzers along with TOF.

The peptides or proteins to be analyzed by ESI are fragmented in an aqueous solution. Peptides exist as ions in the solution due to their functional groups whose ionization is dependent on the pH of the solution. Carboxylic acids are unionized (protonated) at a pH less than 3.0 and ionized at pH above 5.0. In contrast, N-terminal amines are ionized below pH 7.0. Overall, protonation of the amines takes place at pH 3.5 or even less conferring overall net positive charge to peptides and proteins. Deprotonation of the amines and carboxyl groups occur at a pH above 8, conferring overall negative charge. Fragmentation of peptide ions is favored by positive charges on the peptide ions. The sample enters source through HPLC and passes through a stainless-steel cone or needle maintained at a high voltage. As the flow stream exits the needle, it sprays out as a fine mist of droplets, containing peptide ions, components of the HPLC mobile phase, which are further dissolved either by heating or by passing through nitrogen gas. The peptide ions are then passed from the source into the tandem mass analyzer. Three types of tandem mass analyzers are commonly used for proteomics work. These are the quadrupole time of flight (Q-TOF), triple quadrupole and the ion trap, which differs in the working but performs a similar type of analysis. With a mixture of peptide ions generated from the ESI, tandem mass analyzers select a species with single mass/charge (m/z) at one time and the corresponding ions will be subjected to collision-induced dissociation (CID), resulting in the fragmentation of the peptides into ions, analyzed according to their m/z to produce an ion spectrum.

7.3.4.6 Peptide Mass Finger Printing

Identification of protein sequenced by the mass spectrometry is generally done by using the Peptide Mass Fingerprinting (PMF) (also known as Peptide Mass Mapping). PMF is an in silico technique, in which the mass spectrum obtained by the mass spectrometer is used to measure the masses of the proteolytic peptide fragments. The standard technique is to separate the proteins either by 2D gel electrophoresis or HPLC. The separated protein is cleaved with specific endoproteases and their molecular masses are determined by subjecting to either MALDI-TOF or ESI-Tandem Mass spec. Separated proteins can be identified by comparing the experimentally determined masses obtained through the mass spectrum with that of the theoretical peptide masses obtained from the databases such as Swissprot.

7.3.4.7 Advanced Proteomics Methods

1. Isotope-coded affinity tags (ICAT): Is a gel-free method for quantitative proteomics that relies on chemical labeling reagents. ICAT consists of three chemical probes, (1) defined amino acid side chain, (2) an isotopically coded linker and (3) Tag for the affinity isolation of labeled proteins/peptides. Two protein samples with one isotopically light labeled with deuterium and the other one is heavy labeled with ^{13}C. Both the samples are combined with isotope-coded tagging reagents and are analyzed on LC-MS. Tags. This technique is mainly used for the relative quantification of proteins present in two or more biological samples. Visible isotope-coded affinity tags are the additional method in ICAT—Visible tag that allows the electrophoresis position of tagged peptides to be easily monitored.

2. Isobaric Tags for Relative and Absolute Quantification (iTRAQ): Is another non-gel-based technique used to quantify the changes in a proteome, based on the covalent labeling of the N-terminus and side-chain amines of protease digested peptides. 4-plex and 8-plex are the reagents that can be used for the labeling of all peptides from different samples. These samples can be analyzed by using mass spectrometry along with the soft wares such as j-Tracker, j-TraqX 20.
3. Absolute Quantification (AQUA): Studies the absolute quantification and the modification sites of proteins such as covalent modifications, post translational modifications, etc. Covalent modifications can be used for making synthetic proteins, which are chemically identical to the natural posttranslational modifications. Such types of peptides are used to quantify the post translational modifications of proteins using the tandem mass spectrometer.
4. Electrospray ionization-Quadrupole Time of flight mass spectrometer (ESI-QTOF): is the combination of a mass spectrometer with a very good resolution. For ESI, proteins are ionized in a solution, which carries multiple charge state and low mass/charge of values >2000. ESI-QTOF detector has a very good mass accuracy in this scanning range, which produces more accurate mass measurements for proteins.
5. Surface-Enhanced Laser Desorption/Ionization Mass Spectrometer (SELDI-TOF-MS): is used for the detection and analysis of protein mixtures in clinical samples for comparing the protein levels with and without a disease.

7.3.5 Applications of Proteomics

Proteomics comprises of four major applications. (1) Protein mining (2) Protein-expression profiling (3) Protein-network mapping, and (4) Mapping of protein modifications.

1. Protein Mining

Purpose of protein mining is to identify all the proteins in a given tissue or cell culture and catalog the proteome, rather than using the microarrays or expression studies to infer its composition. Mining is the ultimate brute-force exercise of proteomics, which resolves proteins to the possible extent and then uses mass spectrum and associated databases for the identification of the proteins in the simplest way.

2. Protein-Expression Profiling

Protein-expression profiling is the identification of proteins of a sample involved in a function of a particular state of the organism or a cell when exposed to a stimulus such as a drug, chemical, or even a disease. It is a specialized form of mining, practiced for comparing two different states of a particular system, such as healthy and diseased cells can be compared to determine the differentially expressed proteins, which can be used as the potential targets for the therapy.

3. Protein-Network Mapping

It is an approach to understand how different proteins interact with each other in living systems. Most of the proteins perform their functions in a close association with other proteins, which can be used for identifying the function of specific proteins. Much has been studied about the protein-protein interactions of complex proteins involved in complex signaling pathways through in vitro studies involving the yeast two-hybrid system and protein microarrays. Protein-network profiling would offer the ability to assess at once the status of all the participants in the pathway. As such, protein-network profiling represents one of the most ambitious and potentially powerful future applications of proteomics.

4. Mapping of Protein Modifications

Protein modification mapping is applied for identification of the location and the factor responsible for the modification of a protein. Most of the posttranslational modifications such as prenylation, methylation, govern the structure, function, and turnover of proteins. In addition to it, function and turnover of proteins are heavily altered by many environmental factors, chemicals, and drugs. A variety of analytical tools such as Eastern blotting have been developed to identify the post translational modifications and their nature, but their precise sequence sites of a specific modification cannot be identified. Proteomics approaches using protein modification mapping provides the best means of establishing both nature and sequence specificity of posttranslational modifications. Along with this, simultaneous characterization of the modification status of regulated proteins in a network offers a powerful extension to the proteomics technology, which provides a fresh avenue for understanding the changes in the proteome in response to a drug or a chemical stimulus in the living systems.

7.4 Metabolomics

Metabolomics is a new member of the Omics family, involved in the systematic identification and quantification of the small molecule metabolic products in the cell, tissue, organ, biological fluid or an organism, which makes the metabolome of a biological system at a specific point of time. The term metabolome was coined in the year 1998 as a field of study, which uses functional genetics approach for understanding the molecular complexity of life. It is the comprehensive, qualitative and quantitative study of all the small metabolites (less than or equal to 1500 Da) in an organism. Metabolomics focus on the metabolites and the intermediary compounds that form the macromolecular structures and other small molecules that participate in vital cellular functions, such as signaling or secondary metabolites. This study excludes the polymers of amino acids and sugars.

7.4.1 Approaches for Metabolomics

Metabolites are the organic compounds that are either starting materials or intermediates in various reactions that occur in organisms. These are the precursors for constructing more complex molecules, or they can be further broken down into simpler ones. Intermediary metabolites may be synthesized from other metabolites and often release chemical energy. The interaction of the metabolites can further be divided into anabolism, involved in the synthesis of macromolecules and catabolism involved in the release of energy due to the breakdown of complex materials such as proteins, lipids, within an organism. Technically, metabolomics is the measurement of all metabolites that are produced in a living organism, which is not practically possible with the available technology. It might probably require a plethora of complementary technologies as no single technology or a technique can provide comprehensive information about the entire list of metabolites in any specific organism. As metabolomics forms an extensive network of biochemical interactions, its analysis is being done in three major approaches.

1. Metabolite profiling
2. Metabolic fingerprinting
3. Metabonomics

7.4.1.1 Metabolite Profiling

Metabolite profiling is an analytical approach for relative quantification of a number of metabolites from biological samples generated from a specific tissue, organ or organism and analyzing them through highly accurate techniques. Irrelevant of the technique employed, two types of metabolite profiling approaches are followed in the current metabolomics, (1) targeted, and (2) untargeted analyses.

1. A targeted approach is used for monitoring a set of predefined and known substances, which allows absolute quantification and accurate identification.
2. Untargeted analytical methods are applied to find the features of all compounds, which are currently unknown or unidentified at the time of measurement. Untargeted approaches are highly suitable for detecting changes in unexpected parts of the metabolome, leading to new scientific hypotheses.

Metabolite profiling involves the identification and measurement of hundreds or potentially thousands of metabolites. It comprises of a perfectly streamlined pipeline for extraction, separation, and analysis of the metabolites making it feasible for the measurement of large numbers of metabolites in a robust and quantitative manner. The metabolite profiling is performed under two approaches involving Mass Spectrometry and NMR methodologies. Platforms used for the metabolite profiling of the biological samples are as follows:

1. Gas Chromatography Coupled to Mass Spectrometry (GC-MS)
GC-MS is the combination of two complementary technologies, in which the metabolite compounds are Initially separated by the gas chromatography and are then transferred on to a mass spectrometer for the detection. The separation is achieved in six steps.

A. Extraction
Preparation of a metabolite extract should be highly non-selective and as comprehensive as possible, minimizing the specific treatments, which stabilizes one set of metabolites and degrade or modify others. Further, it may be necessary to separate the extracts into fractions for profile tracing of metabolites when the sample is dominated by a small number of highly concentrated metabolites.

B. Derivatization
This step necessary to render the metabolites volatile and amenable to GC-MS. Derivatization is done by using alkylating, acylating and silylating reagents. Presently, trimethylsilylation is the best available choice for the derivatization of the sample. Trimethylsilylation uses the most comprehensive reagent and thus compiles the best with all requirements of a non-biased metabolite profiling. Other reagents are highly specific for a single metabolite class.

C. Separation by GC
Separation of metabolites by GC involves highly standardized conditions. Even a slight changes in the gas-flow conditions, temperature programming and the type of capillary column affect chromatographic retention, and can even alter the order in which compounds are eluted.

D. Ionization
The most widely used ionization method for GC-MS is electron impact (EI) ionization, which is a robust, and reproducible approach that is not subjected to ion suppression effects such as mutual interference between compounds, leading to one or both being underestimated or even undetected. EI transfers a fixed energy load of −70 eV to compounds, which are then converted to molecular ions from the GC outlet into a high vacuum. The energy load exceeds the first ionization energy of all molecules, leading to very efficient generation of molecular ions with almost all molecular ions carrying one positive charge, generating the highly reproducible, compound-specific mass spectral fragmentation pattern, that aids in the identification of compounds.

E. Detection
Three types of mass-detection devices are currently used in GC-MS analysis. Throughput of GC-QUAD-MS systems comprising of Single quadrupole detectors (QUAD) can analyze 10–20 samples per day. GC-TRAP-MS technology comprising of ion-trap and time-of-flight detectors with reaction monitoring capability, in which a pre-defined fragment mass of a parent is sampled and is further subjected to

secondary fragmentation for generating the daughter fragments. This increases selectivity and minimizes the chemical noise especially while analyzing the trace compounds in complex samples. GC-TOF-MS systems can analyze 10–50 scans per second, allowing 30–40 samples per day.

F. Evaluation

Metabolites are identified by matching their chromatographic retention times and mass spectral fragmentation patterns to known and predicted information available in the databases. Typical GC-QUAD-MS software requires expert knowledge about the characteristic fragment masses and retention time of each metabolite and the pitfalls that can lead to misidentification. This manually supported process requires ~2 min per metabolite, which permits around 20 chromatogram files to be evaluated per day. GCTOF-MS systems such as GC-TOF-Pegasus II MS from Leco Corp Inc., St. Joseph, USA are having enhanced software capability, in-built mass-spectral correction for co-eluting metabolites, calculation of retention-time indices, and automated picking of a suitable fragment mass for selective quantification. This machine takes around 2 days for the extraction and derivatization of a batch of 50 samples. Analysis and evaluation of one sample require 35–45 min. Using GC-MS, metabolites with different and almost identical mass spectra such as isomers can also be separated. GC-MS provides quantitative information, which can be widely used for clinical diagnostics and large-scale profiling of complex biological samples.

2. Liquid Chromatography Coupled to Mass Spectrometry (LC-MS)

Liquid chromatography coupled to mass spectrometry (LC-MS), equipped with HPLC has a unique ability to separate the high molecular weight compounds, which cannot be analyzed by gas chromatography. Coupling HPLC to mass spectrometry further provides increased selectivity, unbiased detection, and information about the structures of the separated compounds. Here, metabolites are ionized by ESI through an atmospheric pressure process, transferring the elutes that are emerging from HPLC column into the gas phase, suitable for the mass analysis. Metabolites entered into the mass spectrometer as charged molecules are transported in an electrical field between the end of the column and the entrance of the mass spectrometer. Ionization of the metabolites can occur *via* protonation (+) or de-protonation (−) leading to a single or multiple ions. The presence of multiple ions shifts the mass-to-charge ratio (m/z) into the scanning range of a mass analyzer and the structural information is obtained by collision-induced dissociation (CID). Parent ions produced by ESI are isolated and accelerated inside the mass spectrometer using quadrupole mass analyzer, forcing them to collide with molecules of the bath gas such as helium or argon. The resulting fragment spectrum can be compared with fragmentation libraries for known chemical structures. Depending on the mass analyzer used, several fragment spectra per second can be performed. Using quadrupole ion traps, it is further possible to generate multiple fragment spectra of selected fragments of a parent ion mass. Triple quadrupole instruments allow the quantification by a single-reaction monitoring (SRM) system. A specific mass ion or the metabolite of interest is selected to the first quadrupole, fragmented in the second quadrupole and the corresponding fragment is then selected in the third quadrupole. SRM provides highly specific mass-ion traces for pre-selected metabolites, which can be quantified by peak integration, with high selectivity and sensitivity. LC-MS has enormous potential for metabolite profiling, with high-resolution mass spectrometry, capable of detecting 11,000 mass ions in a single spectrum and high-resolution chromatography will further increase the number of metabolites detected. In spite of so many applications, LCMS has quite a few demerits. It can only be applied to metabolites that have known fragmentation pathways and to a certain number of metabolites per run. The biggest challenge for metabolite profiling is to develop an automated procedure for the evaluation and quantification of metabolites from raw chromatograms and to identify a huge number of unknown metabolites and analytes detected by these powerful analytic platforms, by combining LC-MS with Fourier-transform ion cyclotron resonance mass spectrometry (FTICRMS) and NMR spectroscopy.

3. Fourier-Transform Ion Cyclotron Resonance Mass Spectrometry

In Fourier-transform ion cyclotron resonance mass spectrometry (FTICRMS), extracts are directly infused into the Mass spectrometer using soft ionization techniques, to gain fingerprints of the molecular ions present. It requires a highly accurate mass analyzer for generating the definitive empirical formulae of several hundred ions. Profiling of FTICRMS allows high quality separation of analytes. But, this Mass spectrometer is having three major limitations. (1) lack of chromatography renders the incapability of distinguishing between the isomers of identical molecular masses, making unambiguous discrimination of many metabolites impossible. (2) there is no documentation of vigorous method validation, required to support its utilization for metabolite analyses, (3) Heavy cost for the sample analysis.

4. Nuclear Magnetic Resonance Spectroscopy (NMR)

An entirely different and alternative approach for the detection and quantification of metabolites is the utilization of Nuclear Magnetic Resonance (NMR) spectroscopy for the detection and quantification of metabolites, using the magnetic properties of isotopes of the constituent atoms. As hydrogen, carbon, nitrogen, phosphorus, and oxygen have magnetic isotopes that are detectable by NMR, this approach will help in the detection of a broad range of metabolites. Highly developed computational analysis and chemometric software enable the rapid processing of acquired spectral data and identification of metabolites from the signals. The NMR analysis of metabolites is important for unequivocal determination of metabolite structure, which is one of the major bottlenecks of metabolite profiling. Further, NMR spectroscopy can be applied to unravel the complex metabolic fluxes by following labeled atoms through metabolic intermediates at the atomic level. However, NMR is capable of detecting only fewer metabolites and is having a small dynamic range than that of MS-based technologies. The lower sensitivity of NMR-based techniques restricts their quantification to a very few classes of metabolites.

7.4.1.2 Metabolic Finger Printing

Metabolic fingerprinting is a high-throughput analytical technique, which mostly uses spectroscopic methods for the classification of samples on the basis of their origin or biological relevance. It is the application of a broad analytic technology to discover and discriminate between two biological samples or two different genotypes by making some big differences. It also provides information that helps to orientate a research project. Metabolic fingerprinting helps in the detailed analysis of complex reaction networks and uncovering new drug targets.

Metabolic fingerprinting is currently performed by using two techniques, LC-MS (liquid chromatography-mass spectrometry) and ^{1}H NMR (proton nuclear magnetic resonance) spectroscopy. LC-MS technique is successfully applied in various fields related to plant research such as chemotaxonomy, plant biochemistry, food chemistry, and in the quality control of medicinal plants. The main advantage of using the mass spectrometry is its high sensitivity, which permits the detection of compounds with very low molecular weight, even at the concentrations below nanogram per ml (at optimal conditions).

With its long history in the assessment of plant metabolites, ^{1}H NMR spectroscopy is one the most frequently used technique for metabolic fingerprinting. It is also applied for quality control in food science and technology. NMR techniques are reproducible with rich structure information. The only essential requirement for compound detection in ^{1}H NMR experiments is the availability of observable protons in a molecule, resulting in the applicability of ^{1}H NMR to a wide range of plant metabolites detection that was not detected by LC-MS due to insufficient ionization. ^{1}H NMR spectroscopy has the major advantage of producing the result with predominant reproducibility. However, NMR spectroscopy has a very low sensitivity compared to modern mass spectrometry instrumentations. Compounds of low concentrations may not be detectable with NMR spectroscopy and signal overlapping occurs while analyzing the plant extracts containing more than one compound.

7.4.1.3 Metabonomics

The word metabonomics was coined by Jeremy Nicholson, Elaine Holmes and John Lindon from two Greek words "meta" means change and "nomos" means rules. Metabonomics is defined as the quantitative measurement of the time related to the multi-parametric metabolic response of living systems to pathophysiological stimuli or genetic modification. Metabonomics deals with the response pattern of numerous metabolites or analytes, whose changes are reflected in physiology, as an indicator of efficacy, toxicity, disease, or physiological change. This pattern may be searchable for specific analyte or metabolite information, which might be responsible for the change. Metabonomics is used by the toxicologists for understanding the change made by a specific analyte or metabolite using Biofluid nuclear magnetic resonance spectroscopy (NMR) or Biofluid Mass Spectrometry. Mass spectrometry (MS) and its hyphenated derivations, such as LC-MS, GC-MS, etc. have a sensitivity advantage, which makes it indispensable particularly for the identification of novel biomarkers. On the other hand, the non-selectivity, lack of sample bias, cross-laboratory and cross-platform reproducibility of NMR makes it an extremely important tool for toxicologists for the screening of metabolites.

NMR platform carries a clear advantage while dealing metabonomics. Magic Angle Spinning (MAS NMR) is a technique, which can directly analyze a small quantity of intact tissue. The principle of Magic Angle NMR is when samples are spun rapidly at 54.7° relative to the applied magnetic field (magic angle), line-broadening effects that would ordinarily obfuscate a proton spectra of a solid sample, can be minimized. This technique is carried out by placing a few milligrams of intact tissue into a specially designed rotor and centrifuged at high speed within the bore of the magnet. MAS-NMR is a tool for linking biofluid changes to the mechanism of action in target tissues. This technique is particularly important when products of intermediate metabolism dominate the observed changes in biofluids and are difficult if not impossible to link to any specific target organ. As this technique requires specialized equipment and expertise, it has received much less attention than traditional biofluid NMR for metabonomics studies.

Perhaps, one of the highest impact metabonomics has made is in the area of clinical toxicology. A significant clinical problem with transplant patients is differentiating between patients undergoing graft rejection and those suffering from cyclosporin toxicity. Cyclosporin is an immunosuppressive drug frequently given to transplant patients, whose clinical presentation is very much similar to graft rejection. This problem has been addressed by collecting urine from the patients undergoing kidney transplantation that were given cyclosporin and monitored by NMR spectroscopy. Application of NMR coupled with pattern recognition techniques and trajectory analysis had clearly differentiated a graft rejection patient and patient succumbing to cyclosporine, which is a true clinical metabonomic application, as the spectral pattern and trajectory changes were used to differentiate the cyclosporin toxicity, without any specific biochemical marker. Metabonomics has added value in many areas of biology, particularly in toxicologic sciences. This technology has been most readily accepted and adopted by the industry, whereas the academic community has just now started to recognize its importance. It can be anticipated that metabonomics will be extremely useful in completing the omics circle from genomics to metabolomics.

7.5 Exposomics

Exposome is defined as the measure of all the exposures of an individual in a lifetime and how those exposures are related to health. An individual's exposure begins before birth as it takes the signals from environmental and occupational sources. Understanding how exposures to our environment, diet, lifestyle, etc. will have an impact on our own unique characteristics such as genetics, physiology, and impact of epigenetics on our health can be well articulated. There are two types of exposomes; (1) External exposome and (2) Internal exposome.

7.5.1 External Exposome

The external exposome is the exposure of an individual to external environmental factors such as air pollution, water pollution, dust, UV radiation, etc. Existing population-based studies on environmental exposures, such as the ESCAPE project, along with fully controlled short-term intervention studies, such as the Oxford Street Randomized Trial have shown that acute changes can occur in lung and heart function at low or very low levels of exposure to air pollutants. Analogous studies have been conducted on water contaminants, where genotoxic effects have been observed. One of the main goals of external exposomics is to develop models for short term and long term effects. Development of new land use regression (LUR) models for short term effects such as ultrafine particles (UFP), oxidative potential, personal measurements of particulate matter less than 2.5 μm in diameter (PM2.5) and to investigate the potential for exposure misclassification in using outdoor models of exposure at the residential address as the sole means of exposure assessment. A Detailed 24 h PEM for particulate matter less than 2.5 μm in diameter (PM2.5) and ultrafine particle (UFP) estimates were conducted on approximately 200 participants with data on individual positioning and accelerometry from smartphones in four European countries. UFP exposures for individuals living near to traffic and in background locations suggested that other microenvironments such as journeys, work, home indoors, etc. are important contributors in determining levels of personal UFP exposure. The identification of long-term effects, using GPS-based techniques, smartphones, accelerometers, and PEM to detect pollutants can provide accurate and instant estimates of changes in human exposure and in physical activity.

7.5.2 Internal Expsome

Internal exposomics refers to the changes in the organism with short term exposure to different pollutants. Soterious Kyrtopoulos has identified a few biomarkers related to the exposure in short-term experimental studies and associated metabolic pathways that are potentially linked with health risks. Using high-resolution LC-MS metabolomics, large numbers of small molecules with molecular mass less than 2000 Da, in biological fluids were isolated and characterized. Small molecules include the metabolites that are affected by environmental exposures and are related to inflammation, oxidative stress and various metabolic pathways.

Analogous adductomic technologies will be pioneered for detecting the protein adducts in an untargeted fashion, which will provide a global picture of biomarkers of an individual exposure either to electrophiles or chemicals that are metabolically activated into electrophiles. Because they can directly modify DNA and important proteins, reactive electrophiles are important constituents of an exposome. Due to their greater abundance and residence times in human blood, adducts on the circulating proteins hemoglobin and human serum albumin (HSA) are preferable to those of DNA and glutathione for characterizing adductomes.

Exposomics is one of the rapidly developing fields of exposure assessment, by using omics technologies. It aims to develop a novel approach for the assessment of exposure to high priority environmental pollutants, by characterizing the external and the internal components of the exposome. The primary goals of exposomics are;

1. To provide broad coverage of chemicals, chemical mixtures, outcomes, and life stages.
2. To develop a more robust scientific basis for assessing the health effects of environmental agents.
3. To perform exposure assessment at the personal and population levels by using the tools and methods developed for personal exposure monitoring (PEM).
4. Applying multiple "omics" technologies for the analysis of biological samples, which can be used as internal markers for specific external exposures.

The search for understanding the relationships between external exposures and changes in the profiles of molecular features of same individuals constitutes a novel advancement in the development of "next generation exposure assessment" for pollutants, chemicals with changes in their composition. The linkage of external exposures with disease risks opens the way to the development of exposome-wide association studies (EWAS).

7.6 Connectomics

A connectome is defined as a complete map of all neural connections in the brain of an organism. It is also referred to as a wiring diagram representing the molecular connections between neurons with the brain as an electronic device, axons, dendrites as wires and neuron bodies as components. Connectome comprises of two components. (1) Synaptome, that deals with the specific molecular state of each synaptic connection and (2) Epigenome, which provides the information about any significant changes in the nucleus of each neuron. Connectomes are divided into (1) Whole system connectomes, providing the complete information at the level of whole brains such as fly connectomes, mouse connectomes, human connectomes, whale connectomes and (2) Subsystem connectomes, which provide the neural map of a specific region in the brain such as thalamus (thalamic connectomes), cortex (cortical connectomes).

Connectomics is another omics involved in the study of connectomes made by the comprehensive maps of all neural connections of an organism's nervous system. The principal goal of connectomics is to provide comprehensive mapping and analysis of brain connectivity, from the micro-scale of individual synaptic connections between neurons (Microconnectomics) to the macro-scale of brain regions and interregional pathways (macroconnectomics). Microconnectomics builds the maps of neural circuits in a detailed fashion by including every axonal connection and macroconnectomics attempts to map brain connections at a larger scale.

7.6.1 Need for the Connectome

Since a long time, scientists have been suspecting that striking features of human mental behavior right from general abilities like intelligence to afflictions like depression and schizophrenia are correlated to specific features of the brain. Till date, there is no single precision tool designed for the complete investigation of these hypotheses. With the scope of constructing human connectomes, it will be possible to effectively address the fundamental questions about abilities and the behavior of the human brain. Comparing the connectomes of different human brains will reveal more information about mental exceptionality and pathology, which will increase the scope to the development of advanced and targeted treatments, including better designer drugs, precise surgical interventions, and custom neural prostheses. Another significant reason for the construction of human connectomes is the storage of memories. Many neuroscientists hypothesize that memories are initially stored in the synapses between neurons and new memories are formed when these synapses are strengthened and weakened. Scientists will be able to investigate the neural underpinnings of memory storage and retrieval only with the advancement of connectomics.

Till date, the only organism with a complete connectome is *C. elegans*, a one and a half mM organism with a neural network of 300 neurons and ~7000 synaptic connections. The complete synaptome or epigenome maps for *C. elegans* are not still available. Construction of the *C. elegans* connectome has taken more than 12 years, during which every neuron has been individually identified, its precise location has been determined and its projections to other neurons were traced and cataloged. The only tool available for this work was manual visual recognition and discrimination. *C. elegans* neural pathways were identified manually by tracking their convoluted paths through endless microscopic images.

A human brain is a breathtakingly intricate structure composed of 100 billion neurons with more than 700 trillion synaptic connections, which is 11 times more complex than that of *C. elegans* in its connectivity. Construction of a complete human connectome is inconceivable with existing technology, for two reasons. (1) Electron microscopes currently in existence are not sufficient for imaging the comparatively vast volume of tissue in the human brain. (2) Interpretation of these images and tracing the projections emerging from each neuron is still being done manually by clicking through images in real time.

Invasive techniques for localizing brain regions and tracing anatomical connections are useful in building the human brain connectome. Tracers are injected into a candidate brain region that is taken up inside the cells and transported along the axon. Post-mortem histological staining then reveals the distribution of the tracer stained axons and their connections with distant cells. Tracer techniques are exquisitely precise and accurate for mapping the connections traveling in different pathways or emerging from different cell types or layers. Viral tracers are used for the selective labeling of monosynaptic or multi-synaptic connections.

7.6.2 Measurement of Regional Connections in the Living Human Brain

In-vivo inter-regional connections in the human brain are mapped by using Magnetic Resonance Imaging (MRI), using two different approaches, which rely on very different principles.

7.6.2.1 Diffusion Tractography

This MRI tracker aims to infer the tracks of axon bundles millimeter-by-millimeter as they traverse the brain's white matter. Diffusion tractography studies the anisotropic diffusion of water in and around the axons. Free molecules diffuse equally in all directions. But the presence of semi-permeable boundaries in the tissue will hinder the diffusion rate along some orientations but not others. In the white matter of the human brain, axonal membranes and myelin sheaths hinder the diffusion perpendicular to the axon, permitting the diffusion to occur faster along the axon. Using diffusion-weighted MRI, anisotropy and the peak diffusion orientations in each imaging voxel can be measured. The diffusion orientations are related to the orientations of bundles of coherently oriented axons passing through the voxel. By tracing these orientations voxel-to voxel through the white matter, diffusion tractography reconstructs these bundles and then interregional connections.

7.6.2.2 Resting State Functional MRI (fMRI)

Resting state functional MRI measures spontaneous fluctuations in the blood-oxygenation-level-dependent (BOLD) signal in grey matter regions and estimates statistical dependencies between these BOLD time series, usually expressed as cross-correlations. Instead of measuring the brain connections directly, fMRI expresses these connections as statistical dependencies between the patterns of grey matter activity. fMRI connectivity approaches do not attempt to measure brain connections directly. Instead, they express connectivity as statistical dependencies between patterns of grey matter activity. At rest, regional brain activity exhibits low frequency oscillations that are correlated across distant brain regions. Regions engaging in these oscillations are organized into distinct networks, which are highly consistent across individuals. Exactly similar networks can be reconstructed by considering the co-activate regions across thousands of task related fMRI experiments, suggesting that the underlying functional anatomy measured by resting fMRI plays a central role in the task-engaged brain.

Connectomics attempts to analyze the complex synaptic network in the brain formed by the billions of interconnected neurons. Fine structural details of the human brain can be visualized by thinly slicing neural tissue and imaging each section on a scanning electron microscope at high resolution. Delineation of each neuron and 3-D imaging allows for the high-resolution mapping of all connections made by each cell, which provides the connectome of the entire brain. In reality, the sheer size and complexity of this scientific challenge is a daunting exercise, which needs the collection of more than 10,000 fine sections and recording of over 1,000,000 images, resulting in the generation of terabytes data that require a rapid processing, thus making connectomics a real speed game, necessitating a very high sophistication. Automated sample preparation robots and high throughput electron microscopes have now become available, making 1 mm high-resolution connectome within reach. However, extracting neuronal circuit information from such large datasets is still an intimidating task.

7.7 Microbiomics

The term Microbes is colloquial of microorganisms, comprising of bacteria, fungi, protozoa, which cannot be visualized by the naked eye. Microbiota is a community of all commensal, ammensal, symbiotic and pathogenic microorganisms that are found in and on all multicellular organisms reported till date. The microbiota plays an important role in the induction, training and function of the host immune system, resulting in the evolution of an immune system as a means of maintaining the symbiotic relationship of the host with these microbes. The term microbiome refers to the entire microbial community that is associated with a particular ecological niche or habitat. For example, human gut microbiome refers to the entire group of microorganisms in the human gut.

7.7.1 Human Microbiome

Evolution has rendered humans as one of the most complex living entities on this planet. The human body, as a holobiont is composed of different microbes, which subsist as a single ecological unit. The estimated size of human microbiota is 10–100 trillion, composed of more than 1000 different species level phylotypes. The aggregate size of human microbiome may be equivalent to the human genome and the number of genes in the human microbiome may exceed the total number of human genes by two orders of magnitude. A long co-evolutionary process might have led to mutualistic interactions between the microbiota and humans. For instance, plant polysaccharides that are consumed in the regular diet are rich in xylan, pectin, and arabinose. The human body cannot synthesize the enzymes that can digest these polysaccharides. Plant polysaccharides that are not digestible by humans are the main substrates for microbial growth in their colon. Human microbiome is significantly enriched with the genes involved in the metabolism of glycans, amino acids, xenobiotics, methanogenesis and biosynthesis of vitamins, etc. denoting that genomes of the human microbiome, provide necessary traits to say that humans have not evolved independently on their own and are considered as "superorganisms" with the trillions of associated microorganisms.

Joshua Lederberg, the Nobel Prize winner who is credited with coining the term 'microbiome,' is pioneer in the initiation of the International Human Microbiome Project (HMP), which was started in 2008 by United States National Institute of Health with a budget of 115 million $ with a purpose of identifying and characterizing the microorganisms that makes human microbiome and sequencing of the human microbiota genomes that normally inhabited the skin, mouth, nose, digestive tract, and vagina. It is an interdisciplinary effort made to associate the differences in the microbiome with differences in metabolic function and/or disease.

Human microbiome diversity is studied by using the sequence analysis of 16S rRNA genes which are amplified from the human metagenomic DNA using broad-range PCR primers and used to make libraries of clones. Each clone in a library represents a 16S rRNA gene from a prokaryotic organism. These clones are then differentiated through fingerprinting methods such as Denaturant Gradient Gel Electrophoresis (DGGE) or Amplified Ribosomal DNA Restriction Analysis (ARDRA) and the non-redundant clones are sequenced. After sequencing, 16S rRNA genes are clustered into groups and a threshold of sequence similarity (>97%) is established to distinguish species level phylotypes. Using the 16S rRNA gene sequencing approach, bacterial communities that reside on or in the human body, including skin, mouth, esophagus, stomach, colon, and vagina have been identified and reported. Majority of the human gut microbiome is dominated by two bacterial divisions, the Bacteroidetes, and the Firmicutes, contributing up to 90% of the identified phylotypes. Even though the 16S rRNA analysis reveals the composition microbiomial community, establishing connectivity between microbial diversity and the human physiology, is possible only by understanding their genome content. For instance, many common human pathogens such as *Staphylococcus*, *Streptococcus*, *Neisseria*, *Enterococcus* are closely related to non-pathogenic strains. Genomes of the pathogenic *E. coli* strain O157: H7 and the commensal strain *E. coli* K12 are more different than that of any two mammals.

Another important objective of the Human Microbiome Project is to determine a common core microbiome shared by a group of humans. Comparison of individuals from the same family has revealed that the human gut microbiome is being shared among family members, but variations in the specific bacterial lineages were found in each individual. It has been shown that even identical twins had significant differences at the species level phylotypes.

Regarding the ethnic groups, a Chinese family shared a similar division level phylogenetic relationship with Americans but has shown a clear difference in their microbiomes at the species level. Comparative genomics of human gut bacteria has revealed a large repertoire of genes involved in the acquisition and metabolism of polysaccharides. These genes are organized as polysaccharide utilization loci (PUL) that encode functions necessary to detect, bind, degrade and import carbohydrates in the gut habitat either from the diet or from host glycans associated with mucus and the surfaces of epithelial cells. Genome sequences of the uncultured microbes in the human gut were obtained by single cell genomics approach. In this, single microbial cells are isolated by flow cytometry and their genomes were amplified by Multiple Displacement Amplification (MDA) PCR and are sequenced by the whole genome shotgun approach. Through the microbiome of the human gut, the role of the microbiome in generating age-related diseases such as cancer and cardiovascular diseases and common risk factors such as obesity has been studied to some extent.

However, it is still not clear, which of the hundreds of microbes are important in human health and a very little is known about their host-microbe interactions. The Human Microbiome Project might tell answers to at least a few aspects regarding the infectious as well as non-infectious microbes in human beings.

7.8 Conclusions

It is clearly evident that there is enormous potential in the integration and application of multiple omics data for a better understanding of the molecular mechanisms, processes, and pathways discriminating health and disease. Success in this new dimension of science will solely depends on the shift from a reduction to a global approach, sustained by a lively and proactive flow of data between different fields of expertise and funding programmes that supports this endeavor, which will aid the development of rapid mechanisms for early diagnosis, disease prevention, disease monitoring and treatment. It can be concluded by mentioning that still a lot more is to be done for making the precision medicine possible in this society. It is just in the beginning stage and we are in the process of getting the data. We must make more use of it.

Applications of Genomics

8

Abstract

Applications of genomics in agriculture for the breeding of desired plants with superior traits were enclosed. The critical role of genomics technologies in human genetic testing and in molecular diagnostics for the determination of genetically inherent disorders, in genomic medicine and cancer genomics are discussed. Role of cytogenomics in diagnosing the human brain diseases, the role of genomics in language adaptation of humans and the evolution of the human sex chromosomes are discussed in this chapter.

8.1 Introduction

One of the main objectives of the genomics is to understand the complete genetic component of an organism by applying the technologies for sequencing and its annotation. Till date, most of the genomes related to all model organisms have been elucidated and their details have been maintained in databases. Advances in the genomic technologies have triggered a revolution in the rapid and error free sequencing of the genomes, which has increased our basic understanding of the structure, functions of the most complex biological systems. It has also helped to assess how genetic information is passed down from one generation to another. It has also accelerated the genomics based agriculture system for breeding and increased crop productivity of the plants. The genome sequence is used as an invaluable tool in providing genetic information for understanding the diseased traits, altered genes, mutated genes, which can be targeted for the gene therapy. It is also useful for the identification of short tandem repeats in the DNA fingerprinting profiling of convicted criminals. There are many applications of the genome sequencing, which are discussed in this chapter.

K. V. Chaitanya, *Genome and Genomics*,
https://doi.org/10.1007/978-981-15-0702-1_8

8.2 Application of Genomics in Agriculture

The plant grows and produces only under an optimum range of environmental factors such as water, sunlight, CO_2, etc. Rapidly increasing population, demands more food, fiber, and fuel, generating sufficient pressure on the existing varieties to preform superiorly. The inescapable threat of continuously changing climate and its associated factors such as drought, salinization of the croplands, floods having a check on the global crop productivity. In order to combat these threats, there is an urgent need to identify the novel, high yielding crop plants, which can increase their productivity even under adverse stressful conditions. Advances in genomics offer a promising role in the development of new crop species with superior agronomic traits, improved productivity, and high sustainability.

8.2.1 Plant Breeding

Plant breeding is a part of the human civilization, which is being practiced since prehistoric times for the development of new crops and varieties. In the last century, plant breeding was based on the utilization of natural and mutant-induced genetic variations for the efficient selection, and application of appropriate breeding methods with favorable genetic combinations. Selection and screening for the genetic variants of specific interest were totally based on the phenotypic evaluations. With the improvement of the genomics based technologies at the beginning of the twentieth century, plant breeders are provided with tools that facilitate the study of whole plant genome and its association with the phenotype, shifting the paradigm from phenotypic evaluations to the genotype. The combination of conventional breeding techniques and modern genomic tools has led to the development of genomics based breeding, ideal for the new **greener revolution,** which is very much needed for the world's rapidly growing population without losing natural resources.

Genetic variability is the major pre-requisite for the plant breeding to enhance the frequencies of the desirable alleles and genetic combinations. The main sources for generating the natural genetic variability among plants are landraces and relatives of wild plants. Even though the traditional landraces are substituted by the uniform cultivars, wild plants and their relatives, still many of the vital cultivars are lost because of the genetic erosion. These materials are preserved and maintained by the gene banks, which is the major source of present day's genetic variability. Another vital source of this variability is artificial mutant collections generated by the mutagens such as radiation, chemicals, insertion in the pre-genomics era and methods of generating transgenics on the present day. These artificial mutant collections contain variability that is not found in the natural collections, and so are highly useful in the development of new traits. tDNA tagged lines and transposon tagged lines have been used for the development of mutant collections in several important plant species including Arabidopsis and rice. Gene silencing RNAi technology has also been applied for creating the gene specific mutant collections in Arabidopsis. Accessions of interest in these lines are identified by a reverse genetics approach called Targeting Induced Local Lesions in Genomes (TILLING), which screens and identifies all allelic variants of a specific region in the genome. Similarly, allelic variants for targeting genes in the natural collections can be identified by EcoTILLING. Both these techniques are based on the usage of restriction endonucleases either CEL I or Endo I, for identification and cutting of mismatches in the DNA. Success and identification of the variants for plants breeding depend on the exact selection of target genes that can be achieved by using the next generation sequencing technologies. Sequencing projects of almost all model plants have been completed or in the process of completing. One of the outputs of these genome projects is the generation of all the genetic information needed for a breeder. Gene expression studies have helped in the identification of many candidate genes for TILLING and EcoTILLING studies of rice, barley, maize, wheat, pea, wild peanut, and watermelon. Rice was the first crop plant whose allelic variants were identified by EcoTILLING.

8.2.2 Genome Wide SNP Studies

One of the vital applications of genomics technologies for the improvement of crop plants and their productivity is the identification of genetic variations through next generation sequencing technologies, which has made it possible to conveniently sequence the genomes of multiple individuals of the same species, rapidly at a minimal cost. Parallel development of bioinformatics and computational biology pipeline tools has accelerated the accurate sequencing of these genetic variants, which can be used as genetic markers for screening. Micro Satellite Repeats or Simple Sequence Repeats (SSRs) and Single Nucleotide Polymorphisms (SNPs) are the predominant markers used in the plant genetics.

SNPs are preferred as the markers of choice in genomics due to their abundance, stability, amenability to automation and are increasingly cost-effective. Genome wide SNP discovery has been performed in the model plants by genome re-sequencing. It has begun with the *Arabidopsis thaliana* 1001 genomes project, which was initiated in 2008 for studying the whole genome sequence variation in at least 1001 accessions of *Arabidopsis thaliana*. These accessions are natural inbred lines, produced by natural selection under diverse ecological conditions. Phase I of the project has been completed with the detailed analysis and publication of 1135 genome accessions. Similar re-sequencing projects are being implemented in rice, maize, grape, soybean, poplar, etc. by sequencing their related genotypes by using Roche 454 and Illumina GA next generation sequencers. Re-Sequencing the genomes of selected genotypes is performed for the detection of SNPs in the candidate genes involved in complex biological processes of interest to plant breeders. The transcriptomes of two anthracnose resistant and one susceptible lines of water yam, which is a major staple crop of Africa, were successfully sequenced detecting SNPs in genes putatively involved in the pathogen response and two alfalfa genotypes contrasting for cellulose and lignin content were sequenced, allowing the identification and selection of SNPs useful for the improvement of alfalfa as a forage

Table 8.1 Utilization of SNPs for various agronomic traits and breeding of crop plants

S. No.	Plant Name	No. of SNPs	Application
1	Rice	3.6×10^6	14 agriculturally important traits
2	Barley	1536	*Fusarium* blight resistance
3	Maize	8590	Oleic acid content
4	Grape	97	Muscat flavor gene
5	Pea	384	Linkage map construction
6	Wheat	1536	Diversity studies

crop and cellulosic feedstock. The application of SNPs in the plant breeding of different crops was provided in Table 8.1.

In the species that lack genomic resources, their transcriptomes were re-sequenced as an alternative way to reduce the genome complexity. This technology was initiated by the deep transcriptome sequencing of two maize inbred lines, followed by the rapidly increasing number of transcriptome re-sequencing projects of non-model crop plants comprising of large, complex, uncharacterized genomes such as eucalyptus, oak and polyploidy genomes such as rapeseed, oats, coffee, sweet potato, tomato, chickpea, chickling pea, etc. Targeted amplicon re-sequencing is another strategy being implemented for the identification of novel SNPs in specific genes. The use of genome and transcriptome sequencing for the discovery of the gene specific SNPs, which has resulted in large SNPs collections in most of the crop plants, which are being used in genetic map construction, map saturation, genome-wide diversity studies, and association mapping, etc.

8.2.3 Construction of Genetic Maps

Another important contribution of genomics technology to agriculture is the development of high density genetic maps, by involving thousands of markers from hundreds of various linkage groups. This was achieved by increasing the marker density, integrations of novel markers, identification of SNPs by genome-re-sequencing projects of diploid and polyploid plants. Golden Gate SNP Genotyping platform developed by Illumina followed by Sequenom based SNP typing assays have increased the density of the genetic map 2–3 times at 100-fold faster than that of gel based methods. ~1.016 SNPs were identified and successfully mapped by comparative next generation transcriptome sequencing of 297 maize recombinant inbred lines. Decrease of sequencing cost has also permitted the detection of new genetic markers such as Restriction-site Associated DNA (RAD) for increasing the density of genetic maps. RAD is a marker used for detecting the genetic variations adjacent to restriction enzyme cleavage sites across the target genome. 445 RAD markers useful for linkage map construction were located on all 7 chromosomes of barley. Whole genome re-sequencing of the plants at different coverage levels is another strategy, being successfully applied for the construction of genetic maps. This approach is 35 times more accurate in determining the recombinant breakpoints than PCR based methods. Ultra high quality physical map of rice has been constructed using the whole genome re-sequencing of 128 Chromosome Segment Substitution Lines (CSSLs) and its genetic map has been constructed using the whole genome re-sequencing of 150 Recombinant Inbred Lines (RILs).

8.2.4 Identification of QTL Related Markers

Detection of QTLs is traditionally being done by linkage mapping. But an increase in the accuracy of QTL detection is being done by using the NGS technologies, which permits more number of markers for mapping, thus increasing the mapping resolution enormously. It also facilitates the development of mapping populations such as RILs, CSSLs, and NILs (Near Isogenic Lines) for the detection of appropriate QTLs. A high density map constructed by the whole genome re-sequencing of rice RILs was used for the identification of four QTLs controlling the plant height. The major disadvantage of linkage analysis based QTL detection is the number of recombination events is limited to the generations that are very much needed for developing the mapping population. Linkage disequilibrium mapping is an alternative as well as a novel approach of mapping the complex traits by the identification of genetic loci associated with the phenotypic variation in a collection of individuals.

8.2.5 Association Mapping

Association mapping is another major method of representing the total number of recombination events that occurred in the history of a specific population by using its natural diversity. Currently, two association mapping methods are being practiced. 1. Candidate gene association, which needs a good understanding of the biochemistry and genetics of a trait, 2. Whole genome scan also is known as Genome Wide Association Studies (GWAS), which needs only a few nanograms of good quality DNA from the specific tissue for providing the high density markers required to ensure the sufficient coverage for detecting the linkage between the markers and its causal locus. Association mapping is just being established in model plants as well as in crop plants. Till date, maize is the only crop plant widely studied by association analysis with candidate genes such as *Dwarf8*, *Vgt1,* and *ZmRap2.7*, successfully associated with the flowering time. Other candidate genes have been associated with forage and Kernel quality, oil and carotenoid content. Candidate gene based association studies are under process in grape and a few conifer wood yielding plants. Even though the association mapping is at its fundamental stage, it is considered as a promising tool for the detection of complex traits in crop plants.

8.2.6 Abiotic Stress Tolerance

Plants grow and reproduce only under an optimum range of environmental factors such as radiation, temperature, air humidity, and water supply. Even a slight variation of these factors influences growth and productivity. Changes in the photoperiod, the amount of radiation and its spectral composition, nutrient affluence or starvation, drought or flooding, high-speed wind, UV radiation, anaerobic conditions, extreme temperatures, air and soil pollution, all these abiotic factors can cause stress on the plants. Several abiotic (light, temperature, water, nutrients, and soil structure) factors affect the growth of higher plants. Among these, drought and salinity are the major abiotic factors that limit agricultural crop production. In order to improve agricultural productivity within the limited land resources, it is imperative to ensure higher crop yields against unfavorable environmental stresses. Understanding plant responses to the external environment are of greater importance and also a fundamental part of making the crops stress tolerant. Recent technological advances and aforementioned agricultural challenges have led to the development of high throughput tools for exploring the plant genomes for their abiotic stress tolerance and crop improvement. Genomic research supported by the functional studies that produce the most readily applicable information related to crop improvement. Application of functional genomics techniques such as SAGE, EST libraries helps to unravel the gene functions and interactions between the genes and their regulation.

Application of microarrays will provide the basic knowledge of the transcriptome, which is being analyzed. Extensive microarray expression studies are being employed for the model plants as well as for economically important crop plants such as Arabidopsis thaliana, rice, wheat, barley, soybean, maize, cotton, cassava, and tomato to unravel their stress responsive genes. However, the employment of microarrays for understanding the stress responsive genes suffers from a huge range of disadvantages. Apart from their background and noise, technical limitations such as choice of tissue that is sampled for microarray studies contains complex tissues composed of different cell types, having obscured transcription changes might dilute the overall stress response of the genotype. These difficulties can be avoided by applying Targeting Induced Local Lesions In Genomes (TILLING) arrays for the plants, whose genome sequences are available. TILLING arrays are capable of identifying the novel transcriptional units, alternative splice sites, and methylation sites. TILLING can also be applied to the genes of all species where mutations can be induced. TILLING arrays have already been applied in model species for investigating their responses to abiotic stress. A modified EcoTILLING has been developed for the identification of natural polymorphisms, that are analogous to TILLING-assisted identification of induced mutations. Polymorphisms that demonstrate the natural variations in germplasms are considered as the valuable tools for genetic mapping. EcoTILLING can also be applied to polyploid species, for differentiating the alleles of homologous and paralogous genes.

8.3 Application of Genomics in Genetic Testing and Molecular Diagnostics

Mutations in the human genome are responsible for the development of both single gene disorders and complex multifactorial disorders. Human DNA mutations range from a change in its single nucleotide to the deletion or duplication of a portion or region or entire chromosome. It is estimated that the single base changes in the human genome occur at a rate of 1.7×10^8 per generation. It is also estimated that there are at least two non-silent point mutations in each newborn human and 0.3% of newborn humans are subjected to aneuploidy due to huge alterations in the chromosome number. Every human genome undergoes many chromosomal rearrangements, such as deletions, duplications, leading to the changes in the copy number variations of a gene. The average rate of copy number variations in each human is 1 in 8 for deletion, 1 in 50 for duplication. A high amount of copy number variations has been observed in the human genome, which might be due to inherited or current loss or gain of genome sequences. These alterations in the human genome are responsible for changes in the phenotype and in extreme conditions will lead to the development of abnormal genotypic conditions. Traditionally, point mutations are being detected effectively by Sanger's sequencing and chromosomal aneuploidy, large genome rearrangements are detected by Fluorescence In Situ Hybridization (FISH). Copy number variations and other genomic imbalances are detected by multiplex ligation-dependent probe amplification (MLPA). All these methods suffer from low information content, limited throughput, and the requirement of choosing candidate targets before the test.

8.3.1 Single Gene Disorders

Single gene or monogenic disorders are generated due to mutation at a single genetic locus, located either on autosome or allosome, manifested in a dominant or recessive or X-linked mode. Phenylketonuria (PKU), cystic fibrosis, sickle-cell anemia, and oculocutaneous albinism are examples of the diseases oriented with mutations in a single gene. Huntington's, myotonic dystrophy, polycystic kidney, familial hypercholesterolemia, and neurofibromatosis are autosomal dominant single-gene disorders. Duchenne muscular dystrophy and hemophilia A are X linked recessive, hypophosphatemic rickets, Rett syndrome are inherited by the X linked dominant pattern of inheritance. Non-obstructive spermatogenic failure is a Y-linked disorder arises due to mutations in the ubiquitin-specific protease 9Y gene (USP9Y) on the Y chromosome.

Traditionally, the molecular diagnosis of monogenic disorders was based on Sanger sequencing. The automatic DNA sequencers associated with the capillary sequencing technologies and Sanger sequencing methods were the principal techniques used in the human genome project. This project has greatly accelerated the development of next generation high throughput sequencing technologies, with enhanced speed and accuracy, at low cost and manpower. Identification of causal gene variant in rare genetic diseases is done by applying the next generation sequencing technologies such as Whole Exome Sequencing (WES), methylome sequencing, transcriptome sequencing, etc. Miller syndrome was the first rare Mendelian monogenic disorder whose causal variants were identified by Whole Exome Sequencing. Later, over 100 causal genes responsible for the Mendelian diseases have been identified by whole exome sequencing. In addition to the discovery of genes responsible for the diseases, whole exome sequencing has also been applied for determining somatic mutations and rare mutations in tumors with moderate effect in common disorders as well as clinical diagnoses.

8.3.2 Multifactorial Gene Disorders

Multifactorial disorders or common complex diseases are caused due to a combination of genetics, environment and lifestyle factors, showing their overall effect on the phenotype. Examples of these complex diseases include Alzheimer's, scleroderma, asthma, Parkinson's, multiple sclerosis, osteoporosis, connective tissue diseases, kidney diseases, autoimmune diseases, etc. Most of the factors responsible for these diseases have not been identified. These disorders do not obey the standard Mendelian patterns of inheritance. Majority of the disease development in an individual depends on the environment and his lifestyle. Backgrounds of these complex disorders are studied by population genetic mapping analysis instead of individuals or families. According to the "common disease-common variant" (CDCV) hypothesis, the polymorphisms with an allele frequency of $>1\%$ are associated with common complex diseases. This hypothesis has been tested by haplotype maps and genotyping arrays. After the completion of the Human Genome Project, numerous variations in the human genome sequence were identified and are being considered as the new insights for the identification of multifactorial gene disorders. Post human genome, Genome Wide Association Studies (GWAS) has been designed and initiated by the National Institute of Health, USA for studying the genetic variations across the whole human genome and identification of genetic associations with observable traits, such as the presence or absence of a disease or a condition. GWAS is considered as a major tool for analyzing the complex, multifactorial disorders as they provide a clear glimpse of the molecular pathways that are behind these diseases. GWAS has uncovered novel genes that are associated with a variety of human diseases including obesity, breast cancer, and has identified 1628 SNPs that are associated with 250 loci for common complex diseases. Rapid development in the genome sequencing technologies is permitting to obtain a valid and economical identification of one million SNPs in a single person's genome in a single scan.

8.4 Epigenetics and Epigenomics

Epigenetics is the study of heritable changes in the expression of a gene without any change or alteration in its DNA sequence. To be simpler, it is the study of a change in the phenotype without any change in its genotype. There are four major changes that are regulated by the epigenetics; 1. Histone modifications 2. DNA methylation 3. Small and non-coding RNAs, and 4.Chromatin architecture. These mechanisms, along with transcriptional regulatory events regulate a gene function and its expression either during development or differentiation, or its reply to environmental cues. Epigenetics is one of the most rapidly growing areas of advanced genetics and has become a core aspect of the studies related to development and disease. Epigenetic modifications are crucial for packaging and illustration of a genome and are considered as a relationship that bridges the phenotype with genotype. These epigenetic modifications can serve as markers for demonstrating the gene expression and its activity along with the chromatin state.

The term epigenome refers to the complete epigenetic situation of a genome. It comprises the total description of chemical modifications that are made to the DNA and histone proteins of an organism that results in changes in its chromatin structure and function of its genome. Epigenome represents the total analysis of epigenome markers across the genome of an organism. It is a set of chemical modifications to the DNA and its associated histone proteins in a cell, which alters the gene expression. These modifications also occur due to environmental exposure or disease. When an epigenomic chemical stimulus or environmental signal binds to genomic DNA and modifies its function, it is considered a **marked genome.** These marks will not change the DNA sequence but alter its function. All epigenetic changes that occur to an organism along with the marks are passed down to its offspring by transgenerational epigenetic inheritance (transmission of genetic information from one generation to another that affects the offspring traits without any changes in the DNA sequence).

First major epigenetic mark is the methylation of DNA, which involves the covalent transfer of a methyl group to the C-5 position of DNA cytosine ring catalyzed by methyltransferases (DNMT). In plants, the methylation occurs both in asymmetrical (CHH) and symmetrical (CG or CHG, H refers to A, T, C). In mammalian cells, 98% of DNA methylation occurs at CpG dinucleotides and 2% in non-CpG form. DNA methylation is essential for normal development of an organism during which, it plays a crucial role in many vital processes such as genome imprinting, X-chromosome inactivation, suppression of repetitive element transcription and transposition. Methylation of DNA impedes the binding of transcriptional regulators to the gene. Methylated DNA aids in the chromatin formation by interacting with other epigenetic modifications such as the histone code, polycomb complexes, nucleosome positioning, non-coding RNA, and ATP-dependent chromatin remodeling proteins. Hypomethylation of DNA and locus-specific hypermethylation of CpG islands leads to cancer. Most of the tumors are related to a reduction in DNA methylation (Hypomethylation) at heavily methylated DNA elements such as satellite DNA (SAT2), retrotransposons such as LINES, leading to genome instability and the activation of oncogenes. Locus specific hypermethylation occurs at promoters of tumor suppressor genes resulting in their transcriptional silencing.

Second major epigenetic mark is histone protein modifications, affecting the DNA indirectly. DNA is wrapped in the nucleus by histone proteins, which form spool like structures that enable the long DNA thread to be wound up properly into the chromosomes. Covalent modification of DNA and histone proteins provides a heritable mechanism of gene expression. Tails of histone proteins undergo a variety of covalent modifications such as acetylation, methylation, phosphorylation, ubiquitination, and sumoylation and regulates many key cellular processes including gene transcription, DNA replication, and DNA repair.

The epigenome of an organism along with the post translational modifications of histones, DNA methylation maps can be comfortably studied by using next generation sequencing technologies. Methylation of DNA can be studied by methylated DNA immunoprecipitation (meDIP), whereas the location of transcription factors, post translational modifications of histones at genome level can be identified by ChIP-Seq technology. Interactions of protein with DNA can be evaluated through the combination of chromatin immunoprecipitation with DNA microarray (ChIP-chip).

In 2003, Wellcome Trust Sanger Institute (Hinxton, United Kingdom) and Epigenomics AG (Berlin, Germany) have combinedly announced the launch of the Human Epigenome Project (HEP) to map the DNA methylation sites in the 30,000 genes of the human genome. Phase I of the project has been completed and phase II is on for identifying the acetylation and methylation marks on histone proteins, which alters the chromatin structure, thereby regulating the gene expression. Other human epigenome projects such as ENCODE are currently in the process of cataloging all epigenetic markers across the whole human genome. The resulting reference patterns will usher in epigenetics as an exciting new period of medical science. Epigenomics of Plants International Consortium (EPIC) was formed in 2008 for developing the core ideas related to the plant epigenomes.

8.5 Genomic Medicine

Next generation sequencing technologies for genome sequencing has revolutionized the genomics technology and has also shown certain improvement in the disease diagnosis through the identification of molecular etiologies at minimum cost. Further, advances in the genomics have enabled rational disease prevention strategies with better treatment and development of novel therapies. This application of genomics in medical care is denoted as genomic precision medicine, or genomic medicine, which is defined as the utilization of genomic information and technologies such as whole genome and exome sequencing for determining the disease predisposition, diagnosis, prognosis, selection and prioritization of therapeutic options. Genomic medicine has achieved great success in the diagnosis and prevention of high penetrance monogenic disorders.

Phenylketonuria (PKU) is an autosomal recessive disorder, caused due to mutations in both the alleles of PKU gene, which synthesizes phenylalanine hydroxylase, leading to an increase in the phenylalanine levels in the blood, causing severe health complications. Although PKU is diagnosed by a biochemical test, genomic testing is the most relevant, either as a prenatal or preconception carrier screening test to restore the reproductive confidence. It would also identify the genetic mutation. Genomic testing also has an immediate and effective application in the identification of patients with familial hypercholesterolaemia, which has drastically reduced death to 51%. A similar reduction in the incidents related to fragile X syndrome, cystic fibrosis, and β-thalassemia has been reported as a result of early informed family planning and preconception genetic testing. This is particularly considered as a remarkable step for those with a family history (FH) of a serious genetic disorder. Till date, molecular diagnosis and treatment of rare childhood diseases oriented with more than 1000 monogenic disorders related to intellectual disability have been identified. New therapies such as small molecule substrate inhibition, bone marrow, and hematopoietic stem cell transplantation and gene therapy are in the progress for 81 inborn metabolic disorders.

8.6 Genomics and Cancer Therapy

Cancer is a genetic disease, which has evolved and is progressing by accumulating somatic mutations such as copy number alterations, structural variations, and epigenomic alterations, with or without hereditability. Further, segregation and loss of heterozygosity (LOH) studies on cancer tissues have identified somatic and germline mutations of several tumor suppressor genes, such as *RB1*, *TP53,* and *APC*. Copy-number analysis has found oncogenic activators, such as *HER2/ERBB2* and *MYC*. These oncogenic mutations have been targeted for predicting the sensitivity to the therapy and disease prognosis. Rapid advances in the next generation sequencing technologies and support from the computational technology for handling and analyzing the massive data (~1 TB of raw and sequence data corresponding to cancer and normal DNA) has enabled the sequencing and comprehensive analysis of hundreds of cancer related genes, whole Exome Sequencing, RNA Sequencing and Whole Genome Sequencing, leading to the development of cancer genome profiling. Using these genome based technologies, more than 50,000 cancer genomes have been sequenced, which are deposited in the databases related to The Cancer Genome Atlas (TCGA) and The International Cancer Genome Consortium (ICGC). It is predicted that by the year 2030, hundred millions of the cancer patients will have their genomes sequenced.

Among the technologies developed for cancer genome profiling, Whole Exome Sequencing has been the main platform for cancer profiling and generating vast amounts of data related to mutations in the coding regions of oncogenes. These studies have identified many novel genes and pathways related to regular and rare tumors in humans. Saturation analysis of the data has identified certain driver genes with frequent mutations and rare mutations. Pan cancer analysis of the Whole Exome data has demonstrated that cancer caused by exposure to carcinogens such as melanoma and lung cancer is having the highest number of somatic mutations, whereas leukemia and pediatric tumors have few somatic mutations and more protein altered mutations in their coding regions. COSMIC database has been extensively curating mutations in the coding regions of more than 1,000,000 cancer samples by Whole Exome Sequencing. Several molecular oncology laboratories are now considering the next generation sequencing platforms, related methods, and tools required for quick diagnosis of cancer. The whole transcriptome analysis through RNA-Seq is used for quantifying the gene expression templates, detection of RNA editing, alternative splicing, and fusion transcripts.

Still, limited information is available on somatic mutations such as large deletions, insertions, inversion, duplication, translocation and pathogen (virus) integration in the non-coding untranslated regions (UTR), introns, non-coding functional RNA and repetitive regions, which are spanning up to 98% of the human genome remain to be widely unexplored. These somatic mutations in the non-coding regions of cancer genomes can be further explored by the Whole Genome Sequencing Approaches combined with mathematics and other omics technologies provide a better understanding of the total landscape of cancer genomes and elucidate the functions of these unexplored human genomic regions. The combination of these next generation sequencing technologies gives a high-resolution and global view of the cancer genome, which will definitely provide a better understanding of the disease bridging the new era of genomics technologies to personalized medicine.

8.7 Cytogenomics

Molecular cytogenetics also referred to as cytogenomics is defined as a specific field of biosciences, which targets the study of chromosomes and its molecular resolutions at all stages of the cell cycle. This applied cytogenetics comprises a set of techniques that operate either using the entire genome or with specific DNA sequences for identifying and analyzing genomic, structural and behavioral variations at chromosomal or sub-chromosomal levels. One of the most valuable insights of cytogenomics to the human genome is the identification of exceedingly high inter-individual variations in the human genome, which are important for their diversity and disease resistance. After the elucidation of the human genome, genes corresponding to its 46 chromosomes have been completely mapped according to their position in the specific chromosomes. These chromosomes have been marked and numbered conventionally according to their appearance during the metaphase of the chromosome. During interphase these chromosomes remain indistinct in the nucleus, maintaining their individuality as chromosome territories. The positions of the respective genes in these territories are also important for their expression. A gene might be inactive in an internal position, whereas it might be active at the periphery of a territory. This can be well understood in the case of gene rich chromosome 19, which is internal into the nucleus and the gene poor chromosomes are at the periphery of the nucleus, protecting the gene enriched chromosomes from the mutagens. Three main approaches are followed for identifying the changes at the chromosome level. 1. Cytogenetic approaches applied for detecting the chromosomal abnormalities that are more than 3–5 Mb. 2. Molecular cytogenetic approaches applied for identification of genomic variations at the chromosomal or even at the sub-chromosomal level at a resolution of 1 Kb. 3. Molecular genetic approaches applied for the DNA sequences that are less than 1 Kb. Molecular cytogenetics has evolved as a major subject of studying the structural organization of chromosomes. With the advent of Fluorescent In Situ Hybridization (FISH), Comparative Genomic Hybridization (CGH) and next generation sequencing technologies associated with DNA stretching techniques such as DNA combing and chromosome microdissection have metamorphosed cytogenetics into cytogenomics.

8.7.1 Cytogenomics of Brain Diseases

Many of the brain related diseases in humans are associated with the copy number variations (CNV) or genomic rearrangements. The amount of copy number variations that are related to brain malfunctioning or nervous system disorders grows dynamically. Aneuploidy was the first major genomic imbalance found to be frequently associated with the 0.5% human diseases, which were later named as chromosomal syndromes. It was estimated that these syndromes are the primary causes of pregnancy losses, developmental disabilities, and mental retardation. Another common factor responsible for brain malfunction is intercellular genomic variations. Structural chromosomal abnormalities such as microdeletion syndromes are found among individuals affected by idiopathic mental retardation or major psychiatric and neurological disorders. Application of cytogenomic approaches has allowed the high-resolution screening of the human genome based on array-Comparative Genomic Hybridization technology, resulted in the delineation of numerous microdeletion syndromes associated with intellectual abilities. Sub-chromosomal duplications and triplications will also significantly contribute to the etiology of the brain malfunctioning. Most of the chromosomal imbalances that are directly linked with mental dysfunction are being characterized by the developed and developing cytogenomic approaches.

8.7.2 Cytogenomics of Plants

One of the most important breakthroughs in the history of plant breeding was the development of plant cytogenetics, which has permitted the breeding of selected parental plants. The unprecedented development of plant molecular cytogenetics associated with functional genomics has ushered the development of plant cytogenomics. Development of techniques such as DNA base-specific fluorescence banding, GISH, and FISH has facilitated the identification, localization, and mapping of chromosome-specific markers in the plants of high significance in molecular systematics, species identification, detection of hybrids, alien chromosomes and other chromosomal aberrations. The dynamism of chromatin architecture and cell cycle representing the chromosome function is another vital part of plant cytogenomics, which was applied in the breeding of cereals, legumes, oilseeds, vegetables, horticultural crops, spices and condiments, fiber-yielding plants, medicinal and aromatic plants for diverse types of desirable agronomic and functional traits including abiotic and biotic stress tolerance, herbicide resistance, enhancing the yield and minimizing the anti-nutritional factors, increasing the dietary amount of proteins, minerals, vitamins, essential amino acids, flavonoids, antioxidants and dietary fibers. cytogenomics also plays a pivotal role in the development of functional and therapeutic foods. It is also involved in the identification of diverse traits related to nutritional quality.

8.8 Microarrays

Microarrays are defined as a collection of microscopic features consisting of a picomole concentration of biological molecules (mostly DNA, cDNA or proteins) arranged in a predefined order on glass or a silica substrate. In DNA microarrays, each feature contains predefined short single-stranded DNA oligonucleotides (25–75 b) or large double-stranded DNA (200–800 bp), known as probes. In the microarray experiment, these probes of known DNA sequences will be hybridized with unknown target DNA, whose identity or abundance is being sought. Target DNA will be prepared by embedding fluorescent tags and the microarray is exposed to the conditions that favor hybridization. Unbound DNA fragments will be washed and the microarray positions with DNA–DNA hybrids are then detected by the emission of the fluorescence using a detector. Each microarray experiment produces a vast amount of data, which can be helpful in analyzing the exact nature of the target sample. Various microarray methods are available which differs in type of the probe and the methods applied to detect the target by addressing the probe.

8.8.1 Genomic Microarrays

Genomic microarrays are also termed as molecular karyotyping, used for detecting the gains and losses of bases in the genome by the hybridization of fluorescently labeled patient DNA with the known DNA sequence co-ordinates, spotted on a solid surface. By measuring and comparing the intensity of the signal emitting from the patient DNA with reference DNA, addition and deletions in the genomes can be identified. Two different microarrays are used for genome based studies 1. Single Nucleotide Polymorphism (SNP) array and 2. Comparative Genomic Hybridization (CGH) array.

8.8.1.1 Single Nucleotide Polymorphism (SNP) Array

Single nucleotide polymorphism arrays are designed for detecting the common polymorphisms that are existing in a population. These arrays are mainly used in genome wide association studies of individuals for common multifactorial diseases. An SNP can be identified by calculating the B-allele frequency (BAF), a measurement of the presence of an allele 'B' and allele 'A'. For homozygous AA, BAF is 0, for BB, BAF is 1 and for heterozygous sites AB, BAF is 0.5. Along with the BAF of the respective SNPs, these arrays also detect copy neutral loss of heterozygosity (LOH) (absence of heterozygosity), uniparental isodisomy, uniparental heterodisomy (DNA samples of both the parents are required) and regions that are identical by descent. These platforms are also used for analyzing copy number variations.

8.8.1.2 Comparative Genomic Hybridization (CGH) Array

Comparative genomic hybridization (CGH) is a technique used for detecting and mapping the changes in copy number of DNA sequences. Since it was first reported by Kallioniemi and his colleagues in 1992, this cytogenetic approach has been widely used for the analysis of tumor genomes and constitutional chromosomal aberrations. In CGH, DNA from a test (tumor) and a reference (normal) are differentially labeled and hybridized to a representation of the genome, which was originally a metaphase chromosome spread. Hybridization of repetitive sequences is blocked by the addition of Cot-1 DNA. The fluorescence ratio of the test and reference hybridization signals is determined at different positions along the genome, which provides information on the relative copy number of sequences in the test genome compared with a normal diploid genome. Advances in the microarray technology have converted this technique to a major microarray platform termed as Comparative Genomic Hybridization Array for identifying the changes in the DNA of two contrasting samples.

A Comparative genomic hybridization array (CGH array) is used for comparing the DNA content of two differentially labeled genomes. The two genomes belonging to a patient (considered as a test) and a healthy human being (considered as reference), are co-hybridized on a solid platform, on which cloned or synthesized oligonucleotide DNA fragments have been immobilized. Arrays have been developed with a variety of DNA substrates such as oligonucleotides, cDNAs, and BACs. Resolution of a CGH array depends on the size of cloned DNA targets and the natural distance between these sequences, located on the chromosome. A major application of CGH array is its ability to detect deletions, duplications, amplifications of any locus in a genome simultaneously. These changes are represented by the distance on the array.

8.9 Comparative Genomics

Complete genome sequence of an organism is an ultimate genetic map of a DNA sequence comprising of all heritable characters arranged in an order, along the chromosomes. But, the DNA sequence will not give any direct indication of how this information can be considered as coding or non-coding. Finding the functional part of a genome will tell us how this genetic information will reflect a specific trait and a phenotype, which can be used for improving the health of an individual or for the benefit of the society. The function of a specific genome can be better understood by comparative genomics, in which the genome features of two different organisms such as non-coding sequences, gene sequences, regulatory sequences, and other DNA structural elements are compared for understanding their functions. The basic principle of comparative genomics involves the study of genome conservation within the two organisms. Common features of two organisms are encoded within their DNA, responsible for various cellular functions by generating mRNA and proteins, that are conserved and inherited from their common ancestors. Similarly, regulatory sequences that are involved in controlling the expression of genes in these two organisms are also well conserved. However, the conserved sequence in a set of close and distant species is likely to be constrained due to evolutionary pressure, implying a biological function. Similarly, A DNA sequence can have a specific biological function without being conserved across the species. This aspect is vital for applying comparative genomics, especially when conservation does not necessarily imply any identity.

8.9.1 Comparative Genomics of Human and Mouse

Comparison of genome size and gene density are the two fundamental exercises performed in the comparative genomics by describing the minimal genome for free-living organisms, comparison of non-redundant proteomes, structural and functional genome annotation, and the detection of selection. All these aspects rely on the comparative sequence data and sequencing the genomes of closely related organisms. It was estimated that humans have diverged from mouse roughly 75–80 million years back during which, the large scale gene organization and gene order have been preserved. ~90% of the human genome is in large blocks of homology with the mouse genome. These gene rich regions are arranged on human chromosomes that match with the mouse chromosomes in a similar order. Sequences that were active in the common ancestors of human and mouse such as relics of transposons seems to have lost their functional specificity are also found to be aligned. This alignment has estimated that 5% of the human genome is under purifying selection, which is functional. 99% of the genes in the human genome align with their homologs in the mouse genome. Among them, more than 80% are clear 1:1 orthologs. Intron-exon structures in most of the genes are also highly conserved in these two genomes. At the nucleotide level, 40% of the human genome aligns with the mouse genome. The other 60% is the result of lineage-specific insertions, deletions, and other possible mechanisms, which are divided into two classes of sequences. Class I sequences occupying about 24% of the genome, comprising of repetitive elements that are believed to arise due to transposition only in the human lineage. Class II sequences, comprising of lineage specific and ancestral specific elements have occupied up to 50% of the human genome. The remaining 26% of the human genome cannot be accounted at present, due to limitations in the sensitivity of the alignment. Some of the non-aligning DNA elements could be the orthologous DNA, which might have undergone a lot of changes that current programs could not recognize which sequence has evolved from a common ancestor sequence. Even, the homologs to some of the non-aligning human DNA could be deleted in the mouse genome. Given the large expansion of mammalian genomes by transposable elements, one would expect that a compensatory amount of the ancestral DNA would be deleted from the genome.

8.9.2 Evolution of Sex-Chromosomes in Humans

Evolution of X and Y chromosomes has been studied for several decades, but the major understanding of the process that leads to their evolution was possible through comparative genomics. Human X and Y chromosomes have originated from a homologous pair of chromosomes. The current human Y chromosome has lost many genes in comparison to the formerly identical X chromosome leading to a belief that the human Y chromosome was destined to disappear from the genome forever. Comparison of the human Y chromosome with primate Y chromosomes has shown an entirely different picture of sex chromosome evolution. The iterative mapping and sequencing of the chimpanzee Y chromosome have shown a rapid evolution, not with just the loss of genes, but the addition and duplication of novel sequences. Y chromosome sequencing of the *Rhesus macaque* monkey has shown that no loss of genes either in the human or in the Rhesus Y chromosomes. This has led to a new theory where human Y chromosome lose their gene content quite significantly soon after they lose the ability to cross-over with a homologous chromosome, but the original genes that remain on the Y chromosome exhibit a strict purifying selection, thus continuing its existence.

8.9.3 Language Adaptation in Humans

Evolution of modern human being is studied by comparing its genome with that of extinct ancestors and primates such as the chimpanzee. Comparison of the human genome with that of the Neanderthal genome has revealed that 91% of the human accelerated regions (regulatory regions) have lost their mammalian conservation much earlier than Neanderthal-modern human split. It has been shown that the mutations in FOXP2 gene cause linguistic and grammatical impairment in humans. Comparison of the FOXP2 gene sequence of the modern humans with chimpanzee and orangutan have shown a recent selection of this gene in modern humans resulting in at least two non-synonymous changes responsible for orofacial movement control that has allowed humans to speak, unlike apes. Comparison of modern human genome with Denisovan genome (archaic humans closely related to Neandertals, whose populations overlapped with the ancestors of modern humans) demonstrated that Denisovans possessed an ancestral FOXP2 allele, which shows that occurrence of non-synonymous mutations on FOXP2 gene is a crucial change that had happened in the human lineage in the last 800,000 years, after their divergence from Denisovans and Neanderthals.

8.10 Conclusions

Using powerful next-generation sequencing technologies and microarrays, it is possible to read and understand the genetic changes in organisms at new depths, more rapidly, easily and economically than before, enabling a wide variety of genomics applications. As a result, the discoveries that were unimaginable a few years back are now becoming regular. Crop Genomics is helping to give a major leap in plant breeding by super domestication of economically important crop plants (a process that leads to domestication, with a dramatically increased yield that could not be selected in the natural environments from natural variation without recourse to the new genomics technologies). Completion of the human genome sequencing project has enabled many technological advances in the diagnosis of genetically inherent disorders and diseases. Comparative genomics has paved the path for the understanding of each and every genome in the set and has brought a huge wealth of insights into the evolution of vertebrate genomes. Rapidly advancing genomics technologies is now has the potential to decipher the human evolution and inherent diseases and the link between these two.

Important Databases Related to Genomes 9

Abstract

Genome databases are the locations, which permits storing, sharing, retrieving and comparison of the information related to the genomes of various individuals and organisms. Traditionally the databases were confined to the updated information of certain vital model organisms. Rapid development of technology and high speed internet facilities have created an explosion of databases resulting in the development of specific databases of almost all model organisms and a group of organisms with a common specificity. In the present chapter, details of the databases related to the genomes of viruses, archaea, bacteria, cell organelle, invertebrates, vertebrates, plants, and human beings are provided in a table format which provides an instant information, about different databases and their URLs.

9.1 Virus Databases

Database	Description	URL
CoVDB	Coronavirus genes and genomes	http://covdb.microbiology.hku.hk/
DPV web	Central source of information about viruses, viroids and satellites of plants, fungi, and protozoa	http://www.dpvweb.net/
euHCVdb	European hepatitis C virus database	http://euhcvdb.ibcp.fr/
HCV database	Hepatitis C virus database	http://hcv.lanl.gov/
HERVd	Human endogenous retrovirus database	http://herv.img.cas.cz/
HFV database	Hemorrhagic fever virus sequence database	http://hfv.lanl.gov/
HIV drug resistance DB	Compilation of mutations in HIV genes that confer resistance to anti-HIV drugs	http://www.hiv.lanl.gov/content/sequence/RESDB/
Influenza research DB	Influenza virus	http://www.fludb.org

(continued)

K. V. Chaitanya, *Genome and Genomics*,
https://doi.org/10.1007/978-981-15-0702-1_9

NCBI viral genomes	Viral genomes resource	http://www.ncbi.nlm.nih.gov/genomes/VIRUSES/viruses.html
Papillomavirus episteme	A database of Papillomaviridae family of viruses	http://PaVE.niaid.nih.gov
phiSITE	Gene regulation in bacteriophages	http://www.phisite.org
Poxvirus.org	Poxvirus genomic sequences and gene annotation	http://www.poxvirus.org/
VBRC	Viral bioinformatics resource center	http://www.biovirus.org/
ViTa	microRNAs targets of the influenza virus	http://vita.mbc.nctu.edu.tw/
ViralZone	Molecular and epidemiological information on viral genera and families	http://viralzone.expasy.org/
ViralGenomes	Viral sequence data resource by NCBI	https://www.ncbi.nlm.nih.gov/genome/viruses/

9.2 Archaeal Databases

Database	Description	URL
Ensembl	A browser for archaeal genomes	https://bacteria.ensembl.org/index.html
EZ cloud	EZ cloud for archaea	http://www.ezbiocloud.net/ezgenome
IMG	Integrated microbial genomes and microbiomes	https://img.jgi.doe.gov/
MicrobesOnline resource	A collection of 3707 microbial genomes	http://www.microbesonline.org/
BacDive	The bacterial diversity metadatabase contains detailed information on more than 20,000 strains of bacteria	https://bacdive.dsmz.de/
Wikipedia	List of sequenced archaeal genomes	https://en.wikipedia.org/wiki/List_of_sequenced_archaeal_genomes
UCSC archaeal genome browser	Integrated genome browser for archaeal genomes	http://archaea.ucsc.edu/

9.3 Bacterial Databases

Database	Description	URL
Ensembl bacteria	A browser for bacterial and archaeal genomes	https://bacteria.ensembl.org/index.html
MBGD	Database for comparative analysis of completely sequenced microbial genomes	http://mbgd.genome.ad.jp/
Human microbiome project	Characterisation of microbes in healthy humans	https://hmpdacc.org/

(continued)

GOLD	Genomes online database	https://gold.jgi.doe.gov/
EZ cloud	EZ cloud for bacteria	http://www.ezbiocloud.net/ezgenome
IMG	Integrated microbial genomes and microbiomes	https://img.jgi.doe.gov/
MicrobesOnline resource	A collection of 3707 mmicrobial genomes	http://www.microbesonline.org/
BacDive	The bacterial diversity metadatabase contains detailed information on more than 20,000 strains of bacteria	https://bacdive.dsmz.de/
BIGSdb	Bacterial isolate genome sequence database	https://bigsdb.readthedocs.io/en/latest/
metaMicrobesOnline	Website for browsing and comparing the genomes of 3527 microbial genomes	http://meta.microbesonline.org/
MetaRef	A pan-genomic database for comparative and community microbial genomics	https://omictools.com/metaref-tool
MetaBioME database	MetaBioME is a web resource to find novel homologs for known commercially useful enzymes (CUEs) in metagenomic datasets and completed bacterial genomes	https://metasystems.riken.jp/metabiome/
PATRIC	Pathosystems resource integration center	https://www.patricbrc.org/
TBDB	Tuberculosis database integrates genomic sequences and data for *Mycobacterium* species relevant to drug discovery, vaccines, and biomarkers	http://www.tbdb.org/
EuPathDB	Eukaryotic pathogen database resources is a collection of individual databases, each focusing on specific pathogens, accessible through a common portal	https://eupathdb.org/
coliBase	A database for *E. coli*, *Salmonella* and *Shigella*	http://xbase.warwick.ac.uk/colibase/
DEG	Database of essential genes from bacteria and yeast	http://www.essentialgene.org
Essential genes in *E. coli*	First results of an *E. coli* gene deletion project	http://www.genome.wisc.edu/resources/essential.htm
GenoBase	Escherichia coli K-12 in-frame, single-gene knockout mutants	http://ecoli.naist.jp/
HGT-DB	Horizantal gene transfer database is a genomics database that includes G+C content, codon and amino-acid usage, as well as information on which genes deviate in these parameters for prokaryotic complete genomes	http://genomes.urv.es/HGT-DB/
Pseudomonas genome database	Consists of information about Pseudomonas genome	http://www.pseudomonas.com/
PAIDB	Pathogeneticity islands database	http://www.paidb.re.kr/about_paidb.php

9.4 Cell Organelle Databases

Database	Description	URL
Chloroplast genome database	Chloroplast genome	http://chloroplast.cbio.psu.edu/
OrganelleDB	A web-accessible database cataloging proteins localized to a known organelle	http://organelledb.lsi.umich.edu/
Organelle genomes	Organelle genome resource at NCBI	www.ncbi.nlm.nih.gov/genomes/ORGANELLES/organelles.html
Plant organelles database	Images of plant organelles and protocols for plant organelle research	http://podb.nibb.ac.jp/Organellome/
PLProt	*Arabidopsis thaliana* chloroplast protein database	http://www.plprot.ethz.ch/
Plastid-LCG base	Plastid lineage-based conserved gene-pair database	http://lcgbase.big.ac.cn/plastid-LCGbase/
HMPD	Human mitochondrial protein database	http://bioinfo.nist.gov/hmpd/
HmtDB	Human mitochondrial DataBase	http://www.hmtdb.uniba.it/
MamMiBase	Database of complete mitochondrial genome sequences of over 100 mammalian species	http://www.mammibase.lncc.br/
MitoCarta	Mouse and human mitochondrial proteins	http://www.broadinstitute.org/pubs/MitoCarta
MitoDrome	Annotating D. melanogaster nuclear genes encoding mitochondrial proteins and support studies on human diseases related to mitochondrial dysfunctions	http://mitodrome.ba.itb.cnr.it/
MitoPorteome	Mitochondrial protein sequences encoded by mitochondrial and nuclear genes	http://www.mitoproteome.org/
OGRe	Database containing information on completely sequenced animal mitochondrial genomes of 473 species	http://ogre.mcmaster.ca/
Arabidopsis mitochondrial protein database	Database comprising of experimentally identified proteins in *Arabidopsis* mitochondrial genome	www.plantenergy.uwa.edu.au/applications/ampdb/index.html

9.5 In-Vertebrate Genome Databases

Database	Description	URL
AphidBase	Genome database of pea aphid (*Acyrthosiphon pisum*)	http://www.aphidbase.com/
AppaDB	Database of nematode *Pristionchus pacificus*	http://appadb.eb.tuebingen.mpg.de
BeetleBase	Database of the beetle *Tribolium castaneum*	http://www.beetlebase.org
CeNDR	*C. elegans* natural diversity resource	http://www.elegansvariation.org

(continued)

Comparasite	Database for comparative studies of transcriptomes of parasites	http://comparasite.hgc.jp/
CuticleDB	Structural proteins of arthropod cuticle	http://bioinformatics.biol.uoa.gr/cuticleDB/
FlyBrain	Database of drosophila nervous system	http://flybrain.neurobio.arizona.edu/
Helminth.net	Parasitic nematode sequencing project	http://www.helminth.net
InsectBase	Insect genomes and transcriptomes	http://www.insect-genome.com/
Hymenoptera genome	Genome sequences and annotation for honey bee and the parasitoid wasp *Nasonia vitripennis*	http://HymenopteraGenome.org
InsectBase	Insect genomes and transcriptomes	http://www.insect-genome.com/
SilksatDB	Comprehensive relational database of silkworm microsatellite	http://www.cdfd.org.in/SILKSAT/index.php
SpodoBase	Genomics of the butterfly *Spodoptera frugiperda*	http://bioweb.ensam.inra.fr/spodobase/
VectorBase	Invertebrate vectors of human pathogens	http://www.vectorbase.org
RNAiDB	RNAi phenotypic analysis of *C. elegans* genes	http://www.rnai.org/
NEMBASE	Nematode sequence and functional data database	http://www.nematodes.org/
PLANMINE	Platform for archiving and mining planarian biodiversity	http://planmine.mpi-cbg.de/
WORMBASE	Database of *Caenorhabditis elegans*	http://www.wormbase.org
WorfDB	Worm ORF database which contains predicted proteins from *C. elegans*	http://worfdb.dfci.harvard.edu/

9.6 Vertebrate Genome Databases

Database	Description	URL
FANTOM	Functional annotation of mouse full-length cDNA clones	http://fantom.gsc.riken.jp/
GWIPS-viz	Genomes of human beings and vertebrates	http://gwips.ucc.ie
Xenbase	Human and Vertebrate Genomes	http://www.xenbase.org
Mammalian gene collection	Identify and sequence cDNA clones of FL or ORF for human, mouse and rat genes	http://mgc.nci.nih.gov/
PlasmID	A repository for collection and distribution of plasmid clones for human and vertebrate genomes	http://plasmid.hms.harvard.edu/
Mouse genome database	Mouse genome database for integrated data on the genetics, genomics and biology of the laboratory mouse	http://www.informatics.jax.org

9.7 Human Genome Databases

Database	Description	URL
3DSNP	Database for comprehensively annotating the regulatory function of human noncoding SNPs by exploring their 3D interactions with genes and genetically associated SNPs mediated by chromatin loops	http://biotech.bmi.ac.cn/3dsnp/
ENCODE at UCSC	Encyclopedia of DNA elements an international a consortium of investigators funded to analyze the human genome with the goal of producing a comprehensive catalog of functional elements	http://genome.ucsc.edu/ENCODE
GeneAnnot	Revises and improves the annotation of affymetrix probe sets	http://genecards.weizmann.ac.il/geneannot/
Locus reference genome sequences	Each sequence is a stable genomic DNA sequence for a region of the human genome	http://www.lrg-sequence.org
QTL match maker	Database which integrates QTLs of human mouse and rat genomes to annotate their functional genomes	http://pmrc.med.mssm.edu:9090/QTL/jsp/qtlhome.jsp
TIARA	Total integrated archive of short-read and array (TIARA) accumulates raw-level personal genomic data from whole genome next-generation sequencing (NGS) and comparative genomic hybridization (CGH) arrays	http://tiara.gmi.ac.kr
TRBASE	Database that relates tandem repeats to the gene locations and disease genes in the human genome	http://trbase.ex.ac.uk/
UniSTS	A comprehensive database of sequence tagged sites (STSs), each of which is defined by a distinct pair of primer sequences designed for use in a PCR reaction	http://www.ncbi.nlm.nih.gov/probe
X:MaP	Annotation and visualization of genome structure for Affymetrix exon array analysis	http://annmap.picr.man.ac.uk/
EVOLA	A database of human genes and their vertebrate orthologs	http://jbirc.jbic.or.jp/hinv/evola/
H-InvDB	HumanInvitational database is an integrated database of human genes and transcripts	http://www.h-invitational.jp/
HGPD	Human gene and Protein database is a unique database that stores information from human gateway entry clones and in vitro expression data related to human proteins	http://hupex.hgpd.jp/entrance/index.html

(continued)

HRPD	Human protein reference database that stores the data related to domain architecture, post-translational modifications, interaction networks and disease	http://www.hprd.org/
LIFEdb	A database on the localisation, interaction, functional assays and expression of human proteins	http://www.dkfz.de/gpcf/lifedb.php
NetAffx	A database of public affymetrix probesets and annotations	http://www.affymetrix.com
ORFDB/ORFeome	Invitrogen initiated program intended to provide the open reading frames (ORFs) of all human proteins	http://orf.invitrogen.com/
dbGaP	Database of genotypes and phenotypes	http://www.ncbi.nlm.nih.gov/gap
MSY breakpoint mapper	STSs on human Y chromosome	http://breakpointmapper.wi.mit.edu/
Gene cards	An automated, integrated database of human genes, genomic maps, proteins and diseases	http://www.genecards.org/

9.8 Plant Genome Databases

Database	Description	URL
AgBase	A curated, open-source, web-accessible resource for functional analysis of agricultural plant and animal gene products	http://www.agbase.msstate.edu/
autoSNPdb	Plant SNP discovery database	http://autosnpdb.appliedbioinformatics.com.au/
BarleyBase	Database for barley microarrays holds the data for over 1000 hybridizations from the affymetrix Barley1 GeneChip and genome arrays	http://www.barleybase.org/
Cereal small RNA database	An integrated resource for small RNAs such as miRNAs, siRNAs, ta-siRNAs expressed in rice and maize	http://sundarlab.ucdavis.edu/smrnas/
CR-EST	Crop EST database	https://apex.ipk-gatersleben.de/apex/f?p=116:1
EURISCO	European catalogue for plant genetic resources	http://eurisco.ecpgr.org/
GABiPD	Genome analysis of the plant biological system	http://www.gabipd.org/
GoMapMan	Unified plant-specific gene ontology	http://www.gomapman.org
GrainGenes	Molecular and phenotypic information on grain yielding plants such as wheat, barley, rye, triticale, and oats	http://wheat.pw.usda.gov/

(continued)

GreeNC	Green non-coding database (GreeNC) is a repository of long non-coding RNAs annotated in 37 plant species and six algae	http://greenc.sciencedesigners.com/wiki/Main_Page
MIPS plants database	Plant database resource for integrative and comparative plant genome research	http://mips.gsf.de/proj/plant/jsf/
NIAS	Databases for plant genetic resources and plant disease information	http://www.gene.affrc.go.jp/databases_en.php
P-MITE	Database of plant miniature inverted-repeat transposable elements (MITEs)	http://pmite.hzau.edu.cn/django/mite/
PGDD	A plant genome duplication database	http://chibba.agtec.uga.edu/duplication/
Phytozome	A database for green plant genomics	http://www.phytozome.net/
PIECE	Database of plant intron exon comparison and evolution	https://wheat.pw.usda.gov/piece/
Plant DNA C-values database	A database that provides a one-stop, user-friendly database where the plant genome sizes can be readily accessed and compared	http://www.kew.org/genomesize/homepage.html
Plant ontology database	This database is a collection of plant systematics, botany and genomics involved in developing simple and controlled vocabularies that accurately reflect the biology of plant structures	http://www.plantontology.org/
Plant SnoRNA database	A database of snoRNA genes in plant species	http://bioinf.scri.sari.ac.uk/cgi-bin/plant_snorna/home
Plant stress-responsive gene catalog	Stress-responsive gene in various plant species	http://dayhoff.generationcp.org/
Plant GDB	A database of plant genomic sequences and ESTs	http://www.plantgdb.org/
Plant PAN	Plant promoter analysis navigator database	http://PlantPAN2.itps.ncku.edu.tw
PlantProm	A database of plant promoter sequences	http://www.epd.isb-sib.ch/
PlantsP/PlantsT	PlantsP and PlantsT are plant-specific curated databases that combine sequence derived information with experimental functional genomics data	http://plantsp.sdsc.edu/
Plant tribes	Families of protein-coding genes from five sequenced plant species	http://fgp.huck.psu.edu/tribedb/index.pl
PlantTFDB	Plant transcription factor database provides a comprehensive, high-quality resource of plant transcription factors, regulatory elements and interactions between them	http://planttfdb.cbi.pku.edu.cn
PLAZA	A plant-oriented online resource for comparative, evolutionary and functional genomics	https://bioinformatics.psb.ugent.be/plaza

(continued)

PMDBase	A database of plant microsatellites and marker development	http://www.sesame-bioinfo.org/PMDBase
TIGR database	Plant transcript assemblies database	http://plantta.tigr.org/
TROPGene	A database comprising of genetic and genomic information about tropical crops such as banana, coconut, cotton, cocoa, etc.	http://tropgenedb.cirad.fr/
ForestTree	A resource that centralizes large-scale	http://foresttree.org/ftdb
DataBase	EST sequencing results from several tree species	
LegumeIIP	Model legumes integrative database platform	http://plantgrn.noble.org/LegumeIP/

9.9 Genomes of Economically Important and Model Plants

Database	URL
Medicago truncula ta database	http://www.medicago.org/
Brassica genome	http://www.brassicagenome.net/
Coffee genome hub	http://coffee-genome.org/
CottonGen	http://www.cottongen.org
Maize genetics and genomics database (MaizeGDB)	http://www.maizegdb.org/
The Arabidopsis information resource (TAIR)	https://www.arabidopsis.org/
Oryza sativa (ssp. japonica) genome	http://www.plantgdb.org/OsGDB/
The rice annotation project database	https://rapdb.dna.affrc.go.jp/
PoMaMo—potato maps and more	http://www.gabipd.org/projects/Pomamo/
Shanghai rapeseed database	http://rapeseed.plantsignal.cn/
Genome database for Rosaceae	http://www.rosaceae.org
Solanaceae genomics network	http://solgenomics.net/
SoyBase	http://soybase.org/
SoyKB—soybean knowledge base	http://soykb.org
TED—tomato functional genomics database	http://ted.bti.cornell.edu/
TIGR maize database	http://maize.tigr.org/
TomatEST DB	http://biosrv.cab.unina.it/tomatestdb/
TomatEST DB	http://biosrv.cab.unina.it/tomatestdb/
PANZEA	http://www.panzea.org/
Diatom EST database	http://www.biologie.ens.fr/diatomics/EST/

Glossary

Architectural Proteins: Architectural proteins are those, which mediates the interactions between distant sequences in the genome. The nature of the sequences involved in the genome-wide interaction will be determined by the architectural proteins. In the *Drosophila* genome, 11 different DNA binding architectural proteins have been identified.

Adaptive Mutations: are the mutations that are arising in the non-dividing cells during non-lethal selections. Adaptive mutations have been reported to occur in microorganisms such as bacteria and yeast.

Alphoid DNA (α DNA): is a 170 bp DNA segment arranged as tandem arrays in the centromere of human nuclear chromosomes. These sequences are not transcribed and are hypothesized to play an undefined role in the chromosome cycle.

Bacterial Artificial Chromosome (BAC): is a plasmid used for cloning and stably maintaining the large genomic DNA segments (100-200Kb) inside *E.coli* cells. They also serve as a source of gene specific promoter sequences.

Chromosome Territories: These are the regions in the nucleus, which are specific for a chromosome. The nucleus in most of the eukaryotic cells contains chromosome territories, with very few exceptions such as yeast.

Cleavage Amplified Polymorphic Sequences (CAPS): also referred as PCR-RFLP are the DNA fragments that are amplified by the PCR using specific primers and restriction digestion of the PCR amplified fragments with a restriction endonuclease and identification of the polymorphisms occurring due to the variations in the restriction sites will be tested by agarose gel electrophoresis.

Conjugative Transposons: are the DNA elements that are capable of integrating into the bacterial chromosome, and also have the ability to mobilize the non-transmissible plasmids from cell to cell when they fuse with such plasmids.

K. V. Chaitanya, *Genome and Genomics*,
https://doi.org/10.1007/978-981-15-0702-1

Contig: is a contiguous DNA sequence generated by assembling the overlapped DNA fragments of a chromosome.

Consensus Sequence: is a calculated DNA, RNA or an amino acid sequence order, after determining their positions in a sequence alignment.

Copia Elements: Copia is one of the three repeated sequence family members that are scattered into 30 different locations in the genome of *Drosophila melanogaster*, whose size is up to 5Kb. The sequence of these elements are highly conserved and are non-permutated at each location of the Drosophila genome. These elements show terminal redundancy with direct repeats of 0.5Kb.

CPG Islands are the DNA stretches of length 500–1500 bp with a high CG sequence. In this DNA stretch, C and G are connected by a phosphodiester bond. CPG islands are found near the regulatory regions of the housekeeping genes and the genes whose expression is high in a cell.CPG sequences near inactive genes are densely methylated to prevent their expression.

***Cos* Site:** abbreviated as Cohesive End Site is the sequence of nucleotides that are needed for the recognition and packaging of a lambda (λ) phage into its capsid.

Cosmid: is an engineered vector that combines the characteristics of both plasmids and phages or is a plasmid with a lambda (λ) phage *cos* site. Cosmids are used for constructing the genomic DNA libraries.

Direct Repeat Sequence: is an identical sequence, which is repeated in the same pattern multiple times into the downstream.

Domain Shuffling: is a phenomenon, where the gene segments that codes for a specific domain are shuffled between the genes during their evolution.

Enhancer: is a regulatory sequence that enhances the transcription of its associated gene, when bound by a transcription factor.

Episomes: An extrachromosomal genetic element of bacteria made up of DNA, capable of replicating in the cytoplasm or can be inserted into the bacterial chromosome and can replicate with it. Plasmids are the best examples of episomes in bacteria.

Euchromatin: is the loosely packed and less condensed form of DNA present in the nucleus, rich in the coding regions. It is the transcriptionally active form of the chromatin present in the distal arms of the chromosomes and is replicated during the S phase of the cell cycle.

Expressed Sequence Tags (ESTs): is a Sequence Tagged Site (STS) of a cDNA. It is a unique coding region of a gene used for identification of full length genes and also serves as landmarks for the gene mapping.

Exon: is the coding section of an mRNA that contains the information to encode into a protein.

Exosome: Both prokaryotic and eukaryotic cells harbor a highly conserved RNA processing and degrading protein complex called exosome, which plays a central role in the maturation of rRNA, small nucleolar RNA (snoRNA), small nuclear RNA (snRNA) and in the decay of mRNA.

Gene: is a unit of heredity information that occupies allocation on a chromosome called locus. Genes achieve their functions by directing the synthesis of proteins by undergoing transcription and translation.

Genome Mapping: is a spatial creation for the determination of a gene position and arrangement of the DNA fragments on the chromosomes.

Genome: complete set of an organism's chromosomal, mitochondrial and chloroplast (in the case of plants) DNA including its coding and non-coding regions.

Genome Duplication: is a process by which the additional copies of the entire genome are generated in the gametes during meiosis. This leads to an increase in the copy number (Ploidy).

Genotype: The genetic makeup of an individual, which determines its phenotype.

Gypsy Elements: Gypsy is the class I type of mobile retrotransposable elements that are present in the repetitive portions of the Drosophila genome. These elements are 7. 3 Kb long consisting of well-conserved long terminal repeats of 0.5Kb.

Hairpin Sequence: A region of DNA or RNA containing a short inverted repeat, which can form a short hair pin like structure, when two complementary sequences in a single molecule bind together.

hAT Elements: hAT transposable elements (derived from *hobo* elements of Drosophila, *Activator* elements from maize and *Tam3* elements from the snapdragon) are the group II DNA transposons that are commonly found in the genomes of plants animals and fungi. They are up to the size of 2.5 – 5Kb with short terminal inverted repeats.

Helitrons: are the eukaryotic transposable elements that move by rolling circle replication mechanism *via* a single stranded DNA intermediate. Three coding and structural variants of helitrons have been identified; Helentrons, Proto-Helentron, and Helitron2. Helitrons and Helentrons make up a substantial fraction of many high eukaryotic genomes.

Heteroplasmy: is the presence of two types of mitochondrial DNAs within a cell, one is mutated and one is not. As the mtDNA replicates and sorts randomly among the daughter cells during the cell division, each of the daughter cells will receive different proportions of normal and mutated mtDNA during heteroplasmy.

Heterochromatin: is a tightly packed or highly condensed form of DNA present in the nucleus, with a very low gene density and contains most of the non-coding regions such as repetitive sequences and transposable elements in the genome.

Heteroduplex: is a double stranded DNA molecule originated by the virtue of recombination of single complementary DNA strands belonging to different sources such as homologous chromosomes or different organisms.

Homopolymer Tail: is the function of terminal transferase enzyme to add the nucleoside triphosphate such as A or C to the DNA fragment when provided with dATP or dCTP. This feature enables the addition of two complementary cohesive homopolymer termini or the joining of two different DNA fragments with incompatible termini. Poly A tail found in the mRNA is an example of a homopolymer tail.

Homing Endonucleases: Double stranded DNAses with large asymmetric recognition sites of 12–40 bp and coding sequences embedded either in introns or inteins. They are found inserted in the genes of viruses, archaea, bacteria, and eukaryotes. Due to their activity, introns are spliced out of precursor RNA and inteins are spliced out of precursor proteins. They are named with similar conventions of restriction endonucleases with intron-encoded endonucleases containing the prefix, "I-" and intein endonucleases containing the prefix, "PI-". Recognition sites of Homing endonucleases are extremely rare, with a recognition sequence occur only once in every 7x 10^{10} base pairs of random sequence, which is equivalent to the size of 20 mammalian genomes.

Homoplasmy: refers to cell consisting of a uniform type of mtDNA, either completely normal or fully mutated.

Horizontal Gene Transfer: also called lateral gene transfer is the movement of genetic material between the organisms other than that of transmission from parents to offsprings through reproduction.

Indels: are the insertions and deletion mutations of less than 1Kb in length that is the most common class of mutations in the eukaryotic genomes. They are considered to be responsible for many constitutional and oncologic diseases.

Insertion Sequence Elements (IS Elements) are short DNA segments of 1–2 kb in length having an ability to translocate within and among replicons, leading to numerous molecular and genetic changes including mutations.

Interferons: are a group of soluble glycoproteins that are secreted from the cells in response to a viral infection. They are induced by the stimulation of Toll like receptors. They act by paracrine and autocrine modes and stimulate intra and intercellular networks for upregulating the innate and acquired immune resistance to viral infections.

Inteins: Inteins are the proteins with auto-processing domains found in all living organisms. They carry out protein splicing, a multi-step biochemical reaction comprised of both the cleavage and formation of peptide bonds. Inteins are also functional in exogenous reactions, which can be used to chemically manipulate any polypeptide backbone.

Integrons are highly versatile genetic elements that are most commonly found in the Gram-negative bacterial genomes, involved in the acquisition, expression, and dissemination of antibiotic resistance genes.

Intron: is a non-coding nucleotide sequence found in the ORF of a gene, which will be removed during the post transcriptional splicing of the mRNA.

Invertrons are the DNA elements with inverted terminal repeats at 5′ and 3′ ends, covalently bonded to the terminal proteins. They are involved in the replication of bacterial linear chromosomes. They also function as viruses, linear plasmids, transposons, etc.

Inverted Repeat Sequence: is a single stranded DNA sequence followed by a reverse complement in its downstream. Inverted repeat sequences are abundantly found to be present in the genomes of both prokaryotes and eukaryotes.

Linear Plasmids: are the extrachromosomal linear DNA elements that are found to be present both in Gram-positive and Gram-negative bacteria, Archaea and in the specific strains of *Neurospora intermedia, Podospora anserine*. In eukaryotes, they are mainly found in the mitochondrial DNA. They can be isolated and separated on Pulse Field Gel Electrophoresis (PFGE). Microbial linear plasmids carry either hairpins or proteins at their 5′ ends similar to viruses.

LINEs: Long interspersed nuclear Elements are the non-LTR retrotransposons of the size up to several Kb. They contain an internal promoter for RNA polymerase II, a 5' UTR region, two ORFs with ORF 1 coding for an RNA binding protein and ORF II coding for reverse transcriptase and a restriction endonuclease, a UTR region at 5′ terminal and a 3' Polyadenylation site.

LTR Retro Transposons: are the Class I transposons of size 5–7 Kb, characterized by the presence of long terminal repeats of a few hundred base pairs at each end. These elements were identified in the genome of *Drosophila melanogaster*.

Messenger RNA (mRNA): is the production of the DNA transcription, which undergoes translation on a ribosome for the generation of proteins.

Metagenomics: Metagenomics is also referred to as environmental and community genomics is the genomic analysis of the microorganisms directly from the environment, by extraction cloning and assemblage of DNA. This concept was developed from analyses of 16S rRNA gene sequences amplified directly from the environment, an approach that avoided the bias imposed by culturing and led to the discovery of vast new lineages of microbial life.

Microsatellite Repeats are the di, tri, and tetranucleotide tandem repeats in the DNA sequences of the human genome. The number of repeats is highly variable in the population, which varies from one individual to another. These repeats are evolutionarily relevant due to their high instability. They mutate at a rate of 10^3–10^6 per cell generation, which is 10 orders higher than the magnitude of point mutations.

MITEs: Miniature Invert Repeat Transposable Elements are a group of non-autonomous transposable elements that cannot code for their own transposase. These elements are short (500 bases) consisting of terminal inverted repeats up to 15 bp and two flanking target site duplications. They are reported in the genomes of plants, animals, and fungi.

Monocistronic mRNA: is the property of a eukaryotic mRNA that encodes only one protein. Development of a mature monocistronic mRNA involves the post transcriptional 5'capping, 3'polyadenylation tailing, splicing of introns and joining of exons.

Mosaic: is the presence of two different genotypes in an individual, developed from a single fertilized egg. A mosaic individual will have two or some times more genetically different cell lines in spite of being derived from a single zygote.

Mosaic Gene: A mosaic gene is composed of alternate sequencing polymorphisms that either belongs to the host or the DNA derived from a donor through horizontal gene transfer. Often, the integrated sequence contains a selectable genetic marker, which provides an antibiotic resistance. Most of the bacteria and viruses adapt to varying environmental conditions by acquiring mosaic genes.

Motif: is a sequence of a DNA, RNA or a protein with a specific biological significance.

Multipartite Genome: If the entire genome is split between two or more large DNA fragments, the architecture is referred to as a multipartite genome. The multipartite organization is found in the nitrogen-fixing rhizobia, pathogenic microbes such as Brucella, Vibrio, and Burkholderia.

***Mutator*–Like Transposable Elements (MULEs):** are Class 2 transposable elements that are widely distributed in the genomes of plants, animals, protozoans, and fungi. There are both autonomous and non-autonomous. Transposase of MULEs harbors a zinc finger binding domain and a catalytic domain. These elements consist of a long terminal inverted repeats of 100 bp and 8–10 bp of the flanking target site duplications.

Nucleosome: is a fundamental unit of chromatin, comprising of a section of DNA wrapped by histone proteins.

Open Reading Frame (ORF): is a part of the gene that gets transcribed into mRNA and translated into a protein. It is a sequence of codons that begins with AUG, the start codon and ends at UAA or UAG or UGA, the stop codons.

Operator: is a segment of DNA, which binds to a repressor protein for down regulating the expression of the associated gene. Mostly, an operator is usually found adjacent to a promoter.

Operon: A set of genes that are transcribed and expressed under the control of a single promoter gene. An operon is a DNA sequence consisting of a structural gene, an operator gene and a regulator gene as its adjacent genes, responsible for the transcription and gene regulation under a commonly shared promoter. Operons are found in the bacteria and archaea but not in eukaryotes.

P1 Derived Artificial Chromosome: is an artificial plasmid developed for carrying large fragments of the vertebrate genomes up to the size 300 Kb in an *E.coli* strain. PACs have superior cloning capacity when compared with YAC and BAC.

Palindromic Sequence: A double helix DNA or RNA sequence that is same when reading from 5′ to 3′ on one strand and 3′–5′ on the other complementary strand. It is also known as an inverted-reverse sequence.

Pathogenicity Islands: are the genetic elements present on chromosomes of several pathogenic bacteria, encoding the virulence factors that are normally absent in the non-pathogenic bacterial population. These genomic islands are acquired through horizontal gene transfer.

Physical Map: is the low resolution map that determines the physical structure of a genome such restriction sites, position specific clones and other elements for marking the position of genes.

Plasmid: Plasmids are double stranded circular, self replicating DNA molecules found in most of the bacteria. Presence of plasmid 2 μ circle has also been reported in most of the *Saccharomyces cerevisiae* strains. Plasmids carry the genes which provide resistance to naturally occurring antibiotics or the proteins which may act as toxins in a specific environment. Plasmids also provide elemental nitrogen fixing ability to bacteria and capability to degrade recalcitrant organic compounds under nutrient deprivation. Within a population of microbes, plasmids provide a mechanism for horizontal gene transfer thereby possessing certain selective advantage under certain environmental conditions. Plasmids have a wide range of applications in the molecular biology and genetic engineering, serve as tools for multiplying the copies or for the expression of specific genes.

Polar Mutation: is a mutation that affects the transcription or translation of part of the gene or an operon located in the downstream of the mutant site. Mutations such as nonsense, frameshift, and insertion sequence(IS)-induced mutations are examples of polar mutations.

Point Mutation: is the alteration of a single base in the DNA sequence of an organism, either through addition or deletion. Point mutations are occurred mostly during the DNA replication or by the exposure to mutagens.

Polyadenylation: is an essential feature of Eukaryotic pre-m-RNA processing catalyzed by an enzyme poly(A)polymerase, resulting in the formation of a polyadenylated tail of ~100–200 bases long before the mRNA leaves the nucleus for the translation.

Polymorphism: is a genetic variation resulting in the occurrence of two or more different forms of an organism in a population. A most common source of polymorphism is Single Nucleotide Polymorphism (SNP).

Polycistronic mRNA: is the characteristic feature of bacterial and chloroplast mRNAs, which encodes several proteins from a single mRNA chain. It consists of a leader sequence, which precedes the first translating gene sequence and an intercistronic region that separates two gene sequences. The last gene sequence will be followed by a trailer sequence.

Promoter: is a sequence of the genomic DNA located either directly on the upstream or at the 5′ end of the transcription initiation site of a gene. Most of the eukaryotic genes have a well conserved promoter sequence known as TATA box, situated around 35 bases upstream to the transcription initiation site, for binding with the RNA polymerase and transcription factors and initiation of transcription.

Proteome: Total number of proteins that are expressed by tissue, organ, organism at a specific point of time.

Pseudogenes: are the non-functional genes present in the genome, which might have originated by the duplication of an ancestral gene. The first pseudogene reported was 5S DNA of *Xenopus laevis*, in 1977. Pseudogenes are categorized into the duplicate and processed. ~4000 duplicated pseudogenes and ~8000 processed pseudogenes have been identified in the human genome.

Retro Transposons: Resembles human endogenous retroviruses in their DNA sequence organization and protein coding capacity. They are also called as LTR transposons due to the presence of direct Long Terminal Repeats (LTR) of a few hundred bases at each end. They are found in all eukaryotes and not reported in the prokaryotes. Retrotransposons comprise up to 3% of the yeast genome, 30% of the human genome and up to 75% of the maize genome. Long interspersed nuclear elements (LINEs) are the autonomous retrotransposons, predominantly represented by LINE-1 and small interspersed nuclear elements (SINEs) are non-autonomous retrotransposons, represented by ALUs.

Replisome: It is a large protein complex that carries out DNA replication. Replisome consists of several enzymes, such as helicase, primase, and DNA polymerase and creates a replication fork to duplicate both the leading and lagging strand starting at the origin.

Repressor: is a regulatory protein that binds to a specific DNA sequence for controlling the expression of a specific gene or an operon.

Replicon: A replicon is either a DNA or an RNA molecule, which replicates from a single origin point. Archaeal Sulfolobus contain three replicons. Bacteria such as Rhodobacter sphaeroides and Eukaryotes are known to contain multiple replicons.

Retron is a DNA sequence present in the genomes of bacteria, codes for a reverse transcriptase similar to that of retroviruses and retroelements. Retron reverse transcriptase is used in the synthesis of a msDNA (a complex of DNA, RNA, and a protein).

Retro Transposon: are the transposable elements containing the gene for reverse transcriptase. They can be the long terminal repeat retrotransposons (LTR) and non-long terminal repeat retrotransposons (LINEs, SINEs).

Satellite Chromosomes: The terminal segment of a chromosome that is separated from the main body of the chromosome by a secondary constriction. Part of the chromosome present beyond the secondary constriction is called satellite body (Trabant). Chromosomes bearing satellite bodies are called SAT- chromosomes Atleast 2–4 satellite chromosomes are present in each diploid nucleus, which play a vital role in the formation of the nucleolus in the daughter cell after the completion of cell division. In humans, chromosomes 13, 14, 15, 21 and 22 are SAT-chromosomes.

Satellite DNA: are a large array of tandemly (non-inverting) repeating non-coding DNA found in the human genome. Satellite DNA comprises of minisatellite DNA (5 bp), microsatellite DNA (2 or 3 bp). They form the main constituents of heterochromatin and are important centromere building DNA elements.

Segmental Duplications: are the long stretches of DNA (>1Kb in length), with highly identical sequences (90–100% identity), occurs in multiple locations of the human genome as a result of duplication events. Segmental duplications can be either tandem or interspersed, often leads to chromosomal rearrangement and can cause genome instability.

Sense Strand: The promoter sequence of a gene determine the direction of the transcription and the strand which undergoes transcription. Such strand is called sense strand.

SINEs: Short Interspersed Nuclear elements are the non-LTR retrotransposons of the size 80–400 bp in length. SINEs have an internal promoter for RNA polymerase III and a $3'$ polyadenylation site. Most prominent SINEs are Alu elements in the Human genome and B1 elements in rodents genome.

Shuttle Vectors: Vectors that can propagate in two different host species are called shuttle vectors. Using shuttle vectors, the inserted DNA can be tested in two different cell types such as prokaryotes and eukaryotes. They are frequently used for making the multiple copies of the inserted gene and also used in the Invitro experiments such as mutagenesis. One of the most common shuttle vectors is the yeast shuttle vector that contains components for the replication and selection in both *E. coli* cells and yeast cells. The *E. coli* component of a yeast shuttle vector includes an origin of replication and an antibiotic selection marker, such as β-lactamase. The yeast component of the vector includes an autonomously replicating sequence, a yeast centromere, and a yeast selection marker.

Telomere: is a heterochromatin region found in the end part of a chromosome that constitutes a hexamer repetitive sequence TTA GGG. Telomeres have two functions; (1) They distinguish the true chromosomal ends and prevents the chromosome breakage and (2) Helps in the completion of the DNA replication.

Transposon: Also called as Jumping genes, are the DNA sequences which frequently changes their position with in the genome, responsible for the creation of mutations in the DNA and altering the genetic identity of that organism. Transposons were first discovered by Barbara Mc Clinktock in 1940s in Maize. Transposons are also responsible for the bulking of the eukaryotic genomes.

Trans Splicing: It is a primitive type of RNA processing, which ligates exons from two different primary RNA transcripts end to end. This fundamental mechanism parallels cis-splicing which removes introns in eukaryotes.

Transcriptome: is a complete set of transcript of a cell or the total number of mRNAs that are produced in a cell, tissue, organ at a specific physiological stage of an organism.

Transformation-Competent Artificial Chromosome (TAC): is a vector used for cloning and transforming the plant DNA fragments of size 40 Kb with relevant importance in various agronomic traits.

Twintrons: A special type of spatial arrangement in which two alternatively spliced introns occupy the same position on a genome within a gene and require two different spliceosomes for their processing. Twintrons have been reported in the organellar genomes and in insects.

Whole Genome Duplication: is a common feature found in the higher plants and cancer cells. It involves the duplication of a complete set of chromosomes. Genome doubling is one of the evolutionary factors that facilitate the formation of a new species in plants.

Yeast Artificial Chromosome: Also called as YAC is an artificially constructed system, which can undergo replication. It is a synthetic vector capable of carrying the large DNA fragments of the size up to 2000 Kb. YAC contains a centromere, telomeres, and an autonomously replicating sequence required for the replication and sustenance of YAC inside the yeast cells.

Bibliography

Chapter 1

Agoi VI (1974) Towards the system of viruses. Biosystems 6(2):113–132

Akusjarvi G, Stevenin J (2003) Remodelling of the host cell rna splicing machinery during an adenovirus infection. Curr Top Microbiol Immunol 272:253–286

Amalfitano A, Parks RJ (2002) Separating fact from fiction: assessing the potential of modified adenovirus vectors for use in human gene therapy. Curr Gene Ther 2:111–133

Amiya KB (1987) Transcription and replication of rhabdoviruses. Microbiol Rev 51:66–87

Angelescu DG, Stenhammar J, Linse P (2007) Packaging of a flexible polyelectrolyte inside a viral capsid: effect of salt concentra-tion and salt valence. J Phys Chem B 111:8477–8485

Aparicio O, Razquin N, Zaratiegui M, Narvaiza I, Fortes P (2006) Adenovirus virus-associated RNA is processed to functional interfering RNAs involved in virus production. J Virol 80:1376–1384

Ashelford KE, Day MJ, Fry JC (2003) Elevated abundance of bacteriophage infecting bacteria in soil. Appl Environ Microbiol 69:285–289

Baltimore D (1971) Expression of animal virus genomes. Bacteriol Rev 35(3):235–241

Bearson BL, Allen HK, Brunelle BW, Lee IS, Casjens SR, Stanton TB (2014) The agricultural antibiotic carbadox induces phage-mediated gene transfer in Salmonella. Front Microbiol 5:52. https://doi.org/10.3389/fmicb.2014.00052

Bergh O, Borsheim KY, Bratbak G, Heldal M (1989) High abundance of viruses found in aquatic environments. Nature 340:467–468

Berk AJ (2007) Adenoviridae: the viruses and their replication. In: Knipe DM, Howley PM (eds) Fields virology, 5th edn. Lippincott Williams & Wilkins, Philadelphia, PA, pp 2355–2394

Blinkova O, Joseph V, Yingying L, Brandon FK, Crickette S, Jean-Bosco NN, Martine P, Dominic T, Elizabeth VL, Michael LW, Anne EP, Beatrice HH, Eric LD (2010) Novel circular DNA viruses in stool samples of wild-living chimpanzees. J Gen Virol 91:74–86

Boyd EF (2012) Bacteriophage-encoded bacterial virulence factors and phage-pathogenicity island interactions. Adv Virus Res 82:91–118

Brussaard CP, Wilhelm SW, Thingstad F, Weinbauer MG, Bratbak G, Heldal M, Kimmance SA, Middelboe M, Nagasaki K, Paul JH, Schroeder DC, Suttle CA, Vaque D, Wommack KE (2008) Global-scale processes with a nanoscale drive: the role of marine viruses. ISME J 2:575–578

Cann AJ (1999) DNA viruses: a practical approach. Oxford University Press, Oxford. ISBN 0199637199

Cardinale D, Carette N, Michon T (2012) Virus scaffolds as enzyme nano-carriers. Trends Biotechnol 30:369–376

Casas V, Maloy S (2011) Role of bacteriophage-encoded exotoxins in the evolution of bacterial pathogens. Future Microbiol 6:1461–1473

K. V. Chaitanya, *Genome and Genomics*,
https://doi.org/10.1007/978-981-15-0702-1

Casjens S, Hendrix R (2005) Bacteriophages and the bacterial genome. In: Higgins NP (ed) The bacterial chromosome. ASM Press, Washingtion, DC, pp 39–52

Caspar DLD (1975) Design principles in virus particle construction. In: Horsfall FL, Tamm I (eds) Viral and rickettsial infections in man, 4th edn. JB Lippincott, Philadelphia, PA

Cerritelli ME, Cheng N, Steven AC (1997) Encapsidated conformation of bacteriophage T7 DNA. Cell 91:271–280

Chen J, Novick RP (2009) Phage-mediated intergeneric transfer of toxin genes. Science 323:139–141

Chothia C (1992) Proteins, one thousand families for the molecular biologist. Nature 357:543–544

Christensen JB, Byrd SA, Walker AK, Strahler JR, Andrews PC, Imperiale MJ (2008) Presence of the adenovirus IVa2 protein at a single vertex of the mature virion. J Virol 82(18):9086–9093. https://doi.org/10.1128/JVI.01024-08

Chu FK, Maley GF, Maley F, Belfort M (1984) Intervening sequence in the thymidylate synthase gene of bacteriophage T4. Proc Natl Acad Sci U S A 81:3049–3053

Claverie JM, Abergel C (2009) Mimivirus and its virophage. Annu Rev Genet 43:49–66

Claverie JM, Ogata H, Audic S, Abergel C, Suhre K, Fournier PE (2006) Mimivirus and the emerging concept of "giant" virus. Virus Res 117(1):133–144

Craig NL et al (2002) Mobile DNA. ASM Press, Washington, DC. ISBN 1555812090

Davison AJ, Benko M, Harrach B (2003) Genetic content and evolution of adenoviruses. J Gen Virol 84:2895–2908

Dennis HB, Jonathan MG, David IS (2005) What does structure tell us about virus evolution? Curr Opin Struct Biol 15:655–663

Diemer GS, Stedman KM (2012) A novel virus genome discovered in an extreme environment suggests recombination between unrelated groups of RNA and DNA viruses. Biol Direct 7:13. https://doi.org/10.1186/1745-6150-7-13

Domingo E, Webster RG, Holland JJ (2000) Origin and evolution of viruses. Academic, San Diego, CA. ISBN 0122203607

Dunn JJ, Studier FW (1983) Complete nucleotide sequence of bacteriophage T7 DNA and the locations of T7 genetic elements. J Mol Biol 166:477–535

Elodie G, Naomi AS, Martin S, Jennifer ZF, Vik S, David JS, Jeff S, Hean K, Pavel B, Dmitry D, Tatiana T, Yiming B, Kirsten SG, Jill T, David J, Claire MF, Jeffery KT, Steven LS (2005) Large-scale sequencing of human influenza reveals the dynamic nature of viral genome evolution. Nature 437:1162–1166. https://doi.org/10.1038/nature04239

Eric SM, Elizabeth K, Gisela M, Fumio A, Takashi K, Wolfgang R (2003) Bacteriophage T4 genome. Microbiol Mol Biol Rev 67:86–156

Flint J (1999) Organization of the adenoviral genome. In: Seth P (ed) Adenoviruses: basic biology to gene therapy. R.G. Landes Company, Austin, TX, pp 17–30

Fortier LC, Sekulovic O (2013) Importance of prophages to evolution and virulence of bacterial pathogens. Virulence 4:354–365

Gajdusek DC (1977) Unconventional viruses and the origin and disappearance of kuru. Science 197:943

Gelderblom HR (1991) Assembly and morphology of HIV: potential effect of structure on viral function. AIDS 5:617–637

Ginn SL, Alexander IE, Edelstein ML, Abedi MR, Wixon J (2013) Gene therapy clinical trials worldwide to 2012-an update. J Gene Med 15:65–77

Gott JM, Shub DA, Belfort M (1986) Multiple self-splicing introns in bacteriophage T4: evidence from autocatalytic GTP labeling of RNA in vitro. Cell 47:81–87

Guillaume M, Louis F (2010) Herpesviruses and chromosomal integration. J Virol 84 (23):12100–12109. https://doi.org/10.1128/JVI.01169-10

Harsh BP, Jamie JA, Phillip NW, Michele RSH, Craig EC (2007) Picornavirus genome replication: assembly and organization of the vpg uridylylation ribonucleoprotein (initiation) complex. J Biol Chem 282(22):16202–16213

Hatfull GF (2008) Bacteriophage genomics. Curr Opin Microbiol 11:447–453
Hatfull GF, Sarkis GJ (1993) DNA sequence, structure and gene expression of mycobacteriophage L5: a phage system for mycobacterial genetics. Mol Microbiol 7:395–405
Highton PJ, Chang Y, Myers RJ (1990) Evidence for the exchange of segments between genomes during the evolution of lambdoid bacteriophages. Mol Microbiol 4:1329–1340
Hilleman MR, Werner JH (1954) Recovery of new agents from patients with acute respiratory illness. Proc Soc Exp Biol Med 85:183–188
Hoke CH Jr, Snyder CE Jr (2013) History of the restoration of adenovirus type 4 and type 7 vaccine, live oral (Adenovirus Vaccine) in the context of the department of defense acquisition system. Vaccine 31:1623–1632
Holmes EC (2003) Error thresholds and the constraints to RNA virus evolution. Trends Microbiol 11:543–546
Holmfeldt K, Natalie S, Manesh S, Kristen C, Lasse R, Nathan CV, Matthew BS (2013) Twelve previously unknown phage genera are ubiquitous in global oceans. Proc Natl Acad Sci U S A 110:12798–12803
Horwitz MA (1996) Adenoviruses. In: Knipe DM, Howley PM (eds) Fields virology, vol 2007, 5th edn. Lippincott Williams & Wilkins, Philadelphia, PA, pp 2149–2171
Horwitz MS (2004) Function of adenovirus E3 proteins and their interactions with immunoregulatory cell proteins. J Gene Med 6(Suppl. S1):S172–S183
Hu Y, Zandi R, Gelbart WM (2008) Packaging of a polymer by a viral capsid: the interplay between polymer length and capsid size. Biophys J 94:1428–1436
Ian BD, Justin ER, Leslie RK, Philip MM, Ali Rezaian M (1993) Nucleotide sequence and genome organization of tomato leaf curl geminivirus. J Gen Virol 74:147–151
Jaime I, Susanna CM (2012) Evolutionary dynamics of genome segmentation in multipartite viruses. Proc R Soc B 279:3812–3819
John KR (1996) Positive strands to the rescue again: a segmented negative-strand RNA virus derived from cloned cDNAs. Proc Natl Acad Sci U S A 93(26):14998–15000. https://doi.org/10.1073/pnas.93.26.14998
Juhala RJ, Ford ME, Duda RL, Youlton A, Hatfull GF, Hendrix RW (2000) Genomic sequences of bacteriophages HK97 and HK022: pervasive genetic mosaicism in the lambdoid bacteriophages. J Mol Biol 299:27–51
Karam JD, Drake JW, Kreuzer KN, Mosig G, Hall DH, Eiserling FA, Black LW, Spicer EK, Kutter E, Carlson C, Miller ES (eds) (1994) Molecular biology of bacteriophage T4. American Society for Microbiology, Washington, DC
Kivenson A, Hagan MF (2010) Mechanisms of capsid assembly around a polymer. Biophys J 99:619–628
Koonin EV (2009) On the origin of cells and viruses: primordial virus world scenario. Ann N Y Acad Sci 1178:47–64
Koonin EV, Aravind L, Kondrashov AS (2000) The impact of comparative genomics on our understanding of evolution. Cell 101:573–576
Krupovic M (2013) Networks of evolutionary interactions underlying the polyphyletic origin of ssDNA viruses. Curr Opin Virol 3:578–586
Labonte JM, Suttle CA (2013) Previously unknown and highly divergent ssDNA viruses populate the oceans. ISME J 7:2169–2177
Lamb RA, Krug RM (2001) *Orthomyxoviridae*: the viruses and their replication. In: Knipe DM, Howley PM, Griffin DE et al (eds) Fields virology, 4th edn. Lippincott Williams & Wilkins, Philadelphia, PA, pp 1487–1531
Lee YD, Kim JY, Park JH, Chang H (2012) Genomic analysis of bacteriophage ESP2949-1, which is virulent for *Cronobacter sakazakii*. Arch Virol 157:199–202
Legendre M, Bartoli J, Shmakova L, Jeudy S, Labadie K, Adrait A, Lescot M, Poirot O, Bertaux L, Bruley C, Coute Y, Rivkina E, Abergel C, Claverie JM (2014) Thirty-thousand-year-old distant relative of giant icosahedral DNA viruses with a pandoravirus morphology. Proc Natl Acad Sci U S A 111(11):4274–4279

Linus S, Britt-Marie S (2007) Self-splicing of the bacteriophage t4 group i introns requires efficient translation of the pre-mRNA in vivo and correlates with the growth state of the infected bacterium. J Bacteriol 2:980–990

Mahdi B, Anne B, Guylaine P (2010) Mosaic graphs and comparative genomics in phage communities. J Comput Biol 17(9):1315–1326

Mart K, Patrick F (2015) Single-stranded DNA viruses employ a variety of mechanisms for integration into host genomes. Ann N Y Acad Sci 1341:41–53

Mathews CK, Kutter E, Mosig G, Berget PB (eds) (1983) Bac-teriophage T4. American Society for Microbiology, Washington, DC

Mertens P (2004) The dsRNA viruses. Virus Res 101:3–13

Miller ES, Elizabeth K, Gisela M, Fumio A, Takashi K, Wolfgang R (2003) Bacteriophage T4 genome. Microbiol Mol Biol Rev 67:86–156. https://doi.org/10.1128/MMBR.67.1.86-156.2003

Modi SR, Lee HH, Spina CS, Collins JJ (2013) Antibiotic treatment expands the resistance reservoir and ecological network of the phage metagenome. Nature 499:219–222

Morris SJ, Scott GE, Leppard KN (2010) Adenovirus late-phase infection is controlled by a novel L4 promoter. J Virol 84:7096–7104

Moya A, Holmes EC, González-Candelas F (2004) The population genetics and evolutionary epidemiology of RNA viruses. Nat Rev Microbiol 2:279–288

Musher DM (2003) How contagious are common respiratory tract infections? N Engl J Med 348:1256–1266

Nam KT, Kim DW, Yoo PJ, Chiang CY, Meethong N, Hammond PT, Chiang YM, Belcher AM (2006) Virus-enabled synthesis and assembly of nanowires for lithium ion battery electrodes. Science 312:885–888

Nareerat V, Roongroje T, Alongkorn A, Sanipa S, Sunchai P, Juthatip K, Kanisak O, Piya W, Sukanya P, Apiradee T, Yong P (2004) The genome sequence analysis of H5N1 avian influenza A virus isolated from the outbreak among poultry populations in Thailand. Virology 328:169–176

Oliver KM, Degnan PH, Hunter MS, Moran NA (2009) Bacteriophages encode factors required for protection in a symbiotic mutualism. Science 325:992–994

O'Malley RP, Mariano TM, Siekierka J, Mathews MB (1986) A mechanism for the control of protein synthesis by adenovirus VA RNAI. Cell 44:391–400

Philippe N, Legendre M, Doutre G, Coute Y, Poirot O, Lescot M, Arslan D, Seltzer V, Bertaux L, Bruley C, Garin J, Claverie JM, Abergel C (2013) Pandoraviruses: amoeba viruses with genomes up to 2.5 Mb reaching that of parasitic eukaryotes. Science 341(6143):281–286

Pietilä MK, Roine E, Paulin L, Kalkkinen N, Bamford DH (2009) An ssDNA virus infecting archaea: a new lineage of viruses with a membrane envelope. Mol Microbiol 72:307–319. https://doi.org/10.1111/j.1365-2958.2009.06642

Qazi S, Liepold LO, Abedin MJ, Johnson B, Prevelige P, Frank JA, Douglas T (2013) P22 viral capsids as nanocomposite high-relaxivity MRI contrast agents. Mol Pharm 10:11–17

Raoult D, Forterre P (2008) Redefining viruses: lessons from Mimivirus. Nat Rev Microbiol 6 (4):315–319

Raoult D, Audic S, Rober C, Abergel C, Renesto P, Ogata H, La Scola B, Suzan M, Claverie JM (2004a) The 1.2-megabase genome sequence of Mimivirus. Science 306(5700):1344–1350

Raoult D, Audic S, Robert C, Abergel C, Renesto P, Ogata H, La Scola B, Suzan M, Claverie JM (2004b) The 1.2-megabase genome sequence of Mimivirus. Science 306:1344–1350

Rice G, Tang L, Stedman K, Roberto F, Spuhler J, Gillitzer E, Johnson JE, Douglas T, Young M (2004) The structure of a thermophilic archaeal virus shows a double-stranded DNA viral capsid type that spans all domains of life. Proc Natl Acad Sci U S A 101:7716–7720

Robert B, Oliver G (2007) Pybus and Andrew Rambaut. The evolution of genome compression and genomic novelty in RNA viruses. Genome Res 17:1496–1504

Roberta B, Göran A (2015) Regulation of human adenovirus alternative rna splicing by he adenoviral l4-33k and l4-22k proteins. Int J Mol Sci 16:2893–2912. https://doi.org/10.3390/ijms16022893

Robinson RA, O'Callaghan DJ (1983) A specific viral DNA se-quence is stably integrated in herpesvirus oncogenically transformed cells. Cell 32:569–578

Rocholl C, Gerber K, Daly J, Pavia AT, Byington CL (2004) Adenoviral infections in children: the impact of rapid diagnosis. Pediatrics 113:e51–e56

Rolfsson O, Toropova K, Stockley PG (2010) Mutually-induced conformational switching of RNA and coat protein underpins efficient assembly of a viral capsid. J Mol Biol 401:309–322

Rosario K, Anisha D, Milen M, Jessica W, Simona K, Daisy S, Mya B, Arvind V (2012) Diverse circular ssDNA viruses dis-covered in dragonflies (Odonata: Epiprocta). J Gen Virol 93:2668–2681

Roux S, Enault F, Bronner G, Vaulot D, Forterre P, Krupovic M (2013) Chimeric viruses blur the borders between the major groups of eukaryotic single-stranded DNAviruses. Nat Commun 4:2700. https://doi.org/10.1038/ncomms3700

Rowe WP, Huebner RJ, Gilmore LK, Parrott RH, Ward TG (1953) Isolation of a cytopathogenic agent from human adenoids undergoing spontaneous degeneration in tissue culture. Proc Soc Exp Biol Med 84:570–573

Russell WC (2000) Update on adenovirus and its vectors. J Gen Virol 81(11):2573–2604

Sanger F, Air GM, Barrell BG, Brown NL, Coulson AR, Fiddes CA, Hutchison CA, Slocombe PM, Smith M (1977) Nucleotide sequence of bacteriophage phi X174 DNA. Nature 265:687–695

Sanger F, Coulson AR, Hong GF, Hill DF, Petersen GB (1982) Nucleotide sequence of bacteriophage lambda DNA. J Mol Biol 162:729–773

Shen L, Bao N, Prevelige PE, Gupta A (2010) Fabrication of ordered nanos-tructures of sulfide nanocrystal assemblies over self-assembled genetically engineered P22 coat protein. J Am Chem Soc 132:17354–17357

Sikorski A, Kearvell J, Elkington S, Dayaram A, Arguello-Astorga GR, Varsani A (2013) Novel ssDNA viruses discovered in yellow-crowned parakeet (*Cyanoramphus auriceps*) nesting material. Arch Virol 158:1603–1607

Steinhauer DA, Skehel JJ (2002) Genetics of influenza viruses. Annu Rev Genet 36:305–332

Suttle CA (2005) Viruses in the sea. Nature 437:356–361

Suttle CA (2007) Marine viruses–major players in the global ecosystem. Nat Rev Microbiol 5:801–812

Thomas B, Charles HC, Stephen H (2013) Viruses of the family Bunyaviridae: are all available isolates reassortants. Virology 446:207–216

Tihova M, Dryden KA, Schneemann A (2004) Nodavirus coat protein imposes dodecahedral RNA structure independent of nucleo-tide sequence and length. J Virol 78:2897–2905

Tollefson AE, Ying B, Doronin K, Sidor PD, Wold WS (2007) Identification of a new human adenovirus protein encoded by a novel late l-strand transcription unit. J Virol 81:12918–12926

Tom SYG, Wenjie Z, Yizhi JT (2012) Bunyavirus: structure and replication. Adv Exp Med Biol 726:245–266. https://doi.org/10.1007/978-1-4614-0980-9_11

Van KI, Elliott RM (2015) Flexibility of bunyavirus genomes:creation of an orthobunyavirus with an ambisense S segment. J Virol 89:5525–5535. https://doi.org/10.1128/JVI.03595-14

Venkataraman S, Reddy SP, Reddy VS (2008) Structure of Seneca Valley virus-001: an oncolytic picornavirus representing a new genus. Structure 16:1555–1561

Vijaykrishna D, Mukerji R, Smith GJD (2015) RNA virus reassortment: an evolutionary mechanism for host jumps and immune evasion. PLoS Pathog 11(7):e1004902. https://doi.org/10.1371/journal.ppat.1004902

Volkova VV, Lu Z, Besser T, Grohn YT (2014) Modeling infection dynamics of bacteriophages in enteric *Escherichia coli* estimating the contribution of transduction to antimicrobial gene spread. Appl Environ Microbiol 80:4350–4362

Wagner M, Wagner M, Ruzsics Z, Koszinowski UH (2002) Herpesvirus genetics has come of age. Trends Microbiol 10:318–324

Weitzman MD (2005) Functions of the adenovirus E4 proteins and their impact on viral vectors. Front Biosci 10:1106–1117

Whelan SPJ, Barr JN, Wertz GW (2004) Transcription and replication of nonsegmented negative-strand RNA viruses. In: Kawaoka Y (ed) Biology of negative strand RNA viruses: the power of reverse genetics. Springer, Berlin Heidelberg, pp 62–119

Whitman WB, Coleman DC, Wiebe WJ (1998) Prokaryotes: the unseen majority. Proc Natl Acad Sci U S A 95:6578–6583

Wold WS, Horwitz MS (2007) Adenoviruses. In: Knipe DM, Howley PM (eds) Fields virology, 5th edn. Lippincott Williams & Wilkins, Philadelphia, PA, pp 2396–2436

Wommack KE, Colwell RR (2000) Virioplankton: viruses in aquatic ecosystems. Microbiol Mol Biol Rev 64:69–114

Ying B, Tollefson AE, Wold WS (2010) Identification of a previously unrecognized promoter that drives expression of the UXP transcription unit in the human adenovirus type 5 genome. J Virol 84:11470–11478

Yongwen L, Ying Z, Xiangyin L, Youtian Y, Xianfeng Y, Daiting Z, Xianbo D, Xiaowei W, Xiaofeng G (2012) Complete genome sequence of a highly virulent rabies virus isolated from a rabid pig in south china. J Virol 86:12454–12455

Young PMO, hman MQ, Xu DA, Shub Sjöberg BM (1994) Intron-containing T4 bacteriophage gene sunY encodes an anaerobic ribonucleotide reductase. J Biol Chem 269:20229–20232

Zhao H, Chen M, Pettersson U (2014) A new look at adenovirus splicing. Virology 456–457:329–341

Chapter 2

Ahmed F, Stuart L, Gopi G, Stephen WS (2003) Remarkable sequence signatures in archaeal genomes. Archaea 1:185–190

Anja S, Jimmy HS, Steffen LJ, Katarzyna ZN, Joran M, Anders EL, Roel E, Christa S, Lionel G, Thijs JGE (2015) Complex archaea that bridge the gap between prokaryotes and eukaryotes. Nature 521(7551):173–179. https://doi.org/10.1038/nature14447

Aravind L, Lakshminarayan MI, Vivek A (2003) The two faces of Alba: the evolutionary connection between proteins participating in chromatin structure and RNA metabolism. Genome Biol 4:R64

Berend S, Peer B, Martijn AH (2002) Genomes in flux: the evolution of archaeal and proteobacterial gene content. Genome Res 12(1):17–25

Bult CJ, White O, Olsen GJ, Zhou L, Fleischmann RD, Sutton GG, Blake JA, FitzGerald LM, Clayton RA, Gocayne JD, Kerlavage AR, Dougherty BA, Tomb JF, Adams MD, Reich CI, Overbeek R, Kirkness EF, Weinstock KG, Merrick JM, Glodek A, Scott JL, Geoghagen NS, Venter JC (1996) Complete genome sequence of the methanogenic archaeon, Methanococcus jannaschii. Science 273(5278):1058–1073

Chastain BK, Kral TA (2010) Approaching Mars-like geochemical conditions in the laboratory: omission of artificial buffers and reductants in a study of biogenic methane production on a smectite clay. Astrobiology 10:889–897

Clotilde T, Laurence Z, Sebastian A (2002) Exosomes: composition, biogenesis and function. Nat Rev Immunol 2:569–579

Cox R, Mirkin SM (1997) Characteristic enrichment of DNA repeats in different genomes. Proc Natl Acad Sci 94:5237–5242

David EG, Ross O, Gary J, Carl RW (2000) An archaeal genomic signature. PNAS 97:3304–3308

David SS, Ashley JP, John AT (2014) Archaeal genome guardians give insights into Eukaryotic DNA replication and damage response proteins. Archaea 2014, Article ID 206735, 24 p

Eugene VK, Yuri IW (2008) Genomics of bacteria and archaea: the emerging dynamic view of the prokaryotic world. Nucleic Acids Res 36:216688–216719. https://doi.org/10.1093/nar/gkn668

Francesco C, Juergen W (2014) Anaerobic thermophiles. Life 4:77–104. https://doi.org/10.3390/life4010077

Gary JO, Carl RW (1997) Archaeal genomics: an overview. Cell 89:991–994

Horvath P, Barrangou R (2010) CRISPR/Cas, the immune system of bacteria and archaea. Science 327:167–170

Huber R, Burggraf S, Mayer T, Barns SM, Rossnagel P, Stetter KO (1995) Isolation of a hyperthermophilic archaeum predicted by in situ RNA analysis. Nature 376:57–58

Jerry E (2000) Archaeal protein translocation crossing membranes in the third domain of life. Eur J Biochem 267:3402–3412

Ken FJ, Alison DW, Chitvan B, Juliet MB, Thomas D, James PJC (2011) Major players on the microbial stage: why archaea are important. Microbiology 157:919–936

Kim B, Peter R, Qunxin S, Fabrice C, Yvan Z, Roger AG (2002) Mobile elements in archaeal genomes. FEMS Microbiol Lett 206:131–141

Kira SM, Eugene VK (2003) Comparative genomics of archaea: how much have we learned in six years, and what's next? Genome Biol 4:115

Koonin EV (2015) Origin of eukaryotes from within archaea, archaeal eukaryome and bursts of gene gain: eukaryogenesis just made easier? Philos Trans R Soc B 370:20140333

Marraffini LA, Sontheimer EJ (2010) CRISPR interference: RNA-directed adaptive immunity in bacteria and archaea. Nat Rev Genet 11:181–190

Niels L, Rosalie PCD, Iddo H, Daan V, Maarten CN, Felix JHH, Malcolm FW, Remus TD, Gijs JLW (2012) Alba shapes the archaeal genome using a delicate balance of bridging and stiffening the DNA. Nat Commun 3:1328. https://doi.org/10.1038/ncomms2330

Ogata N, Miura T (2000) Elongation of tandem repetitive DNA by the DNA polymerase of the hyperthermophilic archaeon Thermococcus litoralis at a hairpin–coil transitional state: a model of amplification of a primordial simple DNA sequence. Biochemistry 39:13993–14001

Patrick F (1997) Archaea: what can we learn from their sequences? Curr Opin Genet Dev 7:764–770

Patrick JK, Ford D (1995) Archaea: narrowing the gap between prokaryotes and eukaryotes. Proc Natl Acad Sci U S A 92:5761–5764

Rachel YS, Takayuki O, Stefan MF, Roger LW, Stephen DB (2008) A role for the ESCRT system in cell division in Archaea. Science 322(5908):1710–1713. https://doi.org/10.1126/science.1165322

Shintani M, Sanchez ZK, Kimbara K (2015) Genomics of microbial plasmids: classification and identification based on replication and transfer systems and host taxonomy. Front Microbiol 6:242. https://doi.org/10.3389/fmicb.2015.00242

Simonetta G, Celine B-A (2006) The origin and evolution of Archaea: a state of the art. Philos Trans R Soc B 361:1007–1022

Sofya KG, Marat DK, Mikhail SG (2015) Horizontal gene transfer and genome evolution in *Methanosarcina*. BMC Evol Biol 15:102. https://doi.org/10.1186/s12862-015-0393-2

Stetter KO (2006) History of discovery of the first hyperthermophiles. Extremophiles 10 (5):357–362

Woese CR, Fox GE (1977) Phylogenetic structure of the prokaryotic domain: the primary kingdoms. Proc Natl Acad Sci U S A 74:5088–5090

Xiaoyu X, Lanming C, Xiaoxing H, Yuanmin L, Qunxin S, Li H (2005) Sulfolobus tengchongensis spindle-shaped virus STSV1: virus-host interactions and genomic features. J Virol 79 (14):8677–8686

Chapter 3

Abigail AS, Nadja BS, Ann MS, Hing-yew L (1995) Conjugative transposons: an unusual and diverse set of integrated gene transfer elements. Microbiol Rev 59:579–590

Amitai G, Belenkiy O, Dassa B, Shainskaya A, Pietrokovski S (2003) Distribution and function of new bacterial intein-like protein domains. Mol Microbiol 47:61–73. https://doi.org/10.1046/j.1365-2958.2003.03283

Arber WJ (1991) Elements in microbial evolution. J Mol Evol 33:4–12

Arber W (2000) Genetic variation: molecular mechanisms and impact on microbial evolution. FEMS Microbiol Rev 24:1–7

Arber W (2014) Horizontal gene transfer among bacteria and its role in biological evolution. Life 4 (2):217–224. https://doi.org/10.3390/life4020217

Barzel A, Naor A, Privman E, Kupiec M, Gophna U (2011) Homing endonucleases residing with ininteins:evolutionary puzzles awaiting genetic solutions. Biochem Soc Trans 39:169–173. https://doi.org/10.1042/BST0390169

Bentley SD, Parkhill J (2015) Genomic perspectives on the evolution and spread of bacterial pathogens. Proc R Soc B 282:20150488

Casjens S, Palmer N, Van VR, Huang WM, Stevenson B, Rosa P, Lathigra R, Sutton G, Peterson J, Dodson RJ, Haft D, Hickey E, Gwinn M, White O, Fraser CM (2000) A bacterial genome in flux: the twelve linear and nine circular extrachromosomal DNAs in an infectious isolate of the Lyme disease spirochete Borrelia burgdorferi. Mol Microbiol 35(3):490–516

Claire MF, Jeannine DG, Owen W, Mark DA, Rebecca AC, Robert DF, Carol JB, Anthony RK, Granger S, Jenny MK, Janice LF, Janice FW, Keith VS, Mina S, Joyce F, David N, Teresa RU, Deborah MS, Cheryl AP, Joseph MM, Jean-Francois T, Brian AD, Kenneth FB, Ping-Chuan H, Thomas SL, Scott NP, Hamilton OS, Clyde AH III, Craig Venter J (1995) The minimal gene complement of *Mycoplasma genitalium*. Science 270:397–403

Cole ST, Saint G (1994) Bacterial genomics. FEMS Microbiol Rev 14:139–160

Daniel GG, Gwynedd AB, Cynthia A-P, Evgeniya AD, Holly B-I, Jayshree Z, Timothy BS, Anushka B, David WT, Mikkel AA, Chuck M, Lei Y, Vladimir NN, John IG, Craig VJ, Clyde AH III, Hamilton OS (2008) Complete chemical synthesis, assembly, and cloning of a *Mycoplasma genitalium* genome. Science 319:1215–1220. https://doi.org/10.1126/science.1151721

Dobrindt U, Hochhut B, Hentschel U, Hacker J (2004) Genomic islands in pathogenic and environmental microorganisms. Nat Rev Microbiol 2:414–424. https://doi.org/10.1038/nrmicro884

Dori-Bachash M, Dassa B, Peleg O, Pineiro SA, Jurkevitch E, Pietrokovski S (2009) Bacterial intein-like domains of predatory bacteria: a new domain type characterized in Bdellovibrio bacteriovorus. Funct Integr Genomics 9:153–166. https://doi.org/10.1007/s10142-008-0106-7

Edgell DR, Chalamcharla VR, Belfort M (2011) Learning to live together: mutualism between self-splicing introns and their hosts. BMC Biol 9:22. https://doi.org/10.1186/1741-7007-9-22

Espéli O, Moulin L, Boccard F (2001) Transcription attenuation associated with bacterial repetitive extragenic BIME elements. J Mol Biol 314:375–386. https://doi.org/10.1006/jmbi.2001.5150

Fleischmann RD, Adams MD, White O, Clayton RA, Kirkness EF, Kerlavage AR, Bult CJ, Tomb J-F, Dougherty BA, Merrick JM (1995) Whole-genome random sequencing and assembly of Haemophilus influenzae Rd. Science 269:496–512

Fraser CM, Gocayne JD, White O, Adams MD, Clayton RA, Fleischmann RD, Bult CJ, Kerlavage AR, Sutton G, Kelley JM (1995) The minimal gene complement of Mycoplasma genitalium. Science 270:397–404

Goodner B, Hinkle G, Gattung S, Miller N, Blanchard M, Qurollo B, Goldman BS, Cao Y, Askenazi M, Halling C, Mullin L, Houmiel K, Gordon J, Vaudin M, Iartchouk O, Epp A, Liu F, Wollam C, Allinger M, Doughty D, Scott C, Lappas C, Markelz B, Flanagan C, Crowell C, Gurson J, Lomo C, Sear C, Strub G, Cielo C, Slater S (2001) Genome sequence of the plant pathogen and biotechnology agent *Agrobacterium tumefaciens* C58. Science 294:2323–2328

Guo H, Tse LV, Nieh AW, Czornyj E, Williams S, Oukil S, Liu VB, Miller JF (2011) Target site recognition by a diversity-generating retroelement. PLoS Genet 7:e1002414. https://doi.org/10.1371/journal.pgen.1002414

Hacker J, Carniel E (2001) Ecological fitness, genomic islands and bacterial pathogenicity. A Darwinian view of the evolution of microbes. EMBO Rep 2:376–381. https://doi.org/10.1093/embo-reports/kve097

Haugen P, Simon DM, Bhattacharya D (2005) The natural history of group I introns. Trends Genet 21:111–119. https://doi.org/10.1016/j.tig.2004.12.007

Hayashi T, Makino K, Ohnishi M, Kurokawa K, Ishii K, Yokoyama K, Han CG, Ohtsubo E, Nakayama K, Murata T, Tanaka M, Tobe T, Iida T, Takami H, Honda T, Sasakawa C, Ogasawara N, Yasunaga T, Kuhara S, Shiba T, Hattori M, Shinagawa H (2001) Complete genome sequence of enterohemorrhagic *Escherichia coli* O157:H7 and genomic comparison with a laboratory strain K-12. DNA Res 8(1):11–22

Iwai H, Züger S, Jin J, Tam PH (2006) Highly efficient protein transsplicing by a naturally split DnaE intein from Nostoc punctiforme. FEBS Lett 580:1853–1858. https://doi.org/10.1016/j.febslet.2006.02.045

Jane H, Mohammad H, Ryan RW, David JE, Helen B-J, Ruth MH, Kathryn EH (2015) ISMapper: identifying transposase insertion sites in bacterial genomes from short read sequence data. BMC Genomics 16:667. https://doi.org/10.1186/s12864-015-1860-2

Josephine B, Claire C, Stephen DB (2012) Developing insights into the mechanisms of evolution of bacterial pathogens from whole-genome sequences. Future Microbiol 7(11):1283–1296. https://doi.org/10.2217/fmb.12.108

Kenji S (1990) Invertrons, a class of structurally and functionally related genetic elements that includes linear dna plasmids, transposable elements, and genomes of adeno-type viruses. Microbiol Rev 54(1):66–74

Kleckner N (1981) Transposable elements in prokaryotes. Annu Rev Genet 15:341–404. https://doi.org/10.1146/annurev.ge.15.120181.002013

Lampson BC, Inouye M, Inouye S (2005) Retrons, msDNA, and the bacterial genome. Cytogenet Genome Res 110:491–499. https://doi.org/10.1159/000084982

Lima WC, Paquola AC, Varani AM, Van Sluys MA, Menck CF (2008) Laterally transferred genomic islands in Xanthomonadales related to pathogenicity and primary metabolism. FEMS Microbiol Lett 281:87–97. https://doi.org/10.1111/j.1574-6968.2008.01083.x

Makarova KS, Wolf YI, Snir S, Koonin EV (2011) Defense islands in bacterial and archaeal genomes and prediction of novel defense systems. J Bacteriol 193:6039–6056. https://doi.org/10.1128/JB.05535-11

Mates AK, Sayed AK, Foster JW (2007) Products of the Escherichia coli acid fitness island attenuate metabolite stress at extremely low pH and mediate a cell density-dependent acid resistance. J Bacteriol 189:2759–2768. https://doi.org/10.1128/JB.01490-06

Michael RG (2014) Integrons: past, present, and future. Microbiol Mol Biol Rev 78(2):257–277. https://doi.org/10.1128/MMBR.00056-13

Nicole TP, Guy P III, Valerie B, Bob M, Jeremy DG, Debra JR, George FM, Peter SE, Jason G, Heather AK, György P, Jeremiah H, Sara K, Adam B, Ying S, Leslie M, Erik JGN, Wayne D, Alex L, Eileen D, Konstantinos DP, Jennifer A, Thomas SA, Jieyi L, Galex Y, David CS, Rodney AW, Frederick RB (2001) Genome sequence of enterohaemorrhagic *Escherichia coli* O157:H7. Nature 409(6819):529–533

Ohad G-M, Brett FB (2006) Pathogenicity islands: a molecular toolbox for bacterial virulence. Cell Microbiol 8(11):1707–1719

Que F, Wu S, Huang R (2013) Salmonella pathogenicity island 1 (SPI-1) at work. Curr Microbiol 66:582–587. https://doi.org/10.1007/s00284-013-0307-8

Reznikoff WS (2008) Transposon Tn5. Annu Rev Genet 42:269–286. https://doi.org/10.1146/annurev.genet.42.110807.091656

Schillinger T, Lisfi M, Chi J, Cullum J, Zingler N (2012) Analysis of a comprehensive dataset of diversity generating retroelements generated by the program DiGReF. BMC Genomics 13:430. https://doi.org/10.1186/1471-2164-13-430

Sharp PM, Li WH (1987) The codon daptation index-a measure of directional synonymous codon usage bias, and its potential applications. Nucleic Acids Res 15:1281

Sullivan JT, Trzebiatowski JR, Cruickshank RW, Gouzy J, Brown SD, Elliot RM, Fleetwood DJ, McCallum NG, Rossbach U, Stuart GS, Weaver JE, Webby RJ, De Bruijn FJ, Ronson CW (2002) Comparative sequence analysis of the symbiosis island of Mesorhizobium loti strain R7A. J Bacteriol 184:3086–3095. https://doi.org/10.1128/JB.184.11.3086-3095.2002

Sung JY, Koo SH, Cho HH, Kwon KC (2012) AbaR7, a genomic resistance island found in multidrug-resistant Acinetobacter baumannii isolates in Daejeon, Korea. Ann Lab Med 32:324–330. https://doi.org/10.3343/alm.2012.32.5.324

Tourasse NJ, Kolstø AB (2008) Survey of groupI and group II introns in 29 sequenced genomes of the *Bacillus cereus* group: insights into their spread and evolution. Nucleic Acids Res 36:4529–4548

Virdi JS, Pooja S (2005) Genetic diversity of pathogenic microorganisms: basic insights, public health implications and the Indian initiatives. Curr Sci 89(1):113–123

Wu Y, Richard ZA, Mark MT (2015) Dynamics of bacterial insertion sequences: can transposition bursts help the elements persist? BMC Evol Biol 15:288. https://doi.org/10.1186/s12862-015-0560-5

Yang D, Xuerui B, Lili J, Lei C, Junyan L, Jian M, Dingqiang C, Huawei B, Yanmei L, Guangchao Y (2015) Resistance integrons: class 1, 2 and 3 integrons. Ann Clin Microbiol Antimicrob 14:45–56

Chapter 4

Adrian CB, Christopher JH, Davy PK, Sarah JT (2010) Organization and expression of organellar genomes. Philos Trans R Soc B 365:785–797

Aono N, Shimizu T, Inoue T, Shiraishi H (2002) Palindromic repetitive elements in the mitochondrial genome of *Volvox*. FEBS Lett 521(1–3):95–99

Bender A, Krishnan KJ, Morris CM, Taylor GA, Reeve AK, Perry RH, Jaros E, Hersheson JS, Betts J, Klopstock T, Taylor RW, Turnbull DM (2006) High levels of mitochondrial DNA deletions in substantia nigra neurons in aging and Parkinson disease. Nat Genet 38:515–517

Bennett MS, Wiegert KE, Triemer RE (2012) Comparative chloroplast genomics between *Euglena viridis* and *Euglena gracilis* (Euglenophyta). Phycologia 51:711–718. https://doi.org/10.2216/12-017.1

Bennett MS, Wiegert KE, Triemer RE (2014) Characterization of Euglenaformis gen. nov. and the chloroplast genome of *Euglenaformis* [Euglena] proxima (Euglenophyta). Phycologia 53:66–73. https://doi.org/10.2216/13-198.1

Bitner-Glindzicz M, Pembrey M, Duncan A, Heron J, Ring SM, Hall A, Rahman S (2009) Prevalence of mitochondrial 1555A–NG mutation in European children. N Engl J Med 360:640–642

Copertino DW, Hallick RB (1991) Group II twintron: an intron within an intron in a chloroplast cytochrome *b559* gene. EMBO J 10:433–442

David RS (2012) Updating our view of organelle genome nucleotide landscape. Front Genet 3:175

David RS (2016) The past, present and future of mitochondrial genomics: have we sequenced enough mtDNAs? Brief Funct Genomics 15(1):47–54

David RS, Patrick JK (2015) Mitochondrial and plastid genome architecture: reoccurring themes, but significant differences at the extremes. PNAS 112(33):10177–10184

David RS, Robert WL (2010) low nucleotide diversity for the expanded organelle and nuclear genomes of *Volvox carteri* supports the mutational-hazard hypothesis. Mol Biol Evol 27 (10):2244–2256

Davoodi-Semiromi A, Schreiber M, Nalapalli S, Verma D, Singh ND, Banks RK et al (2010) Chloroplast-derived vaccine antigens confer dual immunity against cholera and malaria by oral or injectable delivery. Plant Biotechnol J 8:223–242

Elliott HR, Samuels DC, Eden JA, Relton CL, Chinnery PF (2008) Pathogenic mitochondrial DNA mutations are common in the general population. Am J Hum Genet 83:254–260

Ellis RJ (1970) Further similarities between chloroplast and bacterial ribosomes. Biochem J 116 (4):28P–29P

Emmanuel DL, Eleftherios Z (2017) Evolution and inheritance of animal mitochondrial DNA: rules and exceptions. J Biol Res (Thessaloniki) 24:2. https://doi.org/10.1186/s40709-017-0060-4

Finsterer J (2008) Leigh and Leigh-like syndrome in children and adults. Pediatr Neurol 39:223–235

Gao L, Su YJ, Wang T (2010) Plastid genome sequencing, comparative genomics, and phylogenomics: current status and prospects. J Syst Evol 48:77–93

Gockel GHW (2000) Complete gene map of the plastid genome of the nonphotosynthetic euglenoid flagellate *Astasia longa*. Protist 151:347–351

Goto IN, Horai S (1990) A mutation in the tRNA(Leu)(UUR) gene associated with the MELAS subgroup of mitochondrial encephalomyopathies. Nature 348:651–653

Green BR (2011) Chloroplast genomes of photosynthetic eukaryotes. Plant J 66:34–44

Hallick RB, Hong L, Drager RG, Favreau MR, Minfort A, Orsat B, Spielmann A, Stutz E (1993) Complete sequence of *Euglena gracilis* chloroplast DNA. Nucleic Acids Res 21:3537–3544

Helen AL, Tuppen EL, Blakely DMT, Robert WT (2010) Mitochondrial DNA mutations and human disease. Biochim Biophys Acta 1797:113–128

Henry D, Choun-Sea LMY, Wan-Jung C (2016) Chloroplast genomes: diversity, evolution, and applications in genetic engineering. Genome Biol 17:134–162

Hrda S, Fousek J, Szabov J, Hampl VV, Vlcek C (2012) The plastid genome of Eutreptiella provides a window into the process of secondary endosymbiosis of plastid in euglenids. PLoS One 7:e33746. https://doi.org/10.1371/journal.pone.0033746

Jan-Willem T (1999) The mitochondrial genome: structure, transcription, translation and replication. Biochim Biophys Acta 1410:103–123

Kaila T, Chaduvla PK, Saxena S, Bahadur K, Gahukar SJ, Chaudhury A, Sharma TR, Singh NK, Gaikwad K (2016) Chloroplast genome sequence of pigeonpea (*Cajanus cajan* L. Millspaugh) and *Cajanus scarabaeoides* (L.) Thouars: genome organization and comparison with other legumes. Front Plant Sci 7:1847. https://doi.org/10.3389/fpls.2016.01847

Kailash CB, Dipnarayan S (2012) Chloroplast genomics and genetic engineering for crop improvement. Agric Res 1(1):53–66. https://doi.org/10.1007/s40003-011-0010-6

Krings M, Stone A, Schmitz RW, Krainitzki H, Stoneking M, Pääbo S (1997) Neandertal DNA sequences and the origin of modern humans. Cell 90:19–30. [PubMed: 9230299]

Kusumi J, Tachida H (2005) Compositional properties of greenplant plastid genomes. J Mol Evol 60:417–425

Lakshmi PS, Verma D, Yang X, Lloyd B, Daniell H (2013) Low cost tuberculosis vaccine antigens in capsules: expression in chloroplasts, bio-encapsulation, stability and functional evaluation in vitro. PLoS One 8:e54708

Maceluch JA, Niedziela M (2006) The clinical diagnosis and molecular genetics of Kearns–Sayre syndrome: a complex mitochondrial encephalomyopathy. Pediatr Endocrinol Rev 4:117–137

Matthew SB, Richard ET (2015) Chloroplast genome evolution in the euglenaceae. J Eukaryot Microbiol 62:773–785

McFarland R, Taylor RW, Turnbull DM (2007) Mitochondrial disease-its impact, etiology, and pathology. Curr Top Dev Biol 77:113–155

Min XJ, Hickey DA (2007) DNA barcodes provide a quick preview of mitochondrial genome composition. PLoS One 2:e325

Monro RE (1967) Ribosome-catalysed peptidyl transfer: effects of some inhibitors of protein synthesis. J Mol Biol 28(1):161–165

Moore MJ, Dhingra A, Soltis PS, Shaw R, Farmerie WG, Folta KM, Soltis DE (2006) Rapid and accurate pyrosequencing of angiosperm plastid genomes. BMC Plant Biol 6:17

Pombert J-F, James ER, Janouskovec JKPJ (2012) Evidence for transitional stages in the evolution of euglenid group ii introns and twintrons in the monomorphina aenigmatica plastid genome. PLoS One 7:e53433. https://doi.org/10.1371/journal.pone.0053433

Rahman S, Poulton J (2009) Diagnosis of mitochondrial DNA depletion syndromes. Arch Dis Child 94:3–5

Richard BH, Ling H, Robert GD, Mitchell RF, Amparo M, Bernard O, Albert S, Erhard S (1993) Complete sequence of *Eugena gracilis* chloroplast DNA. Nucleic Acids Res 21(15):3537–3544

Rotig A, Cormier V, Blanche S, Bonnefont JP, Ledeist F, Romero N, Schmitz J, Rustin P, Fischer A, Saudubray JM et al (1990) Pearson's marrow–pancreas syndrome. Multisystem mitochondrial disorder in infancy. J Clin Invest 86:1601–1608

Salvatore DM (2004) Mitochondrial diseases. Biochim Biophys Acta Bioenerg 1658(1–2):80–88

Santos C, Martinez M, Lima M, Hao YJ, Simoes N, Montiel R (2008) Mitochondrial DNA mutations in cancer: a review. Curr Top Med Chem 8:1351–1366

Schaefer AM, McFarland R, Blakely EL, He L, Whittaker RG, Taylor RW, Chinnery PF, Turnbull DM (2008) Prevalence of mitochondrial DNA disease in adults. Ann Neurol 63:35–39

Serre D, Langaney A, Chech M, Teschler-Nicola M, Paunovic M, Mennecier P, Hofreiter M, Possnert G, Pääbo S (2004) No evidence of Neandertal mtDNA contribution to early modern humans. PLoS Biol 2:E57. [PubMed: 15024415]

Seung BS, Xiangpei Z, Mourad A, Bobby LR, Jonathan K, Antti S, Bruce B (2014) High throughput whole mitochondrial genome sequencing by two platforms of massively parallel sequencing. BMC Genomics 15(Suppl 2):P7

Shinozaki K, Ohme M, Tanaka M, Wakasugi T, Hayashida N, Matsubayashi T et al (1986) The complete nucleotide sequence of the tobacco chloroplast genome: its gene organization and expression. EMBO J 5:2043–2049

Taosheng H (2011) Next generation sequencing to characterize mitochondrial genomic DNA heteroplasmy. Curr Protoc Hum Genet 19. Unit 19.8. https://doi.org/10.1002/0471142905.hg1908s71

Tomohiko K, Kathleen JN (2008) Angiosperm mitochondrial genomes and mutations. Mitochondrion 8:5–14

Valentine N (2013) Mitochondrial genome sequence of the legume *Vicia faba*. Front Plant Sci 4. Article 128/2

Vandebona H, Mitchell P, Manwaring N, Griffiths K, Gopinath B, Wang JJ, Sue CM (2009) Prevalence of mitochondrial 1555A–NG mutation in adults of European descent. N Engl J Med 360:642–644

Wallace DC (1992) Diseases of the mitochondrial DNA. Annu Rev Biochem 61:1175–1212

Weng M-L, Blazier JC, Govindu M, Jansen RK (2014) Reconstruction of the ancestral plastid genome in Geraniaceae reveals a correlation between genome rearrangements, repeats and nucleotide substitution rates. Mol Biol Evol 31:645–659

Wiegert KE, Bennett MS, Triemer RE (2012) Evolution of the chloroplast genome in photosynthetic euglenoids: a comparison of *Eutreptia viridis* and *Euglena gracilis* (Euglenophyta). Protist 163:832–843. https://doi.org/10.1016/j.protis.2012.01.002

Wiegert KE, Bennett MS, Triemer RE (2013) Tracing patterns of chloroplast evolution in euglenoids: contributions from *Colacium vesiculosum* and *Strombomonas acuminata* (Euglenophyta). J Eukaryot Microbiol 60:214–221. https://doi.org/10.1111/jeu.12025

Wong LJ (2007) Diagnostic challenges of mitochondrial DNA disorders. Mitochondrion 7:45–52

Yangrae C, Yin-long Q, Peter K, Jeffrey DP (1998) Explosive invasion of plant mitochondria by a group I intron. Proc Natl Acad Sci U S A 95:14244–14249

Chapter 5

Aimée MD, Adrian B (2011) CpG islands and the regulation of transcription. Genes Dev 25:1010–1022

Alfredo TM (2007) Beyond the sequence: cellular organization of genome function. Cell 128:787–800

Anderson S (1981) Sequence and organization of the human mitochondrial genome. Nature 290:457–465. https://doi.org/10.1038/290457a0

Anthony L, Pierre P (2011) The role of duplications in the evolution of genomes highlights the need for evolutionarybased approaches in comparative genomics. Biol Direct 6:11–23

Brent MR (2007) How does eukaryotic gene prediction work? Nat Biotechnol 25:883–885. https://doi.org/10.1038/nbt0807-883

Brigida G, Jan S, Troels P, Leah S, Veerle S, Beatriz H-M, Adriaan M, Miguel R, Karin V, Loren M, Clotilde T, Brian S, Maryann T, Ariel S, Toby R, Christopher W, Guy B, Steven M, Kevin JV (2016) Domestication and divergence of *Saccharomyces cerevisiae* beer yeasts. Cell 166(6):1397–1410

Burke T, Bruford MW (1987) DNA fingerprinting in birds. Nature 327:149–152

Carlton JM et al (2007) Draft genome sequence of the sexually transmitted pathogen *Trichomonas vaginalis*. Science 31:207–212. https://doi.org/10.1126/science.1132894

Chen M et al (2002) An integrated physical and genetic map of the rice genome. Plant Cell 14:537–545

Cheng Z et al (2002) Functional rice centromeres are marked by a satellite repeat and a centromere-specific retrotransposon. Plant Cell 14:1691–1704

Choi IY et al (2007) A soybean transcript map: gene distribution, haplo type and single nucleotide polymorphism analysis. Genetics 176:685–696

Craig VJ, Mark D, Eugene WM, Li PW, Richard JM et al (2001) The sequence of the human genome. Science 291:1304–1353. https://doi.org/10.1126/science.1058040

Fantom Consortium et al. (2005) The transcriptional landscape of the mammalian genome. Science 309:1559–1563. https://doi.org/10.1126/science.1112014

Feschotte C, Jiang N, Wessler SR (2002) Plant transposable elements: where genetics meets genomics. Nat Rev Genet 3:329–341

Gaut BS, Doebley JF (1997) DNA sequence evidence for the segmental allotetraploid origin of maize. Proc Natl Acad Sci U S A 94:6809–6814

Goffeau A, Barrell BG, Bussey H, Davis RW, Dujon B, Feldmann H, Galibert F, Hoheisel JD, Jacq C, Johnston M, Louis EJ, Mewes HW, Murakami Y, Philippsen P, Tettelin H, Olive SG (1996) Life with 6000 genes. Science 5287:546–567. https://doi.org/10.1126/science.274.5287.546

Gregory TR (2005) Synergy between sequence and size in large-scale genomics. Nat Rev Genet 6:699–708. https://doi.org/10.1038/nrg1674

Jaffe DB et al (2003) Whole-genome sequence assembly for mammalian genomes: Arachne 2. Genome Res 13:91–96

Jeffreys AJ, Wilson V, Thein SL (1985) Hypervariable minisatellite region in human DNA. Nature 314:67–73

Jeremy S, Steven BC, Jessica S, Jianxin M, Therese M, William N et al (2010) Genome sequence of the palaeopolyploid soybean. Nature 463:178–183

Kazazian HH Jr (2004) Mobile elements: drivers of genome evolution. Science 303:1626–1632

Kidwell MG (2002) Transposable elements and the evolution of genome size in eukaryotes. Genetica 115:49–63

Koszul RS, Caburet BD, Fischer G (2004) Eukaryotic genome evolution through the spontaneous duplication of large chromosomal segments. EMBO J 23:234–243

Lin JY et al (2005) Pericentromeric regions of soybean (*Glycine max* L. Merr.) chromosomes consist of retroelements and tandemly repeated DNA and are structurally and evolutionarily labile. Genetics 170:1221–1230

Malenčić D, Cvejić J, Miladinović J (2012) Polyphenol content and antioxidant properties of colored soybean seeds from central europe. J Med Food 15:89–95

McClintock B (1950) The origin and behavior of mutable loci in maize. Proc Natl Acad Sci U S A 36:344–355

Michelmore R, Meyers BC (1998) Clusters of resistance genes in plants evolve by divergent selection and a birth-and-death process. Genome Res 8:1113–1130

Parra G et al (2005) Tandem chimerism as a means to increase protein complexity in the human genome. Genome Res 16:37–44

Reenan R (2005) Molecular determinants and guided evolution of species-specific RNA editing. Nature 434:409–413. https://doi.org/10.1038/nature03364

Scott AJ (2016) Rice: the first crop genome. Rice 9:14–16
Shore D (2001) Transcriptional silencing: replication redux. Curr Biol 11(20):R816–R819
Taie HAA, El-Mergawi R, Radwan S (2008) Isoflavonoids, flavonoids, phenolic acids profiles and antioxidant activity of soybean seeds as affected by organic and bioorganic fertilization. Am Eurasian J Agric Environ Sci 4:207–213
Takuji S, Benjamin B (2000) International rice genome sequencing project: the effort to completely sequence the rice genome. Curr Opin Plant Biol 3:138–141
Van SNI, Roelofs D (2006) Introduction to ecological genetics. Oxford University Press, New York
Venter JC et al (2001) The sequence of the human genome. Science 5507:1304–1351. https://doi.org/10.1126/science.1058040
Villasante A, José PA, Méndez-Lago M (2007) Centromeres were derived from telomeres during the evolution of the eukaryotic chromosome. Proc Natl Acad Sci U S A 104(25):10542–10547
Virginia AZ (2012) Telomeres: the beginnings and ends of eukaryotic chromosomes. Exp Cell Res 318(12):1456–1460. https://doi.org/10.1016/j.yexcr.2012.02.015
Wolfe KH, Shields DC (1997) Molecular evidence for an ancient duplication of the entire yeast genome. Nature 387:708–713
Yandell M et al (2005) A computational and experimental approach to validating annotations and gene predictions in the *Drosophila melanogaster* genome. Proc Natl Acad Sci 102:1566–1571

Chapter 6

Bainbridge MN, Warren RL, Hirst M, Romanuik T, Zeng T, Go A, Delaney A, Griffith M, Hickenbotham M, Magrini V, Mardis ER, Sadar ED, Siddiqui AS, Marra MA, Jones SJ (2006) Analysis of the prostate cancer cell line LNCaP transcriptome using a sequencing-by-synthesis approach. BMC Genomics 7:246
Balasubramanian S (2011a) Sequencing nucleic acids: from chemistry to medicine. Chem Commun 47:7281–7286
Balasubramanian S (2011b) Decoding genomes at high speed: implications for science and medicine. Angew Chem Int Ed 50:12406–12410
Barski AS, Cuddapah KC, Roh TY, Schones DE, Wang Z, Wei G, Chepelev I, Zhao K (2007) High-resolution profiling of histone methylations in the human genome. Cell 129:823–837
Bhinge AA, Ki J, Euskirche GM, Snyder M, Iyer VR (2007) Mapping the chromosomal targets of STAT1 by sequence tag analysis of genomic enrichment (STAGE). Genome Res 17:910–916
Both M, Michael C, Michael PH, Stumpfc Pietro DS (2005) Gene expression profiles of *Blumeria graminis* indicate dynamic changes to primary metabolism during development of an obligate biotrophic pathogen. Plant Cell 17:2107–2122
Brookes KJ (2013) The VNTR in complex disorders: the forgotten polymorphisms? A functional way forward? Genomics 101:273–281
Carrilho E (2000) DNA sequencing by capillary array electrophoresis and microfabricated array systems. Electrophoresis 21:55–65
Chaitanya KV, Akbar Alikhan P, Prasanth Reddy V, Rishabh L, Taraka Ramji M (2010) Role of high-throughput sequencing technologies in genome sequencing. Int J Adv Biotechnol Res 1 (2):120–129
Chien-Yueh L, Yu-Chiao C, Liang-Bo W, Yu-Lun K, Eric YC, Liang-Chuan L, Mong-Hsun T (2013) Common applications of next-generation sequencing technologies in genomic research. Transl Cancer Res 22(1):33–45
Eric LT, Michael SW (1988) Genomic mapping by fingerprinting random clones: a mathematical analysis. Genomics 2:231–239
Ewing B, Green P (1998) Base-calling of automated sequencer traces using phred. II. Error probabilities. Genome Res 8:186–194
Fred Sanger S, Nicklen ARC (1977) DNA sequencing with chain-terminating. Proc Natl Acad Sci 74:5463–5467

Gowda M, Li H, Alessi J, Chen F, Pratt R, Wang GL (2006) Robust analysis of 5¢-transcript ends (5¢-RATE): a novel technique for transcriptome analysis and genome annotation. Nucleic Acids Res 34:e126

Hayden EC (2015) Pint-sized DNA sequencer impresses first users. Nature 521:15–16

Holley RW, Apgar J, Merrill SH, Zubkoff PL (1961) Nucleotide and oligonucleotide com-positions of the alanine-, valine-, and tyrosine-acceptor soluble ribonucleic acids of yeast. J Am Chem Soc 83:4861–4862

Holley RW et al (1965) Structure of a ribonucleic acid. Science 147:1462–1465

Hutchison CA (2007) DNA sequencing: bench to bedside and beyond. Nucleic Acids Res 35:6227–6237

James MH, Benjamin C (2016) The sequence of sequencers: the history of sequencing DNA. Genomics 107:1–8

Korbel JO, Urban AE, Affourtit JP, Godwin B, Grubert F, Simons JF, Kim PM, Palejev D, Carriero NJ, Du L, Taillon BE, Chen Z, Tanzer A, Saunders AC, Chi J, Yang F, Carter NP, Hurles ME, Weissman SM, Harkins TT, Gerstein MB, Egholm M, Snyder M (2007) Paired-end mapping reveals extensive structural variation in the human genome. Science 318:420–426

Lawrence MS, Stojanov P, Mermel CH, Robinson JT, Garraway LA, Golub TR, Meyerson M, Gabriel SB, Lander ES, Getz G (2014) Discovery and saturation analysis of cancer genes across 21 tumour types. Nature 505:495–501

Mardis E (2013) Next-generation DNA sequencing platforms. Annu Rev Anal Chem 6:287–303

Maxam AM, Gilbert W (1977) A new method for sequencing DNA. Proc Natl Acad Sci U S A 74:560–564

Nageswara Rao RN, Challa S (2015) Next generation sequencing and metagenomics, computational biology and bioinformatics: gene regulation. 319–339. ISBN 9781498724975 - CAT K25752. https://doi.org/10.1201/b20026-21

Navin N, Kendall J, Troge J, Andrews P, Rodgers L, McIndoo J, Cook K, Stepansky A, Levy D, Esposito D et al (2011) Tumour evolution inferred by single-cell sequencing. Nature 472:90–94

Nyren PBP, Uhlen M (1993) Solid phase DNA minisequencing by an enzymatic luminometric inorganic pyrophosphate detection assay. Anal Biochem 208:171–175

Sanger F (1980) Frederick Sanger – Biographical. http://www.nobelprize.org/nobel_prizes/chemistry/laureates/1980/sanger-bio.html

Sanger F, Coulson A (1975) A rapid method for determining sequences in DNA by primed synthesis with DNA polymerase. J Mol Biol 94:441–448

Sanger F, Brownlee G, Barrell B (1965) A two-dimensional fractionation procedure for radioactive nucleotides. J Mol Biol 13:373-IN4

Shendure J, Ji H (2008) Next-generation DNA sequencing. Nat Biotechnol 26(10):1135–1144

Tawfik DS, Griffiths AD (1998) Man-made cell-like compartments for molecular evolution. Nat Biotechnol 16:652–656

Venter JC, Remington K, Heidelberg JF, Halpern AL, Rusch D, Eisen JA et al (2004) Environmental genome shotgun sequencing of the *Sargasso Sea*. Science 304(5667):66–74

Zallen DT (2003) Despite Franklin's work, Wilkins earned his nobel. Nature 425:15

Chapter 7

Bernd T, Wolfgang H, Alexander K, Mattowc J, Schmid M, Schmidt F, Jungblut PR (2005) Peptide mass fingerprinting. Methods 35(3):237–247

Carthew RW, Sontheimer EJ (2009) Origins and mechanisms of miRNAs and siRNAs. Cell 136:642–655

Chang SH (1976) High-performance liquid chromatography of proteins. J Chromatogr 125:103–114

Cloonan N et al (2008) Stem cell transcriptome profiling via massive-scale mRNA sequencing. Nat Methods 5:613–619

Donald GR (2005) Metabonomics in toxicology: a review. Toxicol Sci 85:809–822
Dong ZC, Chen Y (2013) Transcriptomics: advances and approaches. Sci China Life Sci 56(10):960–967
Eddy SR (2001) Non-coding RNA genes and the modern RNA world. Nat Rev Genet 2:919–929
Hamilton JP, Buell CR (2012) Advances in plant genome sequencing. Plant J 70:177–190
Helen S, Liselotte K, Renwang J, Ian S, Svetlana I, Andreas M, Xinmiao L, Jandirk S (2012) The potential of metabolic fingerprinting as a tool for the modernisation of TCM preparations. J Ethnopharmacol 140:482–491
Joachim K, Alisdair F, Wolfram W, Yves G, Mark S (2004) Metabolite profiling in plant biology: platforms and destinations. Genome Biol 5(6):109.1–109.9
Lindon JC, Holmes E, Nicholson JK (2004) Metabonomics and its role in drug development and disease diagnosis. Expert Rev Mol Diagn 4:189–199
Lister R et al (2008) Highly integrated single-base resolution maps of the epigenome in Arabidopsis. Cell 133:523–536
Lockhart DJ, Winzeler EA (2000) Genomics, gene expression and DNA arrays. Nature 405:827–836
Marioni J, Mason C, Mane S, Stephens M, Gilad Y (2008) RNA-seq: an assessment of technical reproducibility and comparison with gene expression arrays. Genome Res 18(9):1509–1517. https://doi.org/10.1101/gr.079558.108
Michelle CT, Paolo V, Eduardo S, Michaela D, David B, Roberto B, Marc C-H, Timothy G, John G, Ayoung J, Soterios K, Marco M, Gary WM, Timothy N, Mark N, David HP, Nicole P-H, Jonathan S, Roel V, Jelle V, Martine V, Christopher W, Manolis K, EXPOsOMICS Consortium (2018) EXPOsOMICS: final policy workshop and stakeholder consultation. BMC Public Health 18:260
Morin R et al (2008) Profiling the HeLa S3 transcriptome using randomly primed cDNA and massively parallel short-read sequencing. BioTechniques 245:81–94
Mortazavi A, Williams BA, McCue K, Schaeffer L, Wold B (2008) Mapping and quantifying mammalian transcriptomes by RNA-Seq. Nat Methods 5:621–628
Nagalakshmi U et al (2008) The transcriptional landscape of the yeast genome defined by RNA sequencing. Science 320:1344–1349
Nicholson JK, Lindon JC, Holmes E (1999) 'Metabonomics': understanding the metabolic responses of living systems to pathophysiological stimuli via multivariate statistical analysis of biological NMR spectroscopic data. Xenobiotica 29:1181–1189
Oliver SG, Winson MK, Kell DB, Baganz F (1998) Systematic functional analysis of the yeast genome. Trends Biotechnol 16:373–378
Pongsuwan W, Bamba T, Harada K, Yonetani T, Kobayashi A, Fukusaki E (2008) High-throughput technique for comprehensive analysis of Japanese green tea quality assessment using ultra-performance liquid chromatography with time-of-flight mass spectrometry (UPLC/TOF MS). J Agric Food Chem 56:10705–10708
Rainer S, Rudolf K, Wolfram W, Royston G (2013) Metabolomics and metabolite profiling. Anal Bioanal Chem 405:5003–5004
Rajendhran J,·Gunasekaran P (2010). Human microbiomics. Indian J Microbiol 50:109–112
Stefan S, Serhat SC, Valerio P, Stefan S, Peter S, Volker W, Hermann S (2011) Metabolic fingerprinting of *Leontopodium* species (Asteraceae) by means of ^{1}H NMR and HPLC–ESI-MS. Phytochemistry 72(11–12):1379–1389
Tianniam S, Tarachiwin L, Bamba T, Kobayashi A, Fukusaki E (2008) Metabolic profiling of *Angelica acutiloba* roots utilizing gas chromatography-time-of-flight-mass spectrometry for quality assessment based on cultivation area and cultivar via multivariate pattern recognition. J Biosci Bioeng 105:655–659
Timothy EJ, Behrensa OS (2012) Human connectomics. Curr Opin Neurobiol 22(1):144–153. https://doi.org/10.1016/j.conb.2011.08.005
Urbain A, Marston A, Marsden-Edwards E, Hostettmann K (2009) Ultra-performance liquid chromatography/time-of-flight mass spectrometry as a chemotaxonomic tool for the analysis of Gentianaceae species. Phytochem Anal 20:134–138

Wilhelm BT et al (2008) Dynamic repertoire of a eukaryotic transcriptome surveyed at single-nucleotide resolution. Nature 453:1239–1243

Wright PC, Noirel A, OW S-A, Fazeli A (2012) A review of current proteomics technologies with a survey on their widespread use in reproductive biology investigations. Theriogenology 77:738–765

Chapter 8

Alföldi J, Lindblad-Toh K (2013) Comparative genomics as a tool to understand evolution and disease. Genome 23:1063–1068

Alukdar D, Sinjushin A (2015) Cytogenomics and mutagenomics in plant functional biology and breeding. In: Barh D, Khan M, Davies E (eds) PlantOmics: the omics of plant science. Springer, New Delhi

Atul B (2002) The use and analysis of microarray data 2002. Nat Rev 1:951–960

Bala AA, Stuart JL, Hikmet B (2013) Genomics approaches for crop improvement against abiotic stress. Sci World J 2013. Article ID 361921

Bernhard H, Thomas W (2004) Comparative genomics: methods and applications. Naturwissenschaften 91:405–421

Brennan P, Wild CP (2015) Genomics of cancer and a new era for cancer prevention. PLoS Genet 11(11):e1005522. https://doi.org/10.1371/journal.pgen.1005522

Brett D, Deborah JS, Tony R, John SM (2017) Prioritising the application of genomic medicine. NPJ Genom Med 2:35. https://doi.org/10.1038/s41525-017-0037-0

Craig J (2008) Complex diseases: research and applications. Nat Educ 1(1):184

Donna GA, Daniel P (2003) Genomic microarrays in human genetic disease and cancer. Hum Mol Genet 12(Review Issue 2):R145–R152. https://doi.org/10.1093/hmg/ddg261

Easwar RD, Chaitanya KV (2016) Photosynthesis and antioxidative defense mechanisms in deciphering drought stress tolerance of crop plants. Biol Plant 60(2):201–218

Gincy PT, Kandakumar J, Ahmad SO (2016) Sequencing crop genomes: a gateway to improve tropical agriculture. Trop Life Sci Res 27(1):93–114

Gulzar AN, Riaz-ud-Din S (2006) Biotechnology and genomics in medicine – a review. World J Med Sci 1(2):72–81

Hardison RC (2003) Comparative genomics. PLoS Biol 1(2):156–160

Hidewaki N, Masashi F (2018) Whole genome sequencing analysis for cancer genomics and precision medicine. Cancer Sci 109:513–522

Iourov IY, Vorsanova SG, Yurov YB (2008) Molecular cytogenetics and cytogenomics of brain diseases. Curr Genomics 9:452–465

Ivo DD, Federica T, Fabio M, Petros P, Zhizhong L, Alen Z, Paul E, Jonathan P, Alex G, James AK, Andrew PC, John DVH, Joseph A, Carl K, Arthur WT (2011) Applications of the pipeline environment for visual informatics and genomics computations. BMC Bioinf 12:304–323

Jackson SA, Aiko I, Suk-Ha L, Jeremy S, Randy S (2011) Sequencing crop genomes: approaches and applications. New Phytol 191:915–925

Jane B (2003) Human epigenome project—up and running. PLoS Biol 1(3):316–319

Jin B, Li Y, Keith DR (2011) DNA methylation: superior or subordinate in the epigenetic hierarchy? Genes Cancer 2(6):607–617

Meenakshi D, Ambika R (2018) Application of functional genomics in agriculture. Int J Chem Stud 6(1):462–465

Michal-Ruth S, Christian B, Bernd T (2013) Genomics and epigenomics: new promises of personalized medicine for cancer patients. Brief Funct Genomics 12(5):411–421

Paul DB, Joris RV (2012) Genomic microarrays: a technology overview. Parenat Diagn 32:336–343

Pérez-de-Castro AM, Vilanova S, Cañizares J, Pascual L, Blanca JM, Díez MJ, Prohens J, Picó B (2012) Application of genomic tools in plant breeding. Curr Genomics 13:179–195

Richard R, Nigel PC (2009) Comparative genomic hybridization: microarray design and data interpretation. Methods Mol Biol 529:37–49. https://doi.org/10.1007/978-1-59745-538-1_3

Ridgely FG, David WD, Scott B, Katherine K, Muin JK (2015) Genomics in public health: perspective from the office of public health genomics at the centers for disease control and prevention (CDC). Healthcare 3:830–837. https://doi.org/10.3390/healthcare3030830

Swaroop A, Robert LN (2018) Genetics in mainstream medicine: finally within grasp to influence healthcare globally. Mol Genet Genomic Med 6:473–480

Index

K. V. Chaitanya, *Genome and Genomics*,
https://doi.org/10.1007/978-981-15-0702-1

MIX
Papier aus verantwortungsvollen Quellen
Paper from responsible sources
FSC® C105338

If you have any concerns about our products,
you can contact us on
ProductSafety@springernature.com

In case Publisher is established outside the EU,
the EU authorized representative is:
Springer Nature Customer Service Center GmbH
Europaplatz 3, 69115 Heidelberg, Germany

Printed by Libri Plureos GmbH
in Hamburg, Germany